Delivering Value with BIM

Building Information Modelling (BIM) is a global phenomenon which is gaining significant momentum across the world. Currently there is little information on how to realise and monitor benefits from implementing BIM across the life-cycle of a built environment asset. This book provides a practical and strategic framework to realise value from implementing BIM by adapting Benefit Realisation Management theory. It presents an approach for practitioners aiming to implement BIM across the life-cycle of built environment assets, including both buildings and infrastructure.

Additionally, the book features:

- wide-ranging information about BIM, the challenges of monitoring progress towards benefit goals and the greater context of implementation;
- a set of dictionaries that illustrate: how benefits can be achieved, what the benefit flows are and the enabling tools and processes that contribute to achieving and maximising them;
- a suite of measures that can serve to monitor progress with examples of how they have been used to measure benefits from BIM;
- real-world examples from across the world and life-cycle phases that show how these benefits can be achieved; and
- information on international maturity and competency measures to complement the value realisation framework.

Including a blend of academic and industry input, this book has been developed in close collaborative consultation with industry, government and international research organisations and could be used for industry courses on BIM benefits and implementation for asset management or by universities that teach BIM-related courses.

Adriana X. Sanchez is a Research Associate at Curtin University, Australia.

Keith D. Hampson is CEO of Australia's Sustainable Built Environment National Research Centre (SBEnrc).

Simon Vaux is acting Principal Engineering Manager with Transport for NSW, Australia, and chair of the Transport for NSW BIM Working Group.

Delivering Value with BIM

A whole-of-life approach

Edited by
Adriana X. Sanchez, Keith D. Hampson
and Simon Vaux

LONDON AND NEW YORK

First published 2016
by Routledge
2 Park Square, Milton Park, Abingdon, Oxon OX14 4RN

and by Routledge
711 Third Avenue, New York, NY 10017

Routledge is an imprint of the Taylor & Francis Group, an informa business

British Library Cataloguing in Publication Data
A catalogue record for this book is available from the British Library

Library of Congress Cataloguing in Publication Data
Names: Sanchez, Adriana, editor. | Hampson, Keith (Keith Douglas), editor. | Vaux, Simon, editor.
Title: Delivering value with BIM : a whole-of-life approach / edited by Adriana Sanchez, Keith Hampson and Simon Vaux.
Description: Abingdon, Oxon; New York, NY : Routledge, 2016. | Includes bibliographical references and index.
Identifiers: LCCN 2015040370 (print) | LCCN 2015048209 (ebook) | ISBN 9781138118997 (pbk. : alk. paper) | ISBN 9781315652474 (ebk) | ISBN 9781315652474 () Subjects: LCSH: Building information modeling. | Construction industry–Cost effectiveness. | Economic value added.
Classification: LCC TH438.13 .D45 2016 (print) | LCC TH438.13 (ebook) | DDC 690.068/4–dc23
LC record available at http://lccn.loc.gov/2015040370

ISBN: 978-1-138-11899-7 (pbk)
ISBN: 978-1-315-65247-4 (ebk)

Typeset in Sabon
by Out of House Publishing

Contents

Figures

Tables

Equations

Notes on contributors

Adriana X. Sanchez, MSc
Research Associate, Sustainable Built Environment National Research Centre, Curtin University
Perth, Australia

Adriana holds an MSc in Sustainable Resource Management with honours from the Technische Universität München (TUM). She has been working at the Australian Sustainable Built Environment National Research Centre (SBEnrc) since January 2012.

In this role she has carried out research about different aspects of implementing Building Information Modelling in Australia, such as developing a national strategy, reducing the skills gap, modifying procurement practices and realising benefits. Adriana has also acted as scientific secretary to the International Council for Research and Innovation in Building and Construction (CIB) Task Group 85: R&D Investment and Impact and currently co-coordinates the CIB TG90: Information Integration in Construction.

Prior to this role, Adriana carried out sustainability-related research in different locations around the world.

Dr Keith D. Hampson
CEO, Sustainable Built Environment National Research Centre, Curtin University
Perth, Australia

Keith has more than 30 years of industry, government and research leadership. He has a Bachelor of Engineering with Honours from Queensland University of Technology (QUT), an MBA, and a PhD from Stanford University focusing on innovation and business performance. He is a fellow of the Institution of Engineers Australia, the Australian Institute of Company Directors and the Australian Institute of Management.

Keith serves as CEO of the Sustainable Built Environment National Research Centre, successor to the Australian CRC for Construction

Innovation, for which he led the bid team in 2000 and was CEO for its nine years of operation.

As Professor of Construction Management at Curtin University, he continues to work collaboratively with colleagues across Australia and globally to transform industry performance in sustainability, safety and productivity for a stronger and more competitive industry.

Simon Vaux
Principal Engineering Manager, Transport for NSW
Sydney, Australia

Simon is currently an acting Principal Engineering Manager with Transport for New South Wales (TfNSW), and chair of the TfNSW BIM Working Group. He is a chartered civil engineer with 14 years of experience in managing the delivery of transport infrastructure, having worked in both the private and public sectors in Australia and the UK.

In 2014, Simon led the development of the TfNSW BIM Roadmap, which outlines the long-term strategy for BIM adoption across the Transport cluster. He is also currently leading the first 5D BIM pilot study for TfNSW, testing the application of BIM processes and technologies during the delivery phase of rail infrastructure projects. Simon is a member of the Institution of Civil Engineers (ICE) NSW committee and an active promoter of BIM and civil engineering.

Dr Judy A. Kraatz
Senior Research Fellow, Sustainable Built Environment National Research Centre, Griffith University
Brisbane, Australia

Judy is a Senior Research Fellow working with the Australian Sustainable Built Environment National Research Centre, currently investigating better options for the delivery of social housing in Australia. Previous research with SBEnrc has focused on improving R&D outcomes for the construction industry, especially with regards to gaining productivity benefits from Building Information Modelling systems and approaches.

With more than 25 years as a registered architect, Judy has led the delivery of city-wide solutions for public buildings and open places and integrated sustainability into university curriculum, regional initiatives, and design and business practice. In 2009, Judy completed a PhD (Urban Development) at the Queensland University of Technology addressing issues of corporate responsibility in the delivery of major economic infrastructure projects.

As Group Manager, Architecture, with Brisbane City Council (2001–2005) Judy led a team of up to 40 design professionals delivering urban and social infrastructure across the city. This followed three years as the Program Director for building courses (building design, survey, construction and project management) being delivered in Australia and South-East Asia through the Central Queensland University. Prior to that time, Judy had several roles in the Commonwealth government's building procurement groups, as design architect, change agent and senior manager.

Daniel Månsson, MSc
MPhil Candidate, Curtin University
Perth, Australia

Daniel Månsson holds an MSc in Design and Construction Project Management at Chalmers University of Technology. He is currently pursuing an MPhil by research at Curtin University in Perth, Australia, with a focus on BIM management. Daniel is active in both academia and the industry with a strong interest in innovation-based start-up companies related to the built environment and support functions around it. Daniel's main research area is focusing around processes and management aspects related to innovations in the built environment. Daniel has also been involved with IREC, an international real estate competition for students taking place in Berlin every January, as well as several other industry/education collaborative projects.

Dr Göran Lindahl
Associate Professor, Chalmers University of Technology, Civil and Environmental Engineering
Gothenburg, Sweden

Göran Lindahl is an Associate Professor at the Division of Construction Management at Chalmers. He is also an Adjunct Associate Professor at the Tampere Technical University. Göran is active in both research and education. His focus is primarily on project management and facilities management, often from a client perspective. Topics related to his research are organisation, management and design processes, including issues of participation and stakeholder management. Goran's particular focus in research is on planning of workplaces and healthcare facilities and issues related to usability as well as project development of real estate. He is also active in research related to energy issues and the built environment. In education, his focus is on project management, facilities management, real estate development and governance in and of the construction sector. In addition to his academic work, he also has a background as a specialist in public client organisations and project management consultant.

Dr Sherif Mohamed
Professor, Griffith University, School of Engineering
Gold Coast, Australia

Sherif Mohamed's principal research interests lie in the area of project and construction management. He focuses on the development of theoretical knowledge and operational tools needed for effective management of risk and uncertainty in the context of decision-making, process improvement and project feasibility.

Sherif has developed a particular interest in applying his research skills to the challenges associated with civil infrastructure systems subject to extreme weather events. His ongoing research investigates the economic dimension of climate change at different levels (i.e., system, organisation, industry and region).

He has authored and co-authored more than 150 refereed journal and conference publications, and has supervised 16 postgraduate students to successful PhD completion in the past 14 years. He has also acted as examiner of more than ten PhD theses in different Australian and international universities. He currently holds the Associate Editor role with *Journal of Safety Science*, Elsevier.

Chris Linning
Manager, Building Information, Sydney Opera House
Sydney, Australia

Chris has more than 40 years of industry experience, starting as a cartography manager and later becoming a CAD and IT Manager for a Mineral Exploration company in the pacific basin. Chris has been working at the Sydney Opera House since 2000 and as Manager of Building Information he has been actively involved in and led the development of this institution's BIM journey and implementation.

Chris is often invited to industry conferences to share his extensive knowledge on the development and implementation of BIM for Facility and Asset Management. He has actively collaborated in research into practice efforts with research organisations such as the CRC for Construction Innovation since 2004, as well as carried out his own research for internal purposes.

Will Joske
BIM Development Manager, BIM Academy
Melbourne, Australia

Will Joske is the BIM Development Manager for BIM Academy ANZ with affiliations to Swinburne University and i2C Design & Management; a BIM-enabled architectural design company with offices in Melbourne,

Sydney and Perth. BIM Academy is also a member of the Ryder Alliance; an international move towards integrated multidisciplinary working, with BIM being the catalyst.

Will is a registered architect with 20 years' experience in the industry. Throughout his career, Will has had a passion for learning and a passion for how computers and software are able to not only enhance the practice of architecture, but be a vehicle for innovation and problem solving.

Will's involvement in BIM over the past decade has followed a number of parallel streams. First, as an agent of change, he has worked within organisations to align culture and business practices to derive benefit from BIM. Second, he has led and managed various tasks needed to produce BIM outputs to deliver value to organisations, including content, standards, training and research. Lastly, Will remains a hands-on practitioner and Revit expert who continues to focus on what is possible, practical and delivers results.

With a background including building design, tertiary lectureship, and business owner and visualisation expert, Will is also a committee member of MelBIM, a free, bi-monthly event in Melbourne with guest speakers presenting topics with a broad focus on BIM across the industry.

Preface

It is often said that advanced information and communications technology has the potential to disrupt and transform industries. Building Information Modelling (BIM) is at the forefront of a new digital transformation, promising a step-change in efficiency and productivity. BIM as a socio-technical system may bring about a paradigm shift to the built environment as a whole, driving change in the construction industry, and creating new roles across the supply chain filled by new players and early adopters. Around the globe, BIM uptake is now accelerating as industry and governments are starting to demand and even mandate BIM deliverables. In light of this, however, there is little information on how to realise and monitor the benefits of BIM implementation throughout the life-cycle of a built environment asset.

The benefits from BIM can act as the focal point for the development of metrics and action plans that expand across the life-cycle of an asset. In the past, implementation of these kinds of systems has often been considered only when writing the initial business case, with any ongoing management of benefits rapidly being side-lined. The methodology provided in this book can inform existing strategic management methods and practices to ensure that benefits from BIM are realised. This book provides not only the context and methodology to do this but also a starting point based on evidence from experience reported across the world.

We provide an evidence-based, practical framework for realising the value of BIM through an adaption of Benefit Realisation Management Theory. This presents an approach for practitioners aiming to implement BIM across the life-cycle of built environment assets, including buildings and infrastructure. Additionally, the book features:

- general information about BIM, the challenges of monitoring progress towards benefit goals and the greater context of implementation;
- a set of dictionaries that clearly illustrate how benefits can be achieved, what the benefit flows are and the enabling tools and processes required to achieve them;

- a suite of measures that can serve to monitor progress with examples of how they have been used to measure benefits from BIM;
- real-world case studies that showcase how these benefits can be achieved; and
- information on international maturity and competency measures to complement the value realisation framework.

It is hoped that this book can help public and private sector managers improve their approach to implementing BIM and monitoring their progress towards fully integrated whole-of-life digital asset management systems.

This book is organised in two parts:

- **Part I: Theory and practice** presents the broad context for this book through key definitions and theories, as well as case study examples, industry enablers, performance metrics and the BIM realisation management *how-to* framework.
- **Part II: Dictionaries** is formed by three dictionaries. The *Benefits dictionary* outlines some of the main benefits found in the relevant literature. The *Enablers dictionary* includes definitions of the tools and processes mentioned in the *Benefits dictionary* as enabling specific benefits. The *Metrics dictionary* provides an overview of metrics that can be associated with the benefits outlined in the *Benefits dictionary*.

Adriana X. Sanchez, Keith D. Hampson and Simon Vaux

Acknowledgements

The editors wish to thank all those who have made this publication possible through their contributions and support.

We firstly wish to thank our national and international group of authors who have contributed to the chapters and generously shared the outcomes of many years of research experience in this field. These contributions have been crucial, and without them this book would not have been possible. We would also like to expressly thank those who provided an early review of this book's intent.

The editors received both encouragement and financial support from the Australian Sustainable Built Environment National Research Centre (SBEnrc) and Transport for NSW. Without support from these organisations, together with our global innovation networks, this publication would not have been realised.

Thanks are also extended to representatives of industry associations: Australian Civil Contractors Federation (CCF), Australian Institute of Building (AIB), buildingSMART Australasia, Civil Contractors New Zealand (CCNZ), Engineers Australia (EA), International Facility Management Association (IFMA), NATSPEC and Spatial Industries Business Association (SIBA) as well as representatives of SBEnrc's core members: Aurecon, Curtin University, John Holland, Government of Western Australia, Griffith University, New South Wales Roads and Maritime Services, Queensland Government and Swinburne University of Technology. Construction Skills Queensland also contributed to this initiative as a project partner. These organisations provided guidance and insights to the editors throughout the research activity (primarily through SBEnrc Project 2.34 – Driving Whole-of-life Efficiencies through BIM and Procurement) that formed the genesis of this publication.

Disclaimer

Acronyms

ACIF	Australian Construction Industry Forum
AEC	architecture, engineering and construction
AECO	architecture, engineering, construction and owner-operated/operations
AM	asset management
APCC	Australasian Procurement and Construction Council
API	application programming interfaces
AR	augmented reality
ASHRAE	American Society of Heating, Refrigerating, and Air-Conditioning Engineers
ASTM	American Society for Testing and Materials
BDN	benefits dependency network
BEAM	Building Environmental Assessment Method
BEIIC	Built Environment Industry Innovation Council
BEP	BIM Execution Plan
BIM	Building Information Modelling
BIM(M)	Building Information Modelling and Management
BIMCAT	BIM Competency Assessment Tool
BIMe	BIM Excellence (a commercial online platform for performance assessment)
BIS	UK Department of Business, Innovation and Skills
BMP	BIM Management Plan
BREEAM	Building Research Establishment Environmental Assessment Methodology
BrIM	Bridge Information Modelling
BRM	Benefit Realisation Management
BSI	British Standards Institution
CAAM	computer-aided asset management
CAD	computer-assisted drafting
CAFM	computer-aided facility management
CAVT	computer advanced visualisation tools
CIC	computer integrated construction
CIFE	Center for Integrated Facility Engineering

CII Construction Industry Institute
CMM Capability Maturity Model
CMMS computerised maintenance management system
CNC computer numerical control
CRC Cooperative Research Centre
CRC CI Cooperative Research Centre for Construction Innovation
CRV current replacement value
CSC Construction Specifications Canada
CSI Construction Standards Institute
CURT Construction Users Roundtable
DTI Department of Trade and Industry
EMR experience modification rating
EOI expression of interest
EOT extension of time
ERM exchange requirement model
ERP Enterprise Resource Planning Systems
EUL equipment asset useful life
EUPPD European Union Public Procurement Directive
FF&E furniture, fixtures and equipment
FM facilities management
FMA Facility Management Association of Australia
GDP gross domestic product
GIS geographic information systems
GPS global positioning system
GSA US General Services Administration
HSE Health and Safety Environment
HVAC heating, ventilating, and air conditioning
ICC International Code Council
I-CMM Interactive Capability Maturity Model
IDM Information Delivery Manuals
IFC Industry Foundation Classes
IFR International Roads Federation
IPD integrated project delivery
IPMA International Project Management Association
IRF International Roads Federation
IT information technology
IU Indiana University
KM knowledge management
KPI key performance indicators
KPO key performance outcomes
LEED Leadership in Energy and Environmental Design
MEP Mechanical, Electrical and Plumbing
MMP Model Management Plan
NBCC National Building Code of Canada

NBIMS	United States National Building Information Modelling Standard
NBS	National Building Specification
NIBS	National Institute of Building Science
NIST	US National Institute of Standards and Technology
OCPM	online collaboration and project management
OMM	operating and maintenance manuals
PMI	Project Management Institute
PPC	planned percentage complete
QA	quality assurance
QC	quality control
RFI	request for information
RFID	radio-frequency identification
RIR	recordable/reportable incident rate
ROI	return on investment
RP	rapid prototyping
SBEnrc	Sustainable Built Environment National Research Centre
SMEs	small and medium-sized enterprise
SOH	Sydney Opera House
SSOT	Single Source of Truth
TNO	Netherlands Organisation for Applied Scientific Research
UWB	ultra-wide band
VDC	virtual design and construction
VTT	Technical Research Centre of Finland
WAE	work-as-executed
WLAN	Wireless LAN

Part I

Theory and practice

1 BIM, asset management and metrics

Adriana X. Sanchez and Keith D. Hampson

Introduction

> The global financial recession... has highlighted the importance of improving productivity and finding new ways to do business in the industry, and BIM has contributed to that dialogue by supporting efforts to collaborate and use strategies like prefabrication, as well as to reduce the inefficiencies that continue to plague the design and construction industries.
>
> (McGraw Hill Construction, 2014b)

The construction industry worldwide has been facing significant external pressures such as eroding profit margins, higher owner expectations, rapidly changing technology and a dwindling workforce (Roper and McLin, 2005). Building Information Modelling (BIM) has been identified as a socio-technical system 'that can be used to improve team communication throughout the project life-cycle, produce better outcomes, reduce rework, lower risk, provide better predictability of outcomes and improve operation and maintenance of an asset' (Sanchez et al., 2014b). This has led countries across the globe to start moving towards the implementation of BIM.

In Australia, the Australian Productivity Commission recently highlighted that a more widespread adoption of BIM could enhance productivity across the industry and in turn have a positive impact on the cost structure of infrastructure projects (Australian Government Productivity Commission, 2014). The Commission's findings reinforced the recommendations of an earlier Australian visioning initiative that promoted the use of advanced ICT and virtual prototyping for design, manufacture and operation of constructed facilities (Hampson and Brandon, 2004). In the UK, the government identified construction as an enabling sector for their industry strategy and committed to become a world leader in BIM by: (1) committing to the Department for Business Innovation and Skills (BIS) BIM Programme; (2) aim for growth; and (3) help create their future by continually developing their capabilities (HM Government, 2012). In New Zealand, the BIM Acceleration Committee was established in 2014 with a total initial funding of NZ$250,000 over three years. This committee is based on an alliance between government and industry and aims to coordinate efforts across government, industry and research to increase the use of BIM (Building Performance, 2015; Productivity Partnership, 2014). In Hong Kong, the Construction Industry Council issued a BIM Roadmap in 2014 (HKCIC, 2014) and the Housing Authority has been piloting BIM since 2006 and intends to implement BIM across all its construction projects as of 2015 (HKHA, 2015; Wong and Kuan, 2014). In Singapore, the Building and Construction Authority (BCA) issued a nationwide BIM roadmap in 2010 (Das et al., 2011) and mandated BIM in all new buildings projects larger than 5,000 square metres as of 2015 (BCA, 2013).

In the Nordic Region, Finland was one of the pioneers in this area. The RATAS project (which stands for computer-aided design and buildings) originated from discussions in 1982 about the need to integrate information technology (IT) applications in construction. This was part of a coordinated research, development and standardisation effort to bring computer-integrated construction to Finland (Björk, 1993). This project identified BIM as the central issue in using IT for a more efficient construction industry and brought together most of the Finnish industry key players to develop a roadmap (Björk, 2009). Nowadays, Finland requires the use of BIM for government procurement (Mitchell et al., 2012) and is seen as one of the BIM leaders of Europe (RYM Oy, 2014).

Sweden has followed in the steps of Finland and also initiated concerted efforts to increase a nationwide implementation of BIM. This led to the launch of the non-profit organisation OpenBIM (now BIM Alliance) in 2009 to establish BIM standards in Sweden. Public organisations such as the Swedish Transport Administration also mandated the use of BIM from 2015 (Trafikverket, 2013) as part of their nationwide efficiency programme (Albertsson and Nordqvist, 2013).

In general, there is a great deal of anecdotal and qualitative evidence regarding overall benefits from BIM with some contractors and designer firms stating that they use BIM even if not required by the client as a risk management strategy (Gilligan and Kunz, 2007). However, although some firms are measuring some benefits from using BIM (McGraw Hill Construction, 2014a, 2014b), it is unclear whether they have quantitative benefit measurement that can capture all potential benefits. Unclear business value and return on investment (ROI) have additionally been often identified as a barrier for adoption (Barlish and Sullivan, 2012).

Academic literature acknowledges that identifying, monitoring and managing benefits throughout the life-cycle of a project or asset is a way to ensure success during implementation of new technologies (Yates et al., 2009). Outlining the way in which each benefit will be measured and providing evidence for expected levels of improvement that will result from changes provide a basis for the development of rigorous and realistic business cases and financial arguments for investment (Ward et al., 2007). Capturing and disseminating information to ensure intelligent decision-making can also help reduce risk and deal with the large number of variables characteristic of construction projects (Roper and McLin, 2005).

Additionally, the already fast pace of technology and process development is expected to continue to increase in speed. For example, the report *Built Environment 2050* published by the Construction Industry Council in the UK provides an outlook of the next 35 years and expected evolution of BIM into a digital era (Philp and Thompson, 2014). This socio-technological frontier includes milestones such as self-assembly, industrial 3D printing, autonomous vehicles and advanced robotics. All of these advances will significantly increase productivity and reduce cost, but will also require a high level of digitisation of information and integrated systems.

Within this context, proactively establishing quality improvement cycles based on standardised work processes and corresponding measures of effectiveness will ensure better project outcomes, driven by continuously improving systems and organisational knowledge and understanding. Metrics play a critical role in driving this process (CURT, 2005). There is a great deal of literature on BIM adoption and benefits for specific applications and stakeholders (Bryde et al., 2013; Arayici et al., 2011; Migilinskas et al., 2013; Eadie et al., 2013; Azhar and Brown, 2009; Kasprzak and Dubler, 2012; Teichholz, 2013). However, there is a lack of comprehensive studies that focus on mapping and measuring the benefits of implementing BIM across the whole-of-life of built assets. This book aims to help fill this gap and provide a framework for buildings and infrastructure assets to assess the actual benefits of implementing BIM throughout planning, delivery and management.

Building Information Modelling

> BIM is a verb to describe tools, processes and technologies that are facilitated by digital, machine-readable documentation about a building, its performance, its planning, its construction and later its operation. Therefore BIM describes an activity, not an object.
>
> (Shou et al., 2015)

BIM is often defined by international standards as 'shared digital representation of physical and functional characteristics of any built object [...] which forms a reliable basis for decisions' (Volk et al., 2014). However, BIM can be much more than that. For example, in this book the authors acknowledge that the term 'BIM' also includes a set of interacting policies, processes and technologies generating a 'methodology to manage the essential building design and project data in digital format throughout the building's life-cycle' (Succar, 2009). Mature BIM is a socio-technical system that extends to emerging technological and process changes within the architecture, engineering, construction and operations industry.

In the broadest sense, BIM can also be described as a way of working that:

- allows the development of a strategy for project design, construction, and management 'based on the computer-aided modelling and simulation technologies of the object and its development processes';
- ensures 'the integrated management of graphical and information data flows combined with descriptions of process, all this performing under the integrated software environment';
- transforms 'individual executors into teams and decentralised tools into complex solutions, to integrate individual tasks into processes'; and
- allows performing 'life-cycle operations of a construction project faster, more effective, and with lower costs' (Migilinskas et al., 2013).

At the most basic level, however, a BIM model is characterised by a three-dimensional representation of an asset based on objects that include information about the object beyond the graphical representation (CRC for Construction Innovation, 2009). Here the term 'BIM' refers to a 3D design and modelling technology and database that provides enduring and transferrable digital information for the design, construction, management, logistics and material requirements of built environment assets (BEIIC, 2012). It can also support the use of *4D models*, which include scheduling functions allowing *just-in-time* delivery of information, materials, parts, assemblies and required equipment and resources (3xPT Strategy Group, 2007). The model can be further leveraged to include other layers of information leading to for example *5D models* (includes cost) and *6D models* (includes operations management). BIM therefore promises the ability to create models that combine data that was traditionally spread across multiple documents and databases along with the ability to share information between different models for the production of superior design solutions (McGraw Hill Construction, 2008).

Thus, the use of BIM promotes clearer, more accurate, up-to-date communication by consolidating currently disparate project information. It allows all team members to contribute to the establishment and population of the databases underpinning the planning, design, construction and operation of the asset (APCC and ACIF, 2009).

In asset management, having a data rich model where all components are objectified and have properties and relationships attached to them offers a plethora of opportunities for efficiency gains. For example, information generated automatically as the design model is created can be used for cost estimating, project planning and control, and sustainability and general asset management (Kivits and Furneaux, 2013).

From the procurement point of view, BIM offers the potential for streamlining processes to increase efficiencies. Decision-making can also be made easier by having access to an integrated system that makes apparent abstract ideas as well as more concrete issue such as cost and materials (BSI and buildingSMART, 2010). As it can be observed in Figure 1.1, BIM has the power to encompass numerous functions, from checking planning regulations, through design, fabrication and construction to operations and asset management right through to decommissioning and demolition; 'it can serve a project from cradle to grave' (BSI and buildingSMART, 2010). This figure tries to visually represent the fact that BIM can be the single point of truth that is at the centre of planning and delivering all phases of the life-cycle of an asset, to which all stakeholders can contribute to and rip benefits from, and where all the project functions can be based on.

BIM can therefore be defined as 'a virtual process that encompasses all aspects, disciplines and systems' of an asset within a single digital model, allowing all 'to collaborate more accurately and efficiently than using traditional processes' (Azhar, 2011). BIM can thus be used to carry out a wide range of procedures and processes. For example, Pennsylvania State University's (Penn State) computer integrated construction (CIC) research programme has identified at least 25 BIM uses across the planning, designing, construction and operation phases (Liu et al., 2013).

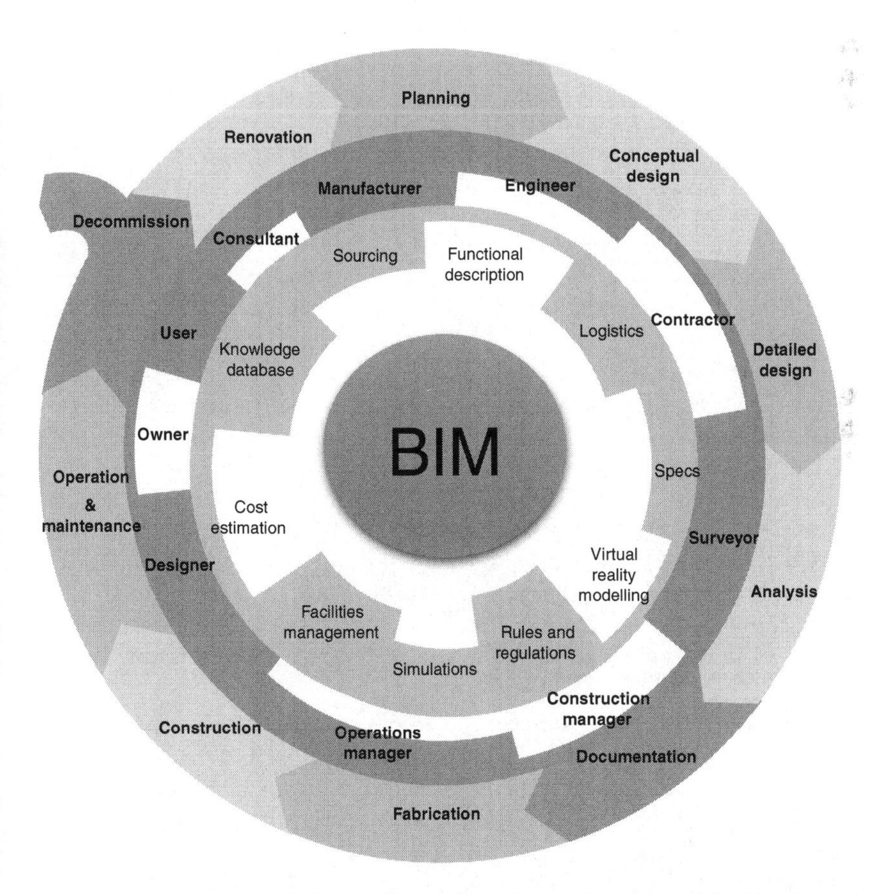

Figure 1.1 Overview based on fully functional BIM by BSI and buildingSMART (2010b)

BIM and procurement

Procurement plays a strong role throughout the life-cycle of a construction project and serves to drive many sustainability outcomes (Hardy, 2013). When referring to procurement, this publication will make use of a somewhat broader definition used in other evaluation frameworks such as that developed by Sanchez et al. (2014c). Under this concept, procurement starts at the strategic planning phase and ends at the project implementation phase, which includes *post-project* activities. Given that about 80 per cent of the asset's cost is commonly incurred during the operational phase (buildingSMART, 2010b), this publication further extends this concept to include operations and decommissioning, providing a whole-of-life costing and asset-management view of procurement.

Government procurement, in particular, often represents a significant share of national GDP, accounting for up to 20 per cent for some countries (Garcia-Alonso and Levine, 2008). Therefore, in industry sectors where government entities commonly constitute the largest client, government procurement practices have a significant impact on the industry (Sanchez et al., 2014c). In Australia, for example, a report issued by Engineers Australia suggested that changes to the public procurement models can have a significant impact on the national budget. They estimate that a 1 per cent cost improvement could generate more than AU$600 million in savings (Engineers Australia, 2013).

BIM can improve procurement processes such as: approval, design, specification and documentation, and tendering, appointment and contract management stages of a project. It can do so by 'increasing data integration and information sharing, as well as reducing design and documentation shortcomings' (Allen Consulting Group, 2010).

There is a close link between procurement and achieving benefits from the implementation of BIM that can lead to a more efficient whole-of-life asset management process. This has led a number of countries to develop procurement frameworks and guidelines that facilitate the adoption of BIM and support the tender and monitoring process (Porwal and Hewage, 2013; NATSPEC, 2011). Chapter 2 will explore the role that agencies delivering public assets can play as industry leaders and agents of change.

BIM for asset management

Numerous reports have highlighted the significant benefits to be gained from implementing BIM in facilities management (FM) (CRC for Construction Innovation, 2007a, 2007b; Allen Consulting Group, 2010; Kivits and Furneaux, 2013; Azhar et al., 2012). FM can be described as 'a business practice that optimises people, process, assets, and the work environment to support delivery of the organisation's business objectives' (Kivits and Furneaux, 2013). However, this concept sometimes excludes other built assets such as infrastructure assets, and efforts to increase the value of such facilities or assets.

The British Standards Institution defines an asset as:

> An item, thing or entity that has potential or actual value to an organization… An asset may be fixed, mobile or movable. It may be an individual item of plant, a system of connected equipment, a space within a structure, a piece of land, or an entire piece of infrastructure or an entire building or portfolio of assets.
>
> (BSI, 2014)

In the context of this book, assets will refer to the physical entities that form the built environment and asset management (AM) refers to the coordinated activity of an organisation to realise value from assets. This value can be tangible, intangible, financial or non-financial, and may vary throughout its life (BSI, 2014). Within this framework, FM is considered part of AM and therefore benefits and uses of BIM for FM are extended to AM as well.

It has also been highlighted by the US General Services Administration (GSA) that FM activities, and therefore AM:

> depend on the accuracy and accessibility of facility data created in the facilities' design and construction phases and maintained throughout the operations and maintenance phase. Lack of this information can result in cost overruns, inefficient building operations, and untimely resolution of client requests.
>
> (GSA, 2011)

BIM has been proposed to be used in three main areas within AM: tracking performance; maintenance; and emergency management (Xu et al., 2014). However, renovation and extension of current assets are also an area where BIM can provide considerable benefits (CRC for Construction Innovation, 2007a). Profiles of benefits, enablers and metrics included in Part II are based on this understanding of AM.

Metrics

> A metric is a recorded piece of data, often captured by a computer system… a measure is a meaningful computation derived from an analysis of the metric, where meaningful means that it is worth reporting to check progress towards the realisation of a benefit and so inform some decision making… the simplest computation is where the measure equals the metric.
>
> (Bradley, 2010)

Some of the main purposes of evaluating benefits are to: justify investments made, compare and rank benefits, provide targets for success and create benchmarks (Bradley, 2010; Costa et al., 2006). BIM has promised many benefits in relation to increasing productivity. However, previous literature has often failed to provide a transferrable, consistent and robust set of metrics that can objectively measure and quantify this productivity improvement, and other tangible and intangible benefits (buildingSMART, 2010a; Becerik-Gerber and Rice, 2010; Tsai et al., 2014; Sebastian and van Berlo, 2010). The benefits and failures of BIM need to be measured throughout the life-cycle of the asset. This guarantees a continual improvement process (Eadie et al., 2013) and supports adoption by new users that depends on how the real benefits of the transition are perceived (Lu et al., 2012). Measuring 'productivity is necessary for assessing if BIM systems provide tangible benefits, and if so, for documenting benefits so as to support effective adoption and implementation' (Sacks et al., 2005).

Although anecdotal evidence has pointed to significant benefits to be gained, the lack of a standard methodology and definition of metrics for BIM benefits represents a challenge for both industry and research to draw conclusions from previous case studies and pilot projects (buildingSMART, 2010a). As a consequence, these types of technologies are often implemented 'as an act of faith, without a full understanding of how business values from investment can be shown' (Becerik-Gerber and Rice, 2010). Nevertheless, without such formal metrics, teams and organisations are unable to consistently measure their own success or failure, and speculation and improper estimation of its benefits are encouraged (Barlish and Sullivan, 2012; Succar et al., 2012). Thus, organisations often have to make the decision of whether or not to utilise BIM based on speculated benefits (Barlish, 2011).

This lack of agreed and readily available metrics for assessing the benefits of BIM implementation has been largely acknowledged in previous research (Sanchez et al., 2014a) and is commonly mentioned as an issue to make the business case for BIM (Barlish and Sullivan, 2012). Furthermore, the use of quantitative metrics related to project process performance, their frequent reporting to the project team, and use in management, contributes towards the success of projects. This is due to the fact that 'achieving these process objectives makes it more likely that projects will reach aggressive overall project objectives' (Kunz and Fischer, 2012).

Management practices that include the use of innovative tools to track and monitor the management process are correlated with higher productivity, profitability and sales growth rates (Kam et al., 2014). The US General Services Administration (GSA), for example, has taken this into consideration and now requires project teams to determine the metrics to measure the success of BIM. These 'stem from the business needs and evaluate how implementation of a given technology provides a value-added service' (GSA, 2007).

Some international reports have reported BIM metrics but rarely include all stakeholders and phases in the life of the asset. For example, McGraw Hill has released a series of SmartMarket Reports on the value of BIM, its global uptake and associated metrics since 2008 (McGraw Hill Construction, 2008, 2009). It recently released *The Business Value of BIM in Australia and New Zealand* (McGraw Hill Construction, 2014b), which provides valuable documentation on the level of use and return on investment for BIM use in this country. However, the surveyed base does not include clients who, as shown in the US, stand to gain the most from implementation (Gallaher et al., 2004). This report also highlights that two-thirds of the surveyed industry is developing and implementing metrics to monitor their ROI, such as process-related metrics (e.g., fewer RFIs and unplanned changes); project schedule metrics (e.g., faster project delivery); and financial metrics (e.g., reduced cost, higher profitability and productivity) (McGraw Hill Construction, 2014b). However, it is unclear whether these metrics would be standard for the industry and could form the basis of industry benchmarks.

There is also a lack of surveys that cover those metrics used by the transport infrastructure industry across the supply chain. For example, in the McGraw Hill (2012) global report that focused on infrastructure, only 13 per cent of the respondents reported working in the transport infrastructure industry and *other*. The *Metrics dictionary* in Part II of this book provides an extensive suite of metrics that can be applied to measure benefits from BIM.

Types of metrics

There are a number of ways of classifying metrics. For example, Pike and Roos (2011) make the distinction between *proper measurement systems* and *indicators*. In this case, indicators are a less rigorous approach that monitor changes as they occur, but cannot be used for decision-making. Indicators can then be divided into leading and lagging where the lagging indicator measures project outcomes and the leading indicator predicts the improvement opportunities (Tsai et al., 2014).

Similarly, indicators can also be classified based on their intended use; that is between key performance outcomes (KPO) and key performance indicators (KPI). KPOs are lagging measures of performance determined at the completion of a process (Rankin et al., 2008). KPIs are often developed to capture the business drivers, to measure and monitor performance (CRC for Construction Innovation, 2007b) and are 'compilations of data measures used to assess the performance of a construction operation' (Cox et al., 2003).

It has been argued that KPIs used for measuring the benefits of information technologies and systems investment: (1) enable stakeholders to assess whether the planned benefits have been delivered; (2) identify the project benefits to measure, and when to measure them; (3) facilitate action based on KPIs measurements; (4) clearly link accountability to measured benefits; and (5) assist the project in being funded (Smith et al., 2008).

At the organisational level, there are also performance factors and performance measures. Performance factors can be either performance driving factors or performance result factors, where the first measures how well the organisation performs, and later measure how much the organisation has achieved (Tsai et al., 2014). Kagioglou et al. (2001) suggest three classifications for performance measures: financial perspective; internal business process perspective (*how are we performing in our key process activities?*); and customer perspective.

Metrics can also be classified in relation to the process level of detail to which they apply; for example, if they are to be used for projects, organisations, industry-wide or even whole economies (Rankin et al., 2008). Metrics can for example be divided into *micro-measures* and *macro-measures*, where the former 'are measures at the individual project level that compare actual project results with expected results as defined in specific project goals and objectives'. Macro-measures then compare and analyse results on a broader scale and are often used for benchmarking (CURT, 2005).

Another parameter for classification is the nature of data collected to calculate a metric. In such cases, metrics are defined as quantitative or qualitative (Rankin et al., 2008). Qualitative metrics can be defined as those indicators that have the potential for measuring behaviours (Cox et al., 2003). Although quantitative metrics are more commonly used and accepted by the construction industry, qualitative measures also provide valuable information for the evaluation of quantitative performance (Kam et al., 2014). Examples of quantitative metrics often suggested by the literature are: units/man-hours, dollars/unit, cost, on-time completion, resource management, quality control, percentage complete, earned man-hours, lost time accounting and punch list (Barlish and Sullivan, 2012).

There is also the distinction between value and cost-based metrics. For example, energy use can be measured on a cost basis (btu/m^2) or based on value (btu/occupant-hour or btu/transaction). The choice might depend on the objective of the measurement. For example, value-based energy measures might provide a much more insightful understanding of the business performance of an asset than cost-based measures (Kunz and Fischer, 2012). Organisations also often develop their own metric classifications. For example, the Construction Industry Institute (CII) in the US has 19 types of metrics (Kang et al., 2008).

This diversity of metric classifications provides some insight into the many ways data can be categorised and may also offer some clues on why industries often find it difficult to create standard metrics.

Development, selection and implementation

Developed metrics should be: relevant; accurate, unambiguous and incorruptible; applicable and inexpensive to track; attainable; consistent and in the appropriate format; cumulative; flexible; informative; neutral; specific; universal; predictable; timely; and, above all, usable (intuitive and able to be easily employed) (Bradley, 2010; Succar et al., 2012). Good metrics should also motivate behaviour that contributes to success, meet the needs of relevant stakeholders and support or at least not undermine the vision or end-goal (Bradley, 2010). Metrics used to form the business case for BIM also need to include vocabulary that is relevant to the upper management (Barlish and Sullivan, 2012).

When used to measure benefits across projects, metrics should:

- focus on critical aspects of outputs or outcomes;
- be maintainable for regular and systematic use;
- be based on a large sample size in order to reduce the impact of project specific variables;
- be designed for use on every project;
- be accepted, understood and owned across disciplines;
- be able to evolve; and
- have a simple graphic display that is easy to update and accessible (Chan and Chan, 2004).

It is also important to determine: (1) whether the most appropriate metrics are comparison metrics (lagging key performance outcomes) or metrics for change (KPI that support improvement); (2) the process level at which they will be used; and (3) the ability to validate the results (data) (Rankin et al., 2008).

Benefits metrics are to measure the value of said benefit, where value is 'the magnitude of the improvement associated with the benefit'. This can be a forecast or predicted value, a target value, or an actual value. For example, the value of the benefit *fewer complaints* might be a 10 per cent reduction of complaints (Bradley, 2010). Additionally, each organisation should identify 'factors to control, process metrics to monitor and use in management and outcomes by which to evaluate project success' (Kunz and Fischer, 2012).

The selection and implementation of metrics should also: (1) focus on critical outcomes or aspects; (2) be formed by a manageable number of indicators; (3) data collection should be made as simple as possible; and (4) indicators must be accepted by users (rigorous consultation is recommended) (Rankin et al., 2008). These metrics should also be based on a small (2–3) set of explicit objectives for each of the following:

- controllable factors, including the BIM modelling and analysis strategy, process objectives to measure and one or two additional factors;
- measurable process performance parameters, such as schedule conformance and response latency; and
- measurable project outcome objectives, such as safety, schedule, cost and functional quality as assessed by post occupancy evaluation (Kunz and Fischer, 2012).

Finally, once the metrics have been developed and selected, they need to be validated in order to be used as effective tools for improvement (Rankin et al., 2008).

Key challenges for BIM benefits metrics development

Metrics developed to assess the return on investment from the implementation of information technologies and systems are often mostly financially based and project reviews do not consistently include assessments of the benefits delivered by the investments (Ward et al., 2007). A common argument in this respect is that the success or failure of every construction project is usually solely measured in terms of four variables: cost, time, quality, and safety (Suermann, 2009; Bassioni et al., 2005). Nevertheless, investment in information technologies, systems and associated requirements are often extremely hard to evaluate in these terms (Becerik-Gerber and Rice, 2010). Financial measures alone have been found to be inadequate to evaluate benefits from this type of technology because they solely rely on the monetary value. These types of measures therefore exclude those benefits that are intangible or *soft*, which are challenging to quantify in monetary terms (Becerik and Pollalis, 2006; Barlish and Sullivan, 2012).

This is in part due to difficulties in quantifying the relevant costs and benefits, as well as due to the high degree of uncertainty with respect to the expected technology value (Becerik-Gerber and Rice, 2010). Determining the cost of implementation is further complicated by infrastructure required by technology (e.g. hardware, servers, etc.) becoming an inextricable part of the organisations' processes and structures. This makes it increasingly difficult to separate its impact from those of other assets and activities (Becerik and Pollalis, 2006).

Using only financial measures has also been criticised because they encourage short-termism; lack broader strategic vision and fail to provide data on quality, responsiveness and flexibility; encourage local optimisation; and do not encourage continuous improvement (Kagioglou et al., 2001). Those measures also tend to be backward-focused, making it difficult to trace operational costs and failing to monitor the ongoing benefits of new investments or applications (Costa et al., 2006).

The challenges associated with measuring benefits from investment in information technology and systems are not unique to the construction industry, but instead are a global issue experienced in all types of business sectors and organisations (Becerik and Pollalis, 2006). This is additionally complicated by the fact that each construction project is different in terms of its parameters and characteristics such as financing and delivery method, inter-organisational relationships and end-user requirements. The construction industry also has a lack of established benchmarks to build on or measure against. Its fragmented supply chain and undercapitalisation further complicate the development and implementation of meaningful metrics (Becerik and Pollalis, 2006; Becerik-Gerber and Rice, 2009).

Additionally, BIM makes some already established productivity measurements less relevant, further adding to the challenges of measuring its impact. Some benefits from BIM are also inherently difficult to measure. For example, BIM is claimed to produce savings through rework avoidance, but 'who can put an accurate price on savings achieved through something that never happened?' (BSI and buildingSMART, 2010). Confidentiality barriers common in the construction industry also remain a large impediment to data collection and therefore to validating developed metrics (Becerik and Pollalis, 2006).

Barlish and Sullivan (2012) further add the following challenges: (1) organisational changes may occur as a result of the introduction of the new system; (2) business benefits are evolutionary over the life-cycle of the system; (3) diverse stakeholders involved will subjectively evaluate the system and may have conflicting opinions; (4) users may feel intimidation or fear of the new system and how it will affect their jobs negatively; and (5) practical difficulties such as improper utilisation, interconnected systems and inability to divide related systems and benefits.

Chapter 4 aims to provide a methodology that will help different stakeholders deal with these challenges while focusing on organisational and project objectives.

BIM performance metrics

> Without such metrics, teams and organizations are unable to consistently measure their own successes and/or failures. Performance metrics enable teams and organizations to assess their own competencies in using BIM and, potentially, to benchmark their progress against that of other practitioners. Furthermore, robust sets of BIM metrics lay the foundations for formal certification systems, which could be used by those procuring construction projects to pre-select BIM service providers.
>
> (ACIF, 2014)

The development of objective metrics that can be used to systematically measure performance improvement are essential to achieving a more efficient industry. Performance metrics can additionally allow projects to be benchmarked against similar projects in order 'to identify standards in the national performance of the construction industry and identify areas for improvement' (Eadie et al., 2013). These metrics enable teams and organisations to assess their own competencies and to benchmark their progress. They can also be used as basis for formal certification systems that facilitate the procurement process (Succar et al., 2012). Measuring BIM performance is important because the more advanced an organisation is in their use of BIM, the greater their ability to receive its benefits and to realise strong returns on investments (McGraw Hill Construction, 2014a).

There have been a number of sets of BIM performance metrics developed across the world. McGraw Hill Construction for example developed a *BIM Engagement Index* to 'measure the level of engagement for every contractor that participated in [their] research, based on their experience, skill and the percentage of BIM projects they work on' (McGraw Hill Construction, 2014a). Kam et al. (2014) developed a BIM [VDC] scorecard to assesses the maturity of implementation of a project across four *areas*, ten *divisions* and 56 *measures*. This method also deploys the *Confidence Level*. This is measured through seven factors that indicate the accuracy of scores (Kam et al., 2014).

The UK-based BIM Industry Working Group uses the *Maturity Level Index* to 'articulate groups of technology and processes and their inherent capabilities' (BIM Industry Working Group, 2011). This index is similar to that developed by the Cooperative Research Centre for Construction Innovation (CRC for Construction Innovation, 2009).

Succar et al. (2012) also introduced a comprehensive set of BIM metrics that is now available through BIMexcellence.[1] This framework is broader and more comprehensive than other BIM assessment methodologies available. However, it has gone through a limited validation process with actual projects, and it has been argued that it lacks a quantifiable scoring mechanism that can account for the project outcome objectives (Kam et al., 2014).

Evaluating BIM performance is considered to be an important, complementary set of metrics to those used for measuring benefits from implementing BIM and will be further explored in Chapter 3. Kam et al. (2014) also provide a useful review of currently available BIM maturity evaluation frameworks.

Conclusions

Implementing BIM in built environment asset management across the life-cycle of buildings and infrastructure promises both significant rewards and challenges. This book aims to provide practitioners with insight into different aspects of implementing BIM, including project examples and a methodology that can easily be applied to an industry setting. In summary, this and the following chapters aim to demystify BIM and provide a practical approach to defining a strategy for implementation in asset management.

Note

1 www.bimexcellence.com.

Bibliography

3xPT Strategy Group, 2007. *Integrated Project Delivery: First Principles for Owners and Teams*, s.l.: 3xPT Strategy Group.

ACIF, 2014. *Project Team Integration – The Key to Optimising BIM*. Available at: www.acif.com.au/acif-news/project-team-integration-the-key-to-optimising-bim [Accessed 5 August 2014].

Albertsson, A.-T. and Nordqvist, H., 2013. *BIM at the Swedish Transport Adminstration*, Stockholm: Trafikverket.

Allen Consulting Group, 2010. *Productivity in the Buildings Network: Assessing the Impacts of Building Information Models*, s.l.: BEIIC.

APCC and ACIF, 2009. *Integrated Project Teams and Building Information Modelling in the Australian Construction Industry. Less Time, Lower Cost, and Better Quality in Construction*, Canberra: Australian Construction Industry Forum.

Arayici, Y., Coates, P., Koskela, L., Kagioglou, M., Usher, C. and O'Reilly, K., 2011. Technology adoption in the BIM implementation for lean architectural practice. *Automation in Construction*, 20(2), pp. 189–195.

Australian Government Productivity Commission, 2014. *Public Infrastructure. Productivity Commission Inquiry Report: Volume 1*, Canberra: Commonwealth of Australia.

Azhar, S., 2011. Building information modeling (BIM): trends, benefits, risks, and challenges for the AEC industry. *Leadership and Management in Engineering*, 11(3), pp. 241–252.

Azhar, S. and Brown, J., 2009. BIM for sustainability analyses. *International Journal of Construction Education and Research*, 5(4), p. 276–292.

Azhar, S., Khalfan, M. and Maqsood, T., 2012. Building information modelling (BIM): now and beyond. *Australasian Journal of Construction Economics and Building*, 12(4), pp. 15–28.

Barlish, K., 2011. *How to Measure the Benefits of BIM: A Case Study Approach*, Phoenix, AZ: Arizona State University.

Barlish, K. and Sullivan, K., 2012. How to measure the benefits of BIM – a case study approach. *Automation in Construction*, 24, pp. 149–159.

Bassioni, H. A., Price, A. D. and Hassan, T. M., 2005. Building a conceptual framework for measuring business performance in construction: an empirical evaluation. *Construction Management and Economics*, 23(5), pp. 495–507.

BCA, 2013. *Building and Construction Authority – 2013 release*. Available at: www.bca.gov.sg/Newsroom/pr17092013_BCM.html [Accessed 30 July 2015].

Becerik, B. and Pollalis, S. N., 2006. *Computer Aided Collaboration in Managing Construction*, Cambridge: Harvard University Graduate Schoolf of Design – Design and Technology Report Series 2006-2.

Becerik-Gerber, B. and Rice, S., 2009. The value of building information modeling: can we measure the ROI of BIM? *AECbytes*, 47.

Becerik-Gerber, B. and Rice, S., 2010. The perceived value of building information modeling in the US building industry. *Journal of Information Technology in Construction*, 15(2), pp. 185–201.

BEIIC, 2012. *Final Report to the Government*, s.l.: The Built Environment Industry Innovation Council.

BIM Industry Working Group, 2011. *A Report for the Government Construction Client Group Building Information Modelling (BIM) Working Party*, London: Department of Business, Innovation and Skills.

Björk, B.-C., 1993. A case study of a national building industry strategy for computer integrated construction, in K. S. Mathur, M. P. Betts and K. W. Tham (eds.), *Management of Information Technology for Construction*, Singapore: World Scientific Publishers, pp. 85–99.

Björk, B.-C., 2009. RATAS, a longitudinal case study of an early construction IT roadmap project. *ITcon*, 14, pp. 385–399.

Bradley, G., 2010. *Benefit Realisation Management: A Practical Guide to Achieving Benefits through Change*, 2nd edn, Surrey: Gower.

Bryde, D., Broquetas, M. and Volm, J. M., 2013. The project benefits of Building Information Modelling (BIM). *International Journal of Project Management*, 31(7), pp. 971–980.

BSI, 2014. *PAS 1192–3:2014: Specification For Information Management for the Operational Phase of Assets Using Building Information Modelling*, London: British Standards Institution.

BSI and buildingSMART, 2010. *Constructing the Business Case: Building Information Modelling*, London: British Standards Institution.

Building Performance, 2015. *Building Information Modelling (BIM) in New Zealand*. Available at: www.building.govt.nz/bim-in-nz#bim-acceleration-committee [Accessed 29 September 2015].

buildingSMART, 2010a. *Investing in BIM Competence*, London: buildingSMART.

buildingSMART, 2010b. *Investors Report: Building Information Modelling (BIM)*, London: buildingSMART.

Chan, A. P. and Chan, A. P., 2004. Key performance indicators for measuring construction success. *Benchmarking: An International Journal*, 11(2), pp. 203–221.

Costa, D. B., Formoso, C. T., Kagioglou, M. and Alarcón, L. F., 2006. Benchmarking initiatives in the construction industry: lessons learned and improvement opportunities. *Journal of Management in Engineering*, 22(4), pp. 158–167.

Cox, R. F., Issa, R. R. and Ahrens, D., 2003. Management's perception of key performance indicators for construction. *Journal of Construction Engineering and Management*, 129(2), pp. 142–151.

CRC for Construction Innovation, 2007a. *Adopting BIM for Facilities Management: Solutions for Managing the Sydney Opera House*, Brisbane: Cooperative Research Centre for Construction Innovation.

CRC for Construction Innovation, 2007b. *FM as a Business Enabler*, Brisbane,: Cooperative Research Centre for Construction Innovation.

CRC for Construction Innovation, 2009. *National Guidelines for Digital Modelling*, Brisbane: Cooperative Research Centre for Construction Innovation.

CURT, 2005. *Construction Measures: Key Performance Indicators*, s.l.: Construction Users Roundtable.

Das, J., Leng, L.E., Lee, P., Kiat, T.C., Palanisamy, L., Leong, N.K., Wee, T.K., Leong, T.K., Kwang, T.W. and Jun, Z.H., 2011. All set for 2015: the BIM roadmap. *Build Smart*, December, p. 2.

Eadie, R., Browne, M., Odeyinka, H., McKeown, C. and McNiff, S., 2013. BIM implementation throughout the UK construction project lifecycle: an analysis. *Automation in Construction*, 36, pp. 145–151.

Engineers Australia, 2013. *Report Puts Spotlight on Government Procurement*. Available at: www.engineersaustralia.org.au/news/report-puts-spotlight-government-procurement [Accessed 8 July 2013].

Gallaher, M. P., O'Connor, A. C., Dettebarn, J. L. and Gilday, L. T., 2004. *Cost analysis of Inadequate Interoperability in the US Capital Facilities Industry, NIST GCR 04-867*. Available at: www.nist.gov/manuscript-publication-search.cfm?pub_id=101287 [Accessed 8 November 2012].

Garcia-Alonso, M. D. C. and Levine, P., 2008. Strategic procurement, openness and market structure, *International Journal of Industrial Organization*, 26, pp. 1180–1190.

Gilligan, B. and Kunz, J., 2007. *VDC Use in 2007: Significant Value, Dramatic Growth, and Apparent Business Opportunity*, Stanford: Center for Integrated Facility Engineering, Stanford University.

GSA, 2007. *GSA Building Information Modeling Guide Series 01 – Overview*, Washington, DC: US General Services Administration.

GSA, 2011. *GSA BIM Guide for Facility Management*, Washington, DC: US General Services Administration.

Hampson, K. D. and Brandon, P. S., 2004. *Construction 2020: A Vision for Australia's Property and Construction Industry*, Brisbane: CRC for Construction Innovation.

Hardy, R., 2013. *The Role of Procurement in Embedding Sustainability Along the Life-Cycle of a Construction Project*, paper presented at CIB World Building Congress, Brisbane, Australia, 5–9 May.

HKCIC, 2014. *Roadmap for Building Information Modelling Strategic Implementation in Hong Kong's Construction Industry*, Hong Kong: Hong Kong Construction Industry Council.

HKHA, 2015. *Hong Kong Housing Authority – Building Information Modelling*. Available at: www.housingauthority.gov.hk/en/business-partnerships/resources/building-information-modelling [Accessed 30 July 2015].

HM Government, 2012. *Building Information Modelling: Industrial Strategy – Government and Industry in Partnership*, London: UK Government.

Kagioglou, M., Cooper, R. and Aouad, G., 2001. Performance management in construction: a conceptual framework. *Construction Management and Economics*, 19(1), pp. 85–95.

Kam, C., Senaratna, D., McKinney, B., Xiao, Y. and Song, M., 2014. *The VDC Scorecard: Formulation and Validation*, Stanford: Center for Integrated Facility Engineering (CIFE), Stanford University.

Kang, Y., O'Brien, W. J., Thomas, S. and Chapman, R. E., 2008. Impact of information technologies on performance: cross study comparison. *Journal of Construction Engineering and Management*, 134(11), pp. 852–863.

Kasprzak, C. and Dubler, C., 2012. Aligning BIM with FM: streamlining the process for future projects. *Australasian Journal of Construction Economics and Building*, 12(4), pp. 68–77.

Kivits, R. A. and Furneaux, C., 2013. BIM: Enabling sustainability and asset management through knowledge management. *The Scientific World Journal*, 2013, pp. 1–14.

Kunz, J. and Fischer, M., 2012. *Virtual Design and Construction: Themes, Case Studies and Implementation Suggestions*, CIFE Working Paper #097, Stanford: Sanford University.

Liu, F., Jallow, A. K., Anumba, C. J. and Wu, D., 2013. *Building Knowledge Modelling: Integrating Knowledge in BIM*, paper presented at 30th CIB W78 International Conference, WQBook, Beijing, China, 9–12 October.

Lu, W., Peng, Y., Shen, Q. and Li, H., 2012. Generic model for measuring benefits of BIM as a learning tool in construction tasks. *Journal of Construction Engineering and Management*, 139(2), pp. 195–203.

McGraw Hill Construction, 2008. *Building Information Modeling Trends SmartMarket Report: Transforming Design and Construction*, Bedford, MA: McGraw Hill Construction.

McGraw Hill Construction, 2009. *The Business Value of BIM in Europe*, Bedford, MA: McGraw Hill Construction.

McGraw Hill Construction, 2012. *The Business Value of BIM for Infrastructure: Addressing America's Infrastructure Challenges with Collaboration and Technology SmartMarket Report*, Bedford, MA: McGraw Hill Construction.

McGraw Hill Construction, 2014a. *The Business Value of BIM for Construction in Major Global Markets: How Contractors Around the World Are Driving Innovation with Building Information Modeling*, Bedford, MA: McGraw Hill Construction.

McGraw Hill Construction, 2014b. *The Business Value of BIM in Australia and New Zealand: How Building Information Modeling is Transforming the Design and Construction Industry: SmartMarket Report*, Bedford, MA: McGraw Hill Construction.

Migilinskas, D., Popov, V., Juocevicius, V. and Ustinovichius, L., 2013. The benefits, obstacles and problems of practical BIM implementation. *Procedia Engineering*, 57, pp. 767–774.

Mitchell, J., Plume, J., Tait, M., Scuderi, P. and Eastley, W., 2012. *National Building Information Modelling Initiative. Volume 1: Strategy. Report to the Department of Industry, Innovation, Science, Research and Tertiary Education*, Sydney: buildingSMART Australasia.

NATSPEC, 2011. *NATSPEC National BIM Guide*, Sydney: Construction Information Systems Limited.

Philp, D. and Thompson, N., 2014. *Built Environment 2050: A Report on our Digital Future*, London: Construction Industry Council.

Pike, S. and Roos, G., 2011. The validity of measurement frameworks: measurement theory, in A. Neely (ed.), *Business Performance Measurement: Unifying Theory and Integrating Practice*, 2nd edn, Cambridge: Cambridge University Press, pp. 220–238.

Porwal, A. and Hewage, K. N., 2013. Building information modeling (BIM) partnering framework for public construction projects. *Automation in Construction*, 31, pp. 204–214.

Productivity Partnership, 2014. *BIM Acceleration Committee: Terms of Reference and Membership*. Available at: www.building.govt.nz/UserFiles/File/Publications/Building/Technical-reports/bac-terms-of-reference.pdf [Accessed 25 September 2015].

Rankin, J., Robinson Fayek, A., Meade, G., Haas, C. and Manseau, A., 2008. Initial metrics and pilot program results for measuring the performance of the Canadian construction industry. *Canadian Journal of Civil Engineering*, 35(9), pp. 894–907.

Roper, K., and McLin, M., 2005. *Key Performance Indicators Drive Best Practices for General Contractors*, Raleigh, US: FMI, Management Consulting, Investment Banking for the Construction Industry, Microsoft Corporation.

RYM Oy, 2014. *BuildingSMART Finland – Development Forum for Infrastructure Modeling*. Available at: http://rym.fi/buildingsmart-finland-development-forum-for-infrastructure-modeling [Accessed 25 May 2014].

Sacks, R., Eastman, C. M., Lee, G. and Orndorff, D., 2005. A target benchmark of the impact of three-dimensional parametric modeling in precast construction. *PCI Journal*, 50(4), pp. 126–138.

Sanchez, A. X., Kraatz, J. A. and Hampson, K. D., 2014a. *Research Report 1 – Towards a National Strategy*, Perth: Sustainable Built Environment National Research Centre.

Sanchez, A. X., Kraatz, J. A., Hampson, K. D. and Loganathan, S., 2014b. *BIM for Sustainable Whole-of-Life Transport Infrastructure Asset Management*, paper presented at IPWEA Sustainability in Public Works Conference, Tweed Heads, Australia, 27–29 July.

Sanchez, A. X., Lehtiranta, L. M., Hampson, K. D. and Kenley, R., 2014c. Evaluation framework for green procurement in road construction. *Smart and Sustainable Built Environment*, 3(2), pp. 153–169.

Sebastian, R. and van Berlo, L., 2010. Tool for benchmarking BIM performance of design, engineering and construction firms in the Netherlands. *Architechtural Engineering and Design Management*, 6(4), pp. 254–263.

Shou, W., Wang, J., Wang, X. and Chong, H. Y., 2015. A comparative review of building information modelling implementation in building and infrastructure industries. *Archives of Computational Methods in Engineering*, 22(2), pp. 291–308.

Smith, D. C., Dombo, H. and Nkehli, N., 2008. *Benefits Realisation Management in Information Technology Projects*, paper presented at Technology Management for a Sustainable Economy Conference, Cape Town, South Africa, 27–31 July.

Succar, B., 2009. Building information modelling framework: a research and delivery foundation for industry stakeholders. *Automation in Construction*, 18, pp. 357–375.

Succar, B., Sher, W. and Williams, A., 2012. Measuring BIM performance: five metrics. *Architectural Engineering and Design Management*, 8(2), pp. 120–142.

Suermann, P. C., 2009. *Evaluating the Impact of Building Information Modelling (BIM) on Construction*, doctoral thesis, Gainesville, US: University of Florida.

Teichholz, P., 2013. *BIM for Facility Managers*, Hoboken: Wiley.

Trafikverket, 2013. *Att införa BIM i Trafikverket (Introducing BIM in Transport Administration)*. Available at: www.trafikverket.se/Foretag/Bygga-och-underhalla/Teknik/Att-infora-BIM-pa-Trafikverket [Accessed 30 April 2014].

Tsai, M.-H., Mom, M. and Hsieh, S.-H., 2014. Developing critical success factors for the assessment of BIM technology adoption: Part I. Methodology and survey. *Journal of the Chinese Institute of Engineers*, 37(7), pp. 845–858.

Volk, R., Stengel, J. and Schultmann, F., 2014. Building information modeling (BIM) for existing buildings – literature review and future needs. *Automation in Construction*, 38, pp. 109–127.

Ward, J., De Hertogh, S. and Viaene, S., 2007. *Managing Benefits from IS/IT Investments: an Empirical Investigation into Current Practice*, paper presented at 40th, IEEE International Conference on System Sciences, Waikoloa, HI, 3–6 January.

Wong, J. K.W. and Kuan, K.L., 2014. Implementing 'BEAM Plus' for BIM-based sustainability analysis. *Automation in Construction*, 44, pp. 163–175.

Xu, X., Ma, L. and Ding, L., 2014. A framework for BIM-enabled life-cycle information management of construction project. *International Journal of Advanced Robotic Systems*, 11(126), pp. 1–13.

Yates, K., Sapountzis, S., Lou, E. and Kagioglou, M., 2009. *BeReal: Tools and Methods for Implementing Benefits Realisation and Management*. Reykjavík: Reykjavik University.

2 Leadership in implementation

Judy A. Kraatz and Adriana X. Sanchez

Introduction

Building Information Modelling (BIM) brings with it a paradigm shift in the way the industry and organisations deliver projects. This requires changes in practice standards, legal arrangements and industry norms. Effective leadership is needed to overcome the challenges brought by this change (Dossick and Neff, 2008, 2010). Such leadership is necessary at project, organisational and industry-wide levels to maximise the benefits and the return on investment.

Effective leadership at an organisational level impacts project and business success and can both guide and inspire delivery teams to collaborate and embrace the new way of working (Dossick and Neff, 2008). This leadership is often embedded in projects and organisations through key champions who map out an incremental learning process. These leaders also often draw on pilot projects to affect the technical and process innovations required to demonstrate and capture the benefits of embracing this new way of working.

At the industry-wide or national level, governments around the world have demonstrated leadership through the development of national strategies to adopt digital technologies and transform the construction industry in their country. Industry-wide collaboration is needed across complex supply chains to achieve this through the development of:

- the case for industry-wide uptake;
- appropriate work process and plans;
- national standards and contracts;
- a skills and training agenda that attracts a new wave of technical expertise to the industry (HM Government, 2012).

The following sections argue these points more extensively and present a series of case studies that showcase different types and scales of leadership for the implementation of BIM in the construction industry.

The need for regional and national leadership

Mature BIM, as both a technology and a process, can bring about wide-scale change in the industrial landscape, including a change of current roles and the creation of new ones. These changes may impact the economic performance of the industry as whole. Uptake should therefore be addressed both as an organisational and industry-wide phenomenon (Crowston and Myers, 2004). In 2012, the Australian Built Environment Industry Innovation Council (BEIIC) also projected that principles of integration and collaboration will come to define how the industry is organised (BEIIC, 2012). The growing adoption of these principles, the up-skilling of the workforce, and new contractual clauses all represent key steps towards more integrated project environments such as BIM. This has the potential to lead to more efficient and profitable infrastructure and building construction projects with less associated risks to the supply chain as a whole.

Within this process, national governments and umbrella organisations can take a leadership role to steer the industry towards specific coordinated goals. These industry stakeholders are already in an advantageous position because they are responsible for providing the overall guidelines and frameworks for the operation of industry sectors. These can underpin a common communication platform around performance and productivity associated with the uptake of new ways of working. Without leadership, however, BIM users may take different approaches to solving issues as they present, resulting in a disjointed approach across an already fragmented industry. This can result in a failure 'to capitalise on the considerable benefits of a coordinated approach based on trust, communication and commitment' (Australian Institute of Architects, 2010).

While any entity involved in this network could take this role, public authorities are already in an empowered position (Hovik and Vabo, 2005). They can reduce tensions, empower particular players and lower the transaction costs associated with the uptake of new technologies (Sørensen and Torfing, 2009). In addition, they are in a position to facilitate and structure policy interactions, as long as the central and regional objectives are broad enough to permit local adjustments and amendments. Public agencies have the power to promote access to a range of skills, and help develop the capacity to learn through collaborative networks, technology diffusion and providing the conditions for organisational change to occur at the firm level (Damgaard and Torfing, 2010). Public authorities are also in a position to 'mobilize the knowledge, resources and energies of a host of public and private actors while retaining their ability to influence the scope, process and outcomes of networked policy-making' (Sørensen and Torfing, 2009).

Public clients therefore have the ability to 'spread and accelerate change to other professional services and contractor firms' (Bonham, 2013). Along with them, other political institutions and umbrella organisations can also influence change. They can, for example, have a significant influence on the speed of diffusion and uptake of new products and services through decisions about: (1) whether to adopt or not; (2) product specification and/or standard setting; and (3) the evaluation of technology once in the market (Howells, 2006).

Industry-wide national case studies

The following section presents four national case studies that consider different approaches to and types of leadership used to promote the industry-wide implementation of BIM. These include the United Kingdom (UK) BIM Task Group, the US General Service Administration, the Swedish Transport Administration and Australia's transport and infrastructure industry. These showcase national or regional leadership by governments or umbrella groups that have led or can lead to the development of a national or regional strategy for the adoption of BIM (Sanchez et al., 2014).

UK BIM Task Group

The UK has a unicameral system of government that has considerable impact on how agreements or relationships between the tiers of government and the sector are played out. The construction industry was identified by the UK government as an enabling sector and BIM as a key agent for economic growth within their industrial strategy. The national government also identified itself as a clear agent of change due to its heavy influence on the construction industry, procuring around 30 per cent of the output of this sector (HM Government, 2012). This led them to take a leadership role in the adoption of BIM in close collaboration with industry umbrella organisations. The resulting BIM implementation strategy has become an exemplar for many countries. Introduced in 2011, they set a five-year implementation plan designed to provide savings in procurement costs, which included a mandate requiring collaborative 3D BIM on all projects as of 2016 (BIM Task Group, 2015). Importantly, the UK government 'established a BIM Task Group to assist both the public sector clients and the private sector supply chain in reengineering their work practices to facilitate BIM delivery' (McGraw Hill Construction, 2014). The BIM Programme, led by this task group, also includes initiatives such as the Government Soft Landings Policy, which aims to, among other things, reduce the cost and improve the performance of asset delivery and operations (HM Government, 2012).

Procurement tools have been used to facilitate a timetable for small and medium-sized enterprises (SMEs) and larger firms to adopt BIM as the technology of choice in design, construction and materials manufacturing (BEIIC, 2012). This has resulted in a journey with industry for sector modernisation, a key objective of which is to reduce capital cost and the carbon burden from the construction and operation of the built environment by 20 per cent (Masterspec, 2012).

This concerted approach between government and industry umbrella organisations has been highlighted in the past as promoting: (1) a series of legal, economic and operational reforms (Melville, 2008); and (2) direct participation by industry in policy development (Strickland and Goodes, 2008). This approach was implemented in the BIM Programme through a national *push–pull* strategy with a number of reforms to be undertaken over the next few years to reach level 2 of their roadmap (BIM Industry Working Group, 2011). This government-led strategy has produced a fast industry response and large scale adoption (HM Government, 2012).

More recently, the UK Cabinet Office has identified BIM as a significant contributor to construction cost savings of £840 million just for the 2013/2014 fiscal year. The delivery of their *Level 2 BIM Programme* helped secure 20 per cent savings on capital expenditures against the 2009/2010 benchmark. In early 2015, the government released a policy paper, *Digital Built Britain: Level 3 Building Information Modelling (BIM) – Strategic Plan*, which goes one step further. This paper envisions a fully computerised construction as the norm and the built environment belonging to a smart and networked world. It also outlines key actions to be delivered in order to achieve their new goals by 2025 (HM Government, 2015).

US General Services Administration (GSA)

The United States is broadly acknowledged as an international leader in BIM (Smith, 2014; Ho and Matta, 2009). McGraw Hill note that industry-wide uptake of BIM in the US increased from 28 per cent in 2007 to 71 per cent in 2012, with investors favouring BIM methods that involve collaboration (McGraw Hill Construction, 2012). This leadership is demonstrated through the implementation of BIM over many years by the GSA. This agency is responsible for new construction and the operations of existing federal properties, including almost 9,000 property buildings and 656 historic buildings. They also deliver annually up to US$50 billion in information technology solutions and telecommunications services through the Federal Acquisition Service (GSA, 2014). As a major public sector client, this organisation has had a significant influence on the adoption of BIM in the US, thus demonstrating the importance of major client and government leadership for the industry (buildingSMART Australasia, 2012).

The GSA first established its *National 3D-4D BIM Program* for buildings in 2003, and by 2007 required 'spatial program BIMs be the minimum requirements for submission' to the Office of Chief Architect (GSA, 2015a). In their BIM Guide Overview, they acknowledged that from all the stakeholders that stand to gain from implementing BIM, owners benefit the most due to the length of the operations phase. This led their move towards BIM and encouraged the development and use of universal, open standards which they view as a governmental imperative (GSA, 2007). The US Army Corps of Engineers (USACE), Air Force and Coast Guard have all moved down the BIM path. USACE, for example, established a Charter in 2006 to 'coordinate capabilities and needs for CADD, GIS and Computer Aided Facility Management (CAFM) technology applications throughout the Department of Defence (DOD)' (US Army Corps of Engineers, 2006). Peggy Ho from GSA and Charles Matta, National Director of the Fellow of the American Institute of Architects (FAIA) note that:

> monitoring and constantly evaluating GSA's design and construction processes while developing and implementing technologies to transform those processes has been one of the cornerstones of GSA's BIM program. Without this understanding and balance, it would be difficult to gain stakeholder buy-in, implement BIM in a way that is least disruptive to the organization's current practices, and measure the benefits of project delivery through BIM.
>
> (Ho and Matta, 2009)

In 2007, the GSA started publishing their BIM Guide Series, covering a range of topics related to the use of BIM for asset management from planning to operations. Although *Guide 08* was published a few years ago, they have recently been working on *BIM Guide 07 – Building Elements*, which will extend previous BIM Guide programmes to include model elements created for the focused areas (GSA, 2015a). They have also published a list of BIM Champions across the GSA offices with their contact information (GSA, 2015b). In addition to their guide series, this organisation is actively involved in conferences and is often nominated for awards such as the American Institute of Architects BIM Awards, which they won in 2011 for their 3D/4D BIM Program in recognition of their exploration into BIM excellence (Hagan, 2011).

Reinforcing this leadership role, they have published facts and statistics to help others build their business case for BIM. For example, they reported that the use of BIM in their Los Angeles Federal Office Building pilot project led to a reduction in the duration of the construction phase of 19 per cent (GSA, 2007). They have also reported that the use of BIM has enabled them to have at least 90 per cent accuracy of space measurements within minutes and reduce operations and management contracting costs by three to six per cent (Suermann, 2009; GSA, 2011). This has resulted in cost savings achieved from using BIM in a single pilot project covering the complete start-up cost of its implementation across the entire pilot programme (GSA, 2007).

Swedish Transport Administration (Trafikverket)

The Nordic countries of Denmark, Finland, Norway and Sweden also provide important examples of BIM adoption and implementation. National governments provide leadership, supported by investment in outcomes in interoperability, open standards and research to support uptake (Smith, 2014).

In Sweden, the uptake of BIM has increased rapidly during the second part of the past decade through the support of organisations such as buildingSMART and OpenBIM (Karrbom-Gustavsson et al., 2012). This activity has included the development of legal guidelines on digital deliveries for construction works in collaboration with the construction industry via the Svenska Byggbranschens Utvecklingsfond (SBUF, Swedish National Construction Industry R&D Organisation) (OpenBIM, 2013). As in the GSA case study, no official mandate has been issued by the national government. Trafikverket, however, the only transport authority in Sweden, has taken a clear lead agent role requiring that all new projects from 2015 use BIM (Trafikverket, 2013a).

Trafikverket oversees all transport infrastructure management across the country, collaborating closely with local councils on specific matters (Trafikverket, 2011). In 2013, Trafikverket stated that a fully integrated system needed to be implemented to maximise the benefits of new digital technologies such as BIM (Trafikverket, 2013b). However, it is not clear whether this level of integration has been achieved yet. This would require all actors to have access to the same input data, with internal information technology (IT) processes being unambiguous, tool-independent and common to all areas (Trafikverket, 2013b).

Interviews conducted that year by Australian SBEnrc researchers showed that international competition from other countries including Germany, France and Poland was leading to a collaborative national approach in the implementation of BIM. Of importance alongside this, however, was the lead role taken by contractors and consultants in developing standards within the context of BIM project deliverables, and actively participating in research in this field. Contractors and designers interviewed expressed their support for Trafikverket in taking responsibility for the model quality and developing appropriate standards with industry (Sanchez et al., 2014).

As with the implementation in the UK, Trafikverket worked across the industry with lead government agencies, academic research institutions, umbrella professional organisations and vendors. They have, for example, been actively involved with the BIM Alliance, created in 2014 when OpenBIM, Fi2 Förvaltningsinformation and buildingSMART Sweden merged (BIM Alliance, 2014a). Together with representatives from some of the largest contractors and consultants in the industry as well as universities, Trafikverket is part of the BIM Alliance board of directors (Trafikverket, 2014) and provides funds to increase uptake through networking and information sharing (BIM Alliance, 2014a).

These efforts have influenced uptake in other sectors of the industry. For example, five major governmental property developers and managers joined forces to develop an overarching BIM strategy that includes a set of BIM and information management guidelines (BIM Alliance, 2014b).

The Australian transport and infrastructure industry

In Australia, there has been a lot said about the potential benefits of BIM to the industry and the need to increase productivity. The Industry Innovation and Competitiveness Agenda, for example, which was announced by the federal government in 2014, highlighted the need to boost productivity across the industry nationally (Australian Government, 2014). Given that the construction industry accounts for more than 10 per cent of Australia's gross domestic product (GDP), and is a major source of work and income (ACIF, 2014), productivity gains can have a significant effect on the overall national economy. This is especially important considering the forecasts for declining or static investment in construction as a percentage of GDP for the next ten years (ACIF, 2014). In this context, increasing the efficiency of investment becomes an important objective to avoid wastage and maximise returns.

It has been argued that the adoption of BIM has the potential to increase productivity in the Australian construction sector by up to 9 per cent with high benefit/cost ratio (Brewer et al., 2012). This would potentially lead to an increase in national GDP of up to AU$7.6 billion for the period 2011–2025, based on the building sector alone (Allen Consulting Group, 2010). Such potential benefits have led many umbrella organisations, comprising some of the most influential industry associations from across the supply chain, to advocate and work towards greater uptake of BIM.

Although some government bodies have advocated for wider implementation of BIM, Little coordination at the national government level has occurred. In 2014, for example, the Australia Productivity Commission inquiry into Public Infrastructure suggested that BIM has the potential to help mitigate risk against inadequate initial specifications, generate efficiencies in construction processes, provide better quality information, reduce whole-of-life cost and increase productivity (Australian Government Productivity Commission, 2014a, 2014b). This commission also made a series of recommendations in which BIM and early contractor involvement were featured. They highlighted that:

- there is a widespread view that there was scope for more innovation and diffusion of new technologies in the industry;
- given the potential savings from BIM, government clients should consider the use of BIM from early design stages; and
- while it is in governments' best interests to pursue these reforms, it can hardly be said that reform has proceeded either apace or uniformly throughout Australia, hinting to the need for a more efficient nationwide strategy (Australian Government Productivity Commission, 2014a).

However, due to Australia's unique form of federalism and the current political context, many argue that a national mandate from the federal government is unlikely in the near future and other avenues should be explored.

In addition to this, although the federal government plays a role in policy development, the majority of services in Australia are procured and delivered by the state governments (Melville, 2008). For example, the responsibility for planning, funding, designing, constructing, maintaining and operating transport infrastructure is shared between the three levels of government (federal, state and local; Figure 2.1). In road transport infrastructure, state governments are also the largest client contributing to 57 per cent of total expenditure in this sector versus 13 per cent invested by the Commonwealth government in 2012/2013 (BITRE, 2014). This means that even if the federal government were to mandate BIM, such leadership could be ineffective without agreement from state and local governments.

Although there is generally good coordination between levels, this results in complex intergovernmental arrangements for infrastructure delivery and management. In addition, there are also several consensus-based umbrella organisations with complex interrelations, often delivering specific services to their members (Sanchez et al., 2014). This creates the potential for problems (Newman et al., 2012) and contributes to a lack of national standard practices and strategies that could support the widespread uptake of BIM.

This led the Australian Sustainable Built Environment National Research Centre (SBEnrc) to carry out research in 2013/2014 to identify a potential leader for the transport infrastructure sector that could take a similar role to that of the GSA and Trafikverket (SBEnrc, 2015). This body had to have significant and appropriate spheres of influence, dissemination strategies and role within the industry for a national strategy to have the most impact on the adoption of BIM (Sanchez et al., 2014).

The research found that a body composed of the federal, state and territory transport infrastructure agencies and industry representatives would be in the best position to provide direction and consistency across state boundaries. This body would also need to be able to establish a close collaboration and dialogue with industry, industry research and umbrella organisations in order to ensure the success of a national strategy.

The Transport and Infrastructure Council is such a body, and importantly also includes the Local Government Association ensuring engagement with other government clients. This body relies on the expertise of the Transport and Infrastructure Senior Officials' Committee (TISOC),[1] which is composed of senior officials from all transport infrastructure agencies in Australia and that of Austroads.[2] In this scenario, Austroads could take a proactive, constructive and collaborative approach informed by its own applied research and strategic partnerships, which has been so successful in other sectors at a state level. TISOC, on the other hand, would serve to reach an agreement between the federal and state governments in order to have a consistent and coordinated national strategy. This process is to be most efficient if a bi-directional line of communication was established between

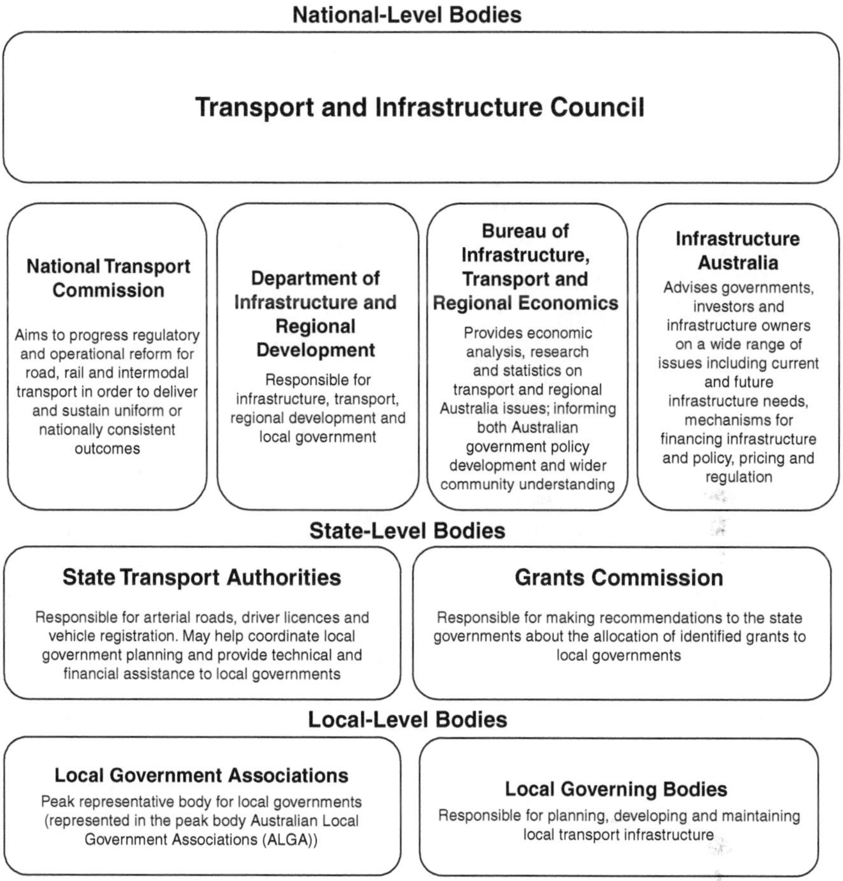

National-Level Bodies

Transport and Infrastructure Council

National Transport Commission
Aims to progress regulatory and operational reform for road, rail and intermodal transport in order to deliver and sustain uniform or nationally consistent outcomes

Department of Infrastructure and Regional Development
Responsible for infrastructure, transport, regional development and local government

Bureau of Infrastructure, Transport and Regional Economics
Provides economic analysis, research and statistics on transport and regional Australia issues; informing both Australian government policy development and wider community understanding

Infrastructure Australia
Advises governments, investors and infrastructure owners on a wide range of issues including current and future infrastructure needs, mechanisms for financing infrastructure and policy, pricing and regulation

State-Level Bodies

State Transport Authorities
Responsible for arterial roads, driver licences and vehicle registration. May help coordinate local government planning and provide technical and financial assistance to local governments

Grants Commission
Responsible for making recommendations to the state governments about the allocation of identified grants to local governments

Local-Level Bodies

Local Government Associations
Peak representative body for local governments (represented in the peak body Australian Local Government Associations (ALGA))

Local Governing Bodies
Responsible for planning, developing and maintaining local transport infrastructure

Figure 2.1 Australian bodies responsible for transport infrastructure within each level of governance (2013)

Note: Based on the Allen Consulting Group, (2009), Australian Government (2013), Infrastructure Australia (2014) and ALGA (2014).

TISOC and Austroads. This approach would actively engage all transport authorities in the development of outputs that would be submitted to the council for discussion and approval. The strategy outlined by the SBEnrc group also highlighted that by establishing a Task Group, TISOC and the council could leverage on:

- their individual members' already invested resources towards BIM pilot projects; and
- their Memorandum of Understanding with Austroads and the ARRB[3] who are already receiving funding to carry out research to support decisions taken by TISOC and council members.

This proposal was published in late 2014 and its reach is yet to be determined, but the outlook is positive. TISOC members such as the Department of Planning, Transport and Infrastructure (DPTI) of South Australia and Queensland Transport and Main Roads have reported research and pilot projects into BIM requirements and procedures (NATSPEC, 2014). Transport for New South Wales, another TISOC member, has also recently recommended that this body takes a leadership role in smart ICT design and planning of infrastructure standards (Transport for New South Wales, 2015). This committee member has also put together a ten-person team to develop a strategy for BIM adoption. The Victoria government has also announced their own BIM pilot programme with the 'potential staged implementation of BIM across infrastructure projects in future' (Hayes, 2015). Talks among the ACIF/APCC[4] BIM Summit Group, formed by the leadership of these organisations and others such as buildingSMART Australasia, have also led to the decision of having a greater focus on infrastructure for future dissemination and advocacy actions (ACIF and APCC, 2015). Most recently, Prime Minister Malcolm Turnbull has been a vocal supporter of technological innovation as a driver for the competitiveness of Australia's economy. This has, for example, led to the establishment of a Digital Transformation Office that follows a UK model (Turnbull, 2015) and may influence the future of BIM in Australia across all sectors.

The case for organisational leadership

Beyond a nationwide and/or regional approach, whether by governments or umbrella groups, there is a need to also acknowledge the very real disincentives to implementing BIM collaboratively on projects because of parent company obligations and the need to manage associated scope (Dossick and Neff, 2008). To address this, inspirational leadership focusing on project-based relationships can enhance the acceptance of new more integrated approaches (ACIF and APCC, 2014). The following section presents an example of such collaborative leadership from one of Australia's earlier government BIM adopters.

Australian early adopter case study: Queensland Department of Public Works (2005–2012)

Leadership was the cornerstone of the implementation of digital technologies in the Queensland government's Department of Public Works (QDPW) project delivery agency until a change of political leadership in the 2012 elections. At the time of this case study, QDPW was a large construction-based client organisation managing the design and construction of public built assets including schools, hospitals and courthouses across the Australian state of Queensland.

The initial implementation of computer-aided design and documentation (CADD) commenced in the 1990s, followed by: experimentation with BIM from the mid-2000s; embedding integrated practice (IP); and finally steps towards integrated project delivery (IPD) with the integration of contractors in the design/delivery process until 2012.

Queensland Project Services BIM implementation was driven by a vision of improved business outcomes, which included:

- Increased production and process efficiency as highlighted by:
  - the Australian Mechanical Contractors Association (AMCA) underlining potential reduction in change orders, reducing from 18.4 per cent on 2D projects down to 2.7 per cent using collaborative BIM (Cannistraro, 2011); and
  - ACIF, who highlighted the potential for 5–12 per cent reductions in rework based on estimates calculated by the Australian Construction Industry Forum (ACIF) (Barda, 2011).
- Better communication and collaboration through:
  - maximising multidisciplinary engagement, facilitated by Project Services being a multidisciplinary design agency;
  - development of effective supply chain networks; and
  - development of a model server.
- Demonstrating added value to government clients through: more effective environmental modelling; potential for whole-of-life asset management; and less waste in the delivery process.
- Creating a stimulating working environment, which helped in maintaining a core team of skilled professionals in a public sector environment.

A research project carried out by SBEnrc identified key elements of this implementation such as having a strong vision, project and organisational champions, an incremental approach to implementation and leadership across the supply chain.

Implementing a vision

Underpinning Project Services' adoption was a strong vision developed across the whole of the delivery supply chain (Figure 2.2). This focused on developing more efficient project delivery mechanisms through the integration of several aspects of process improvement including: the use of new technology enablers, implementing process changes through pilot projects, strong engagement with researchers and targeted industry leadership and partnerships.

Key steps in this vision included:

- Bringing together the multidisciplinary team of professionals within Project Services, to collaborate on the development of the 3D BIM

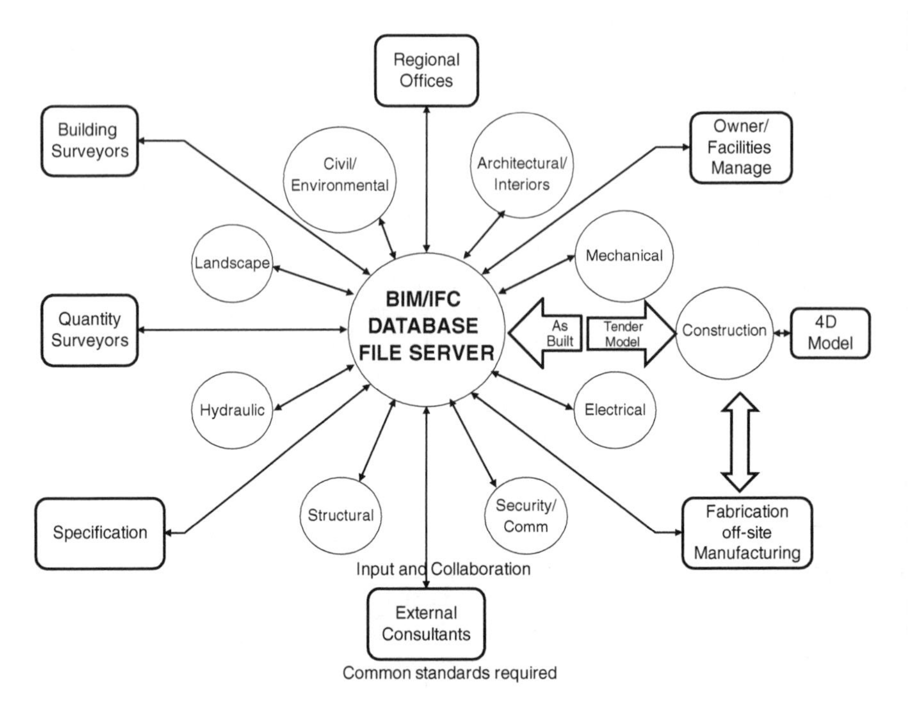

Figure 2.2 QDPW vision for implementation of BIM (adapted for presentation by the authors)

model. This included architects, structural and services engineers, and quantity surveyors.

- The establishment of working relationships with contractors, subcontractors and consultants willing to participate in the implementation of new work processes and practices was considered critical. One example of this was the development of 4D models to assist with construction through high level clash detection, and rehearsals of the construction sequence with both programming and safety benefits.
- Engagement with subcontractors for off-site manufacturing including structural steel and mechanical components.
- Developing common standards to enable the integration of external consultants.
- The development of a BIM/Industry Foundation Classes (IFC) database/file server.

The role of champions

This case demonstrates the importance of *champions* in the successful implementation of new processes and technology. Champions are defined here as high-level individuals whose role is to promote innovation within the organisation. They 'play a critical role in marketing the innovation to the decision makers, developing an implementation plan, facilitating resource allocation, and removing roadblocks to implementation' (Premkumar, 2003). These champions aligned the support from others including:

- strategic support of the organisational hierarchy;
- project delivery support provided to CADD managers, BIM managers, discipline leaders, principal consultants, project directors and superintendents;
- support of strategically selected IT contractors and vendors; and
- buy-in of contractors and subcontractors engaged in pilot projects.

In doing so:

> successfull collaboration leaders leverage the forces of scope and company to bring the teams together. Thus, advocating that group problem-solving collaboration can benefit the individual team member's companies. In this way, they do not put the project at odds with the demands of scope and company, but philosophically align the goals of collaboration with these demands.
>
> (Dossick and Neff, 2008)

This broad support is essential to ensure the necessary significant organisational commitment that is required for learning and capability development over time to ensure success (Doherty et al., 2012). This support was also evidenced in the investment in both training and technology that underpinned this advancement, and engagement with researchers as Project Services moved beyond their own internal proof of concept stage.

An incremental development approach

This approach was led by key champions from both the executive and delivery teams, and was driven by their experience of and sensitivity to the needs of the industry (Figure 2.3). The core BIM team worked to establish a proof of concept for targeted aspects of project delivery on specific projects; for example, energy modelling at the design stage. When satisfactorily proven, this aspect of delivery would be implemented on a pilot project and, subject to benefits, would then become part of their business-as-usual delivery in a continuous process improvement cycle.

BIM was first implemented on selected pilot projects from 2005 (CRC for Construction Innovation, 2009a, 2009b). This involved collaboration with contractors and subcontractors on specific aspects of the work to both test the implementation and, also importantly, as a process of knowledge transfer.

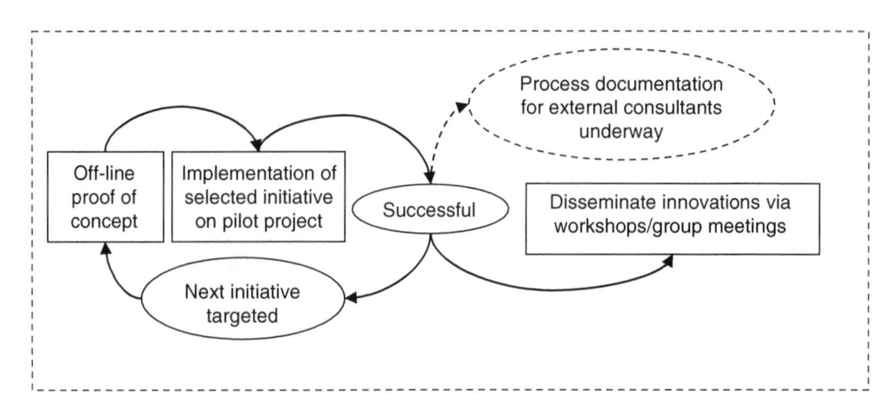

Figure 2.3 Incremental innovation processes

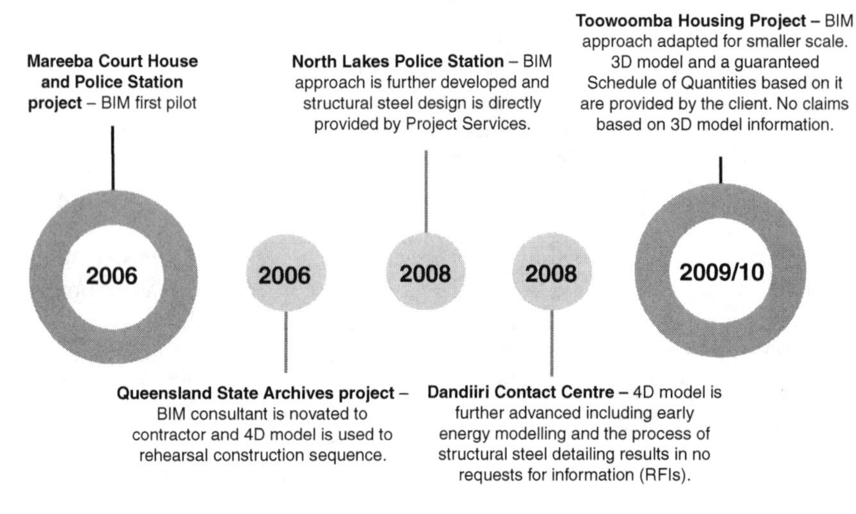

Figure 2.4 Timeline for BIM implementation within Project Services

Practical examples of the integration of these new approaches include:

- Investigating building life-cycle asset management tools and processes, for example with ArtrA software.[5]
- Working with the UK-based subcontractor A3D on 4D model development including modelling for construction rehearsals on construction of the Queensland State Archives project in 2006 by Laing O'Rourke.
- Innovation from suppliers such as that associated with model server development.
- The steel fabrication for the North Lakes Police Station from a model provided directly to the manufacturer.

- Engagement with subcontractors such as the Australian Mechanical Contractors Associations in terms of standard library development.
- Onsite training of workers on the Queensland State Archives Project.

Two key elements in the success of this approach were:

1 the preparedness of Queensland Project Services to take the risk associated with the use of 3D, 4D and 5D models for and by contractors and suppliers; and
2 the establishment of a collaborative environment based on openness and trust.

The strategic national research collaboration with strong international linkages through the Cooperative Research Centre (CRC) for Construction Innovation was also of importance. All these elements contributed to a motivated team of individuals with a commitment to quality outcomes.

Client as leader through supply chain engagement

Strong relationships outside this government agency fundamentally underpinned this implementation. As evidenced in the pilot projects, this occurred with direct industry players including contractors, subcontractors and suppliers. Critical relationships were also established with umbrella organisations and research institutions:

- Key relationships were established with national industry umbrella organisations such as NATSPEC and buildingSMART Australasia and industry sector organisations such as the Australian Institute of Architects (AIA) and AMCA. This engagement was important to raise industry-wide awareness of potential benefits and identify essential areas for collaboration, such as the development of standard digital libraries and addressing intellectual property concerns.
- Strategic research engagement occurred through national research collaboration with strong international linkages through CRC for Construction Innovation, Queensland University of Technology (QUT) and Royal Melbourne Institute of Technology (RMIT), and with engagement through the Australian Research Council (ARC) Linkage scheme. This important engagement helped the agency move beyond proof of concept into project adoption.
- Active engagement with umbrella organisations including: Built Environment Industry Innovation Council (BEIIC) in reporting on productivity impacts of BIM (2010); the Australian Construction Industry Forum (ACIF); the Australian Procurement and Construction Council (APCC); the Australian Institute of Architects (AIA); and the International Alliance for Interoperability (now buildingSMART). This

engagement provided an active industry-driven forum for discussion of methods and tools.

- Significant contribution to NATSPEC through leadership of the development of the National BIM Guidelines (NATSPEC, 2011b) following from the initiative through the CRC for Construction Innovation.

Long-term legacy

This case clearly highlights the role that can be played by clients in introducing new technology and work practices to an industry and partnerships with industry bodies and national research groups. These are important roles in facilitating greater industry-wide productivity improvements and relationship building as a base for ongoing innovation networks (Keast and Hampson, 2007). An informed and involved client with clearly defined needs is of key importance to ensure the benefits of using BIM are maximised.

Queensland Project Services took this key leadership role within an environment where the potential for maximising benefits existed. This experience was carried out into industry more broadly, by the key organisational champion, through strong industry engagement and the use of pilot projects to demonstrate value. At the same time, risk to the organisation was minimised through an incremental approach to learning while enabling benefits to be demonstrated before these lessons learned were passed on to the industry more broadly.

Project Services, like many organisations that were early adopters, were unable to specifically quantify all benefits they derived from the implementation of digital technologies and associated changed work practices. They did, however, deliver quality outcomes and built a cohort of employees and partners with enhanced skills that improved the industry's skill and knowledge base. They also operated in a strategic manner in asking: do the lessons from this pilot project become standard practice in future? What are the target lessons for the next pilot? How to further expand the field of influence of their activities? And, how to partner with research and education partners to disseminate findings at a national level and across the supply chain to maximise impact?

In doing this, key champions within Project Services took a national industry leadership with national research centres such as CRC for Construction Innovation, now SBEnrc, and industry umbrella organisations such as NATSPEC.[6] They also received broad industry recognition for this national leadership role.

Leadership, collaboration and BIM uptake

Leadership and collaboration within organisations and across the supply chain are critical to support successful implementation. International experience highlights three important messages:

1 Industry takes action when the government demonstrates clear leadership.
2 A national strategy facilitates the adoption of new information technologies such as BIM.
3 Collaboration with industry is required to implement this strategy.

Leadership is required within organisations 'to overcome the organizational disincentives inherent in scope and company and to hold the people together and inspire collaboration' (Dossick and Neff, 2008). Key characteristics of organisational champions include:

- communicates a vision;
- develops staff;
- provides support;
- empowers staff;
- is innovative;
- leads by example;
- is charismatic (Carless et al., 2000).

This aligns with the findings of the early adopter case study presented in this chapter. The pivotal role of such leadership should not be underestimated in the efforts to improve industry performance in the effective uptake and implementation of BIM. In the case of QDPW, this leadership was then transferred from the organisation to the industry more broadly through engagement with: (1) contractors, subcontractors, consultants and suppliers in a project environment; (2) industry-focused research; (3) key industry associations and peak bodies; (4) industry standards and related organisations; and (5) research organisations, including the Australian CRC for Construction Innovation, now SBEnrc, through contributions to, for example, the Off-Site Manufacture in Australia project (Blismas, 2007) and the National BIM Guidelines (NATSPEC, 2011a); and Queensland University of Technology (QUT) and RMIT through a series of ARC Linkage projects, among others.

Industry-wide leadership by an entity without direct organisational or project responsibilities is also required to reach beyond the immediate self-interests of project delivery or business responsibilities. Even in collaborative project environments, different stakeholders have internal priorities related to their particular organisation and driven by profit requirements (Dossick and Neff, 2010). Innovation theory notes that innovation brokers are organisations 'acting as a member of a network of firms focused on enabling other organisations to innovate' (Winch, 2005). As such these brokers can also play a key role 'facilitating diffusion' (Winch and Courntey, 2007). It is proposed that like innovation brokers (Kraatz and Hampson, 2013) clients and national umbrella organisations can take on such industry-wide leadership role.

Conclusions

This chapter has highlighted the need for leadership at project, organisational and industry-wide levels to achieve the goal of improving productivity through the uptake of BIM as an approach, process and tool. Significant lessons are available from countries such as Australia, the US, the UK and the Scandinavian region. These lessons can be readily adapted to differing governance arrangements if the energy and expertise of those already engaged in this pursuit can be brought together in a collaborative and trusting environment within a framework of national leadership.

Strong, informed leadership is the cornerstone to improving productivity through the effective implementation of BIM, whether across the industry or within an organisation. Collaborative leadership especially has been demonstrated to be effective when providing direction and combined with active engagement leading to wider acceptance and ownership of the required changes. At a project or organisational level, those implementing BIM need to identify champions at various levels of decision-making to enable the greatest benefits and returns. At a national level, a coordinating lead agent with close ties to industry research and umbrella organisations will help maximise the productivity benefits of BIM implementation by minimising wasteful disjointed efforts by individual firms and organisations.

Acknowledgements

The authors acknowledge the funding and support provided by Australia's Sustainable Built Environment National Research Centre and its partners. Core members across the time-frame of these case studies included the Aurecon, Curtin University, Government of Western Australia, Griffith University, John Holland, New South Wales Roads and Maritime Services, Parsons Brinckerhoff, Queensland Government, Queensland University of Technology, and Swinburne University of Technology. Funding provided by the Australian Research Council (ARC) Linkage program (2011–2015) is also acknowledged.

Notes

1 The Transport and Infrastructure Council is advised by the Transport and Infrastructure Senior Officials' Committee (TISOC) on all non-infrastructure priorities and the Infrastructure Working Group on coordination of infrastructure planning and investment. Meets bi-annually and makes decisions based on consensus.
2 Austroads brings together transport agencies from each state and territory. Programme management is based on strategic priorities set by the Governing Board. Each programme manager and support group identifies outputs which are

developed by task groups as reference documents and published in their portal (Austroads, 2011).

3 ARRB has been providing 'advice, technical expertise and solutions to transport and road agencies across the world for over 50 years. ARRB's member agencies include Australian and New Zealand federal, state and local government bodies responsible for managing the nations transport and road networks' (ARRB, 2014).

4 The Australian Construction Industry Forum (ACIF) and Australasian Procurement and Construction Council (APCC) established a BIM Summit group as a 'knowledge hub to promote the better understanding of BIM activities underway and to avoid duplication of effort' (ACIF, 2013).

5 www.artra.co.uk.

6 NATSPEC is an Australian not-for-profit organisation owned by the design, build, construct and property industry through professional associations and government property groups (NATSPEC, 2015).

Bibliography

ACIF, 2013. *ACIF and APCC PTI and BIM Summit Success*. Available at: www.acif.com.au/acif-news/acif-and-apcc-pti-and-bim-summit-success [Accessed 15 July 2015].

ACIF, 2014. *Policy Compendium: Policy on Design and Construction for a Sustainable Industry*. Canberra: Australian Construction Industry Forum.

ACIF and APCC, 2014. *Project Team Integration Workbook*, Canberra: Australian Construction Industry Forum and Australasian Procurement and Construction Council.

ACIF and APCC, 2015. *ACIF/APCC 4th PTI and BIM summit outcomes and actions*. Sydney: Unpublished.

ALGA, 2014. *About ALGA*. Australian Local Government Association. [Online] Available at: http://alga.asn.au/?ID=42&Menu=41,81 [Accessed 5 May 2014].

Allen Consulting Group, 2009. *Options for Improving the Integration of Road Governance in Australia*, s.l.: Infrastructure Australia.

Allen Consulting Group, 2010. *Productivity in the Buildings Network: Assessing the Impacts of Building Information Models*, s.l.: BEIIC.

ARRB, 2014. *About ARRB*. Available at: www.arrb.com.au/Home/About-ARRB.aspx [Accessed 20 February 2013].

Australian Government, 2013. *Transport*. Available at: http://australia.gov.au/topics/transport [Accessed 5 May 2014].

Australian Government, 2014. *Industry Innovation and Competitiveness Agenda: An Action Plan for a Stronger Australia*, Canberra, ACT: Australian Government.

Australian Government Productivity Commission, 2014a. *Public Infrastructure. Productivity Commission Inquiry Report: Volume 1*, Canberra: Commonwealth of Australia.

Australian Government Productivity Commission, 2014b. *Public Infrastructure. Productivity Commission Inquiry Report: Volume 2*, Canberra: Australian Government.

Australian Institute of Architects, 2010. *BIM in Australia. Report on BIM/IPD Forum*, s.l.: Australian Institute of Architects.

Austroads, 2011. *About Austroads*. Available at: www.austroads.com.au/about-austroads/about-austroads [Accessed 12 February 2014].

Barda, P., 2011. Integrated project delivery – creating added value from construction. In *MESH – Sharing Information, Building Innovation*. Sydney: buildingSMART Australasia.

BEIIC, 2012. *Final Report to the Government*, s.l.: Built Environment Industry Innovation Council.

BIM Alliance, 2014a. *Enad kraftsamling för BIM med ny förening (United mobilization for BIM with new compound)*. Available at: www.bimalliance.se/aktuellt/press/pressmeddelanden/130930_pressmeddelande [Accessed 12 January 2015].

BIM Alliance, 2014b. *BIM I staten (BIM in the state)*. Available at: www.bimalliance.se/natverk_och_utveckling/projekt/bim_i_staten [Accessed 12 January 2015].

BIM Industry Working Group, 2011. *A Report for the Government Construction Client Group Building Information Modelling (BIM) Working Party*. Strategy Paper, Department of Business, Innovation and Skills.

BIM Task Group, 2015. *Building Information Modelling (BIM) Task Group*. Available at: www.bimtaskgroup.org [Accessed 15 July 2015].

BITRE, 2014. *Yearbook 2014 – Australian Infrastructure Statistics*, Canberra: Australian Government Department of Infrastructure and Regional Development, Bureau of Infrastructure, Transport and Regional Economics.

Blismas, N., 2007. *Off-Site Manufacture in Australia: Current State and Future Directions*, Brisbane: CRC for Construction Innovation.

Bonham, M. B., 2013. Leading by example: new professionalism and the government client. *Building Research & Information*, 41(1), pp. 77–94.

Brewer, G., Gajendran, T. and Le Goff, R., 2012. *Building Information Modelling (BIM): Australian Perspectives and Adoption Trends*, Callaghan: Centre for Interdisciplinary Built Environment Research (CIBER).

buildingSMART Australasia, 2012. *National Building Information Modelling Initiative: Vol.1*, Sydney: Department of Industry, Innovation, Science, Research and Tertiary Education.

Cannistraro, M., 2011. *JC Cannistraro Improves Job-site Productivity Using TSI BIM Solutions*, s.l.: Smart Solutions.

Carless, S. A., Wearing, A. J. and Mann, L., 2000. A short measure of transformational leadership. *Journal of Business and Psychology*, 14(3).

CRC for Construction Innovation, 2009a. *National Guidelines for Digital Modelling*, Brisbane: CR for Construction Innovation.

CRC for Construction Innovation, 2009b. *National Guidelines for Digital Modelling: Case Studies*, Brisbane: CRC for Construction Innovation.

Crowston, K. and Myers, M. D., 2004. Information technology and the transformation of industries: three research perspectives. *Journal of Strategic Information Systems*, 13(1), pp. 5–28.

Damgaard, B. and Torfing, J., 2010. Network governance of active employment policy: the Danish experience. *Journal of European Social Policy*, 20(3), pp. 248–262.

Doherty, N. F., Ashurst, C. and Peppard, J., 2012. Factors affecting the successful realisation of benefits from systems development projects: findings from three case studies. *Journal of Information Technology*, 27(1), pp. 1–16.

Dossick, C. and Neff, G., 2008. *How Leadership Overcomes Organizational Divisions in BIM-enabled Commercial Construction*, paper presented at

LEAD Conference, the Engineering Project Organisation Society (EPOS), Lake Tahoe, CA.

Dossick, C. S. and Neff, G., 2010. Organizational divisions in BIM-enabled commercial construction. *Journal of Construction Engineering and Management*, 136(4), pp. 459–467.

GSA, 2007. *GSA BIM Guide Overview*, Washington, DC: US General Services Administration.

GSA, 2011. *GSA BIM Guide for Facility Management*, Washington, DC: US General Services Administration.

GSA, 2014. *Agency Financial Report*, Washington, DC: US General Services Administration.

GSA, 2015a. *3D-4D Building Information Modelling*. Available at: www.gsa.gov/portal/content/105075 [Accessed 15 July 2015].

GSA, 2015b. *BIM Champions*. Available at: www.gsa.gov/portal/content/102497 [Accessed 15 July 2015].

Hagan, S. R., 2011. *Award Winning BIM: Lessons Learned from GSA's 3D/4D BIM Program and a Case Study of 2011 AIA BIM Award Winner Chicago Federal Centre*. Available at: http://network.aia.org/viewdocument/?DocumentKey=205f368e-2e54-407c-bdaa-8bd159bae1d4 [Accessed 14 July 2015].

Hayes, G., 2015. *June 2015 Victoria Division President Report*. Available at: https://www.engineersaustralia.org.au/portal/news/june-2015-victoria-division-president-report [Accessed 15 July 2015].

HM Government, 2012. *Building Information Modelling: Industrial Strategy: Government and Industry in Partnership*, London: UK Government.

HM Government, 2015. *Digital Built Britain: Level 3 Building Information Modelling (BIM) – Strategic Plan*, London: HM Government.

Ho, P. and Matta, C., 2009. Building better: GSA's national 3D-4D-BIM program. *Design Management Review*, 20(1), pp. 39–44.

Hovik, S. and Vabo, S. I., 2005. Norwegian local councils as democratic meta-governors? A study of networks established to manage cross-border natural resources. *Scandinavian Political Studies*, 28(3), pp. 257–275.

Howells, J., 2006. Intermediation and the role of intermediaries in innovation. *Research Policy*, 35, pp. 715–728.

Infrastructure Australia, 2014. *About Infrastructure Australia*. Available at: www.infrastructureaustralia.gov.au/about [Accessed 5 May 2014].

Karrbom-Gustavsson, T., Samuelsson, O. and Wikfo, Ö., 2012. Organizing IT in construction: present state and future challenges in Sweden. *Journal of Information Technology in Construction (ITCon)*, 17, pp. 520–534.

Keast, R. and Hampson, K. D., 2007. Building constructive innovation networks: the role of relationship management. *Journal of Construction, Engineering and Management, American Society of Civil Engineering*, 133(5), pp. 364–373.

Kraatz, J. A. and Hampson, K. D., 2013. Brokering innovation to better leverage R&D investment. *Building Research & Information*, 41(2), pp. 187–197.

Masterspec, 2012. *New Zealand National BIM Survey 2012*, Auckland: Construction Information Ltd.

McGraw Hill Construction, 2012. *The Business Value of BIM in North America*, Bedford, MA: McGraw Hill Construction.

McGraw Hill Construction, 2014. *The Business Value of BIM for Construction in Global Markets*, Bedford, MA: McGraw Hill Construction.

Melville, R., 2008. 'Token participation' to 'engaged partnerships': lessons learnt and challenges ahead for Australian non-for-profits, in *Strategic Issues for the Non-for-Profit Sector*, Sydney: University of NSW Press, pp. 103–124.

NATSPEC, 2011a. *NATSPEC National BIM Guide and Reference Schedule*, Canberra: NATSPEC.

NATSPEC, 2011b. *NATSPEC National BIM Guide*, Sydney: Construction Information Systems Limited.

NATSPEC, 2014. *BIM R&D Projects*. Available at: http://bim.natspec.org/index.php/research-development/bim-r-d-projects [Accessed 15 July 2015].

NATSPEC, 2015. *About NATSPEC*. Available at: www.natspec.com.au/index.php/about-us/about-natspec1 [Accessed 8 April 2015].

Newman, P., Hargroves, C., Desha, C., Kumar, A., Whistler, L., Farr, A., Wilson, K., Beauson, J., Matan, A. and Surawski, L., 2012. *Reducing the Environmental Impact of Road Construction*, Brisbane: Sustainable Built Environment National Research Centre.

OpenBIM, 2013. *About OpenBIM*. Available at: www.openbim.se/OpenBIM/Om_OpenBIM/About_OpenBIM.aspx [Accessed 24 June 2013].

Premkumar, G., 2003. A meta-analysis of research on information technology implementation in small business. *Journal of Organizational Computing and Electronic Commerce*, 13(2), pp. 91–121.

Sanchez, A. X., Kraatz, J. A. and Hampson, K. D., 2014. *Research Report 1: Towards a National Strategy*, Perth: Sustainable Built Environment National Research Centre.

SBEnrc, 2015. 2.24 *Integrated Project Environments – Leveraging Innovation for Productivity Gain through Industry Transformation*. Available at: www.sbenrc.com.au/research-programs/2-24-integrated-project-environments-leveraging-innovation-for-productivity-gain-through-industry-transformation/ [Accessed 8 April 2015].

Smith, P., 2014. *BIM Implementation: Global Initiatives & Creative Approaches*, Prague: Creative Construction Conference.

Sørensen, E. and Torfing, J., 2009. Making governance networks effective and democratic through metagovernance. *Public Administration*, 87(2), pp. 234–258.

Strickland, M. and Goodes, K., 2008. *Review of Tasmanian DHHS-Funded Peak Bodies and the Development of a Peak Body Strategic Framework*, Hobart: Office for the Community Sector Department of Health and Human Services.

Suermann, P. C., 2009. *Evaluating the Impact of Building Information Modelling (BIM) on Construction*, doctoral thesis, Gainesville: University of Florida.

Trafikverket, 2011. *The Swedish Transport Administration Annual Report 2010*, Borlänge: Trafikverket.

Trafikverket, 2013a. *Att införa BIM I trafikverket (Introducing BIM in transport administration)*. Available at: www.trafikverket.se/Foretag/Bygga-och-underhalla/Teknik/Att-infora-BIM-pa-Trafikverket [Accessed 30 April 2014].

Trafikverket, 2013b. *Stockholm Bypass Becomes Pilot for Epochal BIM*. Available at: http://translate.googleusercontent.com/translate_c?depth=1&hl=en&prev=_dd&rurl=translate.google.com&sl=auto&tl=en&u=http://www.anpdm.com/article/5D415D4A764743/6193550/841782&usg=ALkJrhh4R8enV4vhXU41F6d_FRpuP_AXVQ [Accessed 4 July 2013].

Trafikverket, 2014. *Trafikverket I styrelsen för BIM Alliance Sweden (Transport Administration on the Board of BIM Alliance Sweden)*. Available at: www.trafikverket.se/Foretag/Bygga-och-underhalla/Teknik/Att-infora-BIM-pa-Trafikverket/Nyheter/Trafikverket-i-styrelsen-for-BIM-Alliance-Sweden [Accessed 27 January 2015].

Transport for New South Wales, 2015. *Transport for New South Wales Submission: Inquiry into the Role of Smart ICT in the Design and Planning of Infrastructure*, Canberra: Department of Infrastructure and Regional Development.

Turnbull, M, 2015. *Supporting Innovation by Transforming Government Service Delivery*. Available at: www.malcolmturnbull.com.au/media/supporting-innovation-by-transforming-government-service-delivery [Accessed 28 September 2015].

US Army Corps of Engineers, 2006. *Charter*.

Winch, G., 2005. Managing complex connective processes: innovation brokering, in A. Manseau and R. Shields (eds.), *Building Tomorrow: Innovation In Construction And Engineering*, London: Ashgate, pp. 81–100.

Winch, G. M. and Courntey, R., 2007. The organization of innovation of innovation brokers: an international review. *Technology Analysis and Strategic Management*, 19 (6), pp. 747–763.

3 BIM performance and capability

Daniel Månsson and Göran Lindahl

Starting point

A concept that is constantly evolving and adapting to new ways of usage within the architecture, engineering, construction and operations (AECO) industry is Building Information Modelling (BIM). BIM is defined in varied ways across literature; from being solely perceived as a software tool for visual information-sharing by project participants; to an elaborate methodology for integrating data-rich, object-based models across the whole project life-cycle (Succar, 2009). An elaboration of this definition has been provided in Chapter 1.

When using BIM software tools, data and information associated with an object upon its creation can be later referenced across several views or representations. When the source information is later altered, objects linked to the source are changed and are instantly visible. Through such interconnectedness, design problems and rework are greatly reduced thus saving time and money (Lee and Sexton, 2007). By adding time and cost attributes to 3D elements – referred to as 4D and 5D respectively – organisations can reduce resources needed and streamline project execution. In addition to 4D and 5D, there are other dimensions, that help extend the benefits of using BIM tools and methods. Also, multidimensional modelling or 'nD' provide avenues for including additional data (Jung and Joo, 2011) and can integrate several additional benefits such as using models to assess energy use, materials recycling and operations logistics. Using this multidimensional understanding, the term 'BIM' can be expanded to cover additional analysis areas (Lee and Sexton, 2007).

From a different perspective, BIM represents the future of integrating processes through model-based information sharing. As BIM software tools evolve, the boundaries of BIM will continuously expand and allow the creation of new applications and utilisations. However, in such a rapidly expanding technological environment, many questions arise including how best to measure the levels of BIM use and establish stakeholders' capability beyond the simple use of BIM software tools.

As technologies continuously evolve and authorities release more detailed BIM directives, the AECO industry thus faces significant challenges. These challenges are exacerbated by the absence of an agreed BIM framework for assessing and comparing basic BIM capabilities across organisations and measuring their respective BIM maturity. Developing such a framework is not a simple undertaking due to the wide range of construction industry stakeholders, their multitude of disciplines and specialties, and their varied perceptions of expected BIM benefits. To date, very little effort has been exerted to develop formal guides and tools that can be used to establish and compare organisational BIM capability and maturity.

Requirements and drivers: authorities and procurement

National guidelines and incentives affect how performance and capability must be addressed. For example, by 2016, the European Union Public Procurement Directive (EUPPD) will require all 28 European Union (EU) member states to encourage, specify or mandate the use of BIM for publicly funded construction and building projects (Travaglini et al., 2014). While it remains to be seen how this directive will be enforced and followed up, the EUPPD signals how access to public funding will require higher efficiency in the form of better software tools, process transparency, information-sharing and data integration. Such directives also highlight the need for establishing common BIM performance criteria for the procurement of services. They also focus attention on the challenges facing organisations, private and public alike, as they implement BIM internally or participate in collaborative BIM projects.

This directive is not unique to the EU and, as discussed in previous chapters, has been preceded by the UK, the Netherlands and Nordic countries, which already require the use of BIM on publicly funded building projects.

Since the 1980s, Nordic countries have been leading the world in information management research. Major research efforts are currently underway in Denmark (EU social funds), Norway and Finland (under the direction of public sector clients) responding to government initiatives to digitise construction processes. In the UK, the government developed a national BIM strategy for the AECO industry in partnership with the private sector and academia. The declared aim of the government is to enable the UK construction industry to become world leaders in BIM utilisation. As a first step, the government is currently mandating Level 2 as illustrated in Figure 3.1 (Hooper, 2015).

The application of BIM tools and workflows are not exclusive to buildings and infrastructure, but also applies across the larger urban setting. In this respect, Germany has been leading the development of standardised 3D city models (Kolbe, 2009), and the UK Ordnance Survey has been working on linking BIM models to their national survey map (Morin et al., 2014). In Sweden, the National Board of Housing, Building and Planning (Boverket)

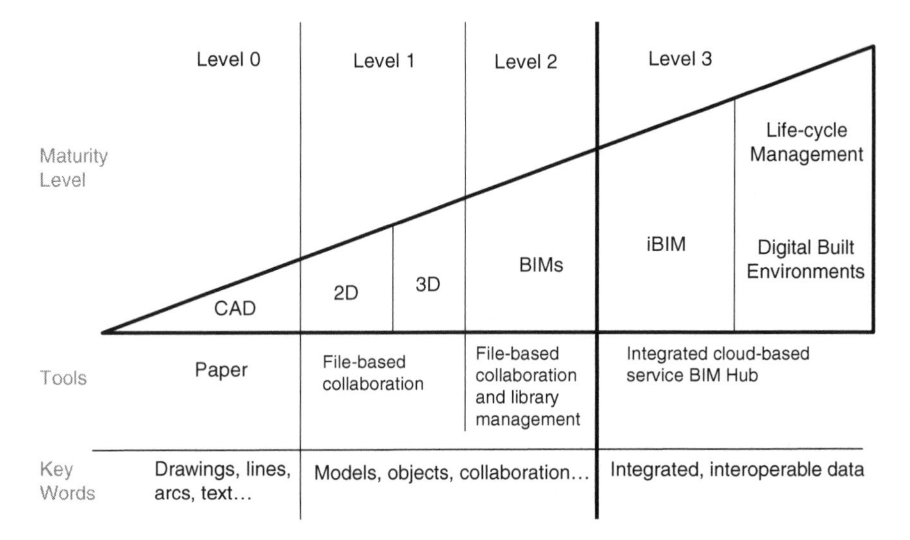

Figure 3.1 Illustration of the UK iBIM model based on the adaptation made by Barlish and Sullivan (2012) of the original Bew and Richards model

have also started to integrate geographical information systems (GIS) with the building permit process (Boverket, 2012).

In Hong Kong, the Housing Authority has required the use of BIM on all new projects since 2014. In South Korea, the Public Procurement Service currently requires the use of BIM on projects over SKW50 million, and on all public sector projects by 2016. Starting in 2010, Singapore developed an e-submission system that streamlines the building permissions process by requiring the submission of models for use by planning authorities (Wong and Fan, 2013).

Many authorities around the world are thus driving BIM implementation by encouraging industry stakeholders to investigate and/or adopt new technologies and processes, but even with such encouragements, many challenges remain. From the supply chain perspective, AECO organisations need to both manage their typical workloads and adhere to additional requirements for model generation and information management. From the authorities' perspective, policymakers need to identify how best to prescribe BIM use within tender documents and measure actual BIM use on projects. It is certainly not enough to mandate the adoption of certain software tools and workflows. That is, even if all companies across the supply chain adopted identical tools and workflows, clients/employers will still need clear metrics to identify and compare the BIM performance of different stakeholders.

To address supply chain BIM capability issues, it is important to acknowledge that organisations manage their performance, both internally and as part of a project team, in different ways. Some organisations opt not to adopt BIM software tools and workflows but prefer to implement workarounds to avoid the costs and challenges of adoption. These workarounds, according to Merschbrock and Figueres-Munoz (2015), reflect insufficient financial resources or inadequate know-how and may lead to messy information exchanges between project participants. Other organisations opt to develop their own workflows, protocols and standards in complete isolation from others. Such an approach is not only costly and time-consuming for these organisations, but signals their hesitance to openly share improvements and allow the spread of incompatible protocols across the industry.

To improve performance across the AECO industry, stakeholders must understand the importance of adopting common BIM tools and workflows as well as sharing information and best practices.

Organisational performance measurement

Organisational performance measurement refers, in this chapter, to measuring the effect of adopting BIM technologies and their respective processes on the performance of an organisation. While organisations within the construction industry vary greatly based on the type of business they conduct, it is important to apply unified, reliable and valid metrics that allow the measurement and comparison of their performance. Organisational performance measurement will also need to report on whether an organisation actually increases its overall productivity, delivers better outputs and meets or even surpasses its target performance improvement objectives. However, the term *performance* is relatively generic and is thus often replaced with *BIM capability* and *BIM maturity*.

As a term, *capability* describes what an organisation is able to do and what actions it can take based on its organisational competence. It combines the aspects of knowing what to do with the ability to do it. In other words, capability measurements identify what an organisation can deliver in comparison to what is specified or expected, and reflect its dynamic ability to rethink existing know-how and develop new solutions.

In the information technology (IT) sector, the development of the Capability Maturity Model (CMM) has spawned a number of similar models applied in varied industries including manufacturing, healthcare and construction (Curtis et al., 2002, 2009). As a performance measurement approach, CMM was originally developed to identify the strengths, weaknesses and risks of software projects (Paulk et al., 1993). Later model adaptations were intended to assess infrastructure projects yet proved less able to address the specific challenges of the construction industry (Jia et al., 2011).

In addition to *capability*, the term *maturity* is often used to describe how well an organisation or its members can manage their processes and tools (Andersen and Jessen, 2003; Yazici, 2009). In more general terms, maturity identifies the performance criteria to be fulfilled by assigning a maturity level or milestone that best describes the utilisation of available knowledge, best practices and innovative techniques within a business or across the whole market.

There are multiple approaches to maturity measurement, including those that yield a formal certificate. For example, in the field of project management, there are a number of specialised entities that offer to certify project managers according to well-defined and commonly used guidelines. The Project Management Institute (PMI) and the International Project Management Association (IPMA) are two such global organisations providing certification and delivering courses through a network of national associations (IPMA, 2015; PMI, 2015).

Nonetheless, there are a number of ongoing efforts to measure how construction organisations perform. With increasing demands for construction efficiency and the availability of more detailed procurement protocols, the development of metrics to establish and compare organisational performance is now needed. As mentioned earlier, the development of the EUPPD will necessarily exert transformational pressures on the industry to develop such metrics, and may well drive the generation of common guidelines for assessing and improving organisational performance within and beyond the EU.

BIM performance frameworks

There is an increasing number of frameworks that can be referred to as BIM performance frameworks. These include the Interactive Capability Maturity Model (I-CMM, 2009); BIM Proficiency Matrix (Indiana University, 2009); BIM Maturity Levels (Bew and Richards model, see BIM Industry Working Group, 2011); BIM QuickScan (Sebastian and Berlo, 2010); BIM Maturity Matrix (Succar, 2010a); Vico BIM Score (Vico, 2011); CPIx-BIM Assessment Form (CPI, 2011); BIM Excellence (BIMe, 2013); bimSCORE (2013); BIM Planning Guide for Facility Owners (CIC, 2013); and BIM Competency Assessment Tool (Giel and Issa, 2015). A short description of each is provided as follows:

Interactive Capability Maturity Model

This Interactive Capability Maturity Model (I-CMM) is part of the United States National Building Information Modeling Standard (NBIMS). I-CMM was first published in 2007, slightly modified in 2009 and released as v1.9. However, it has not been updated since then. The I-CMM establishes the maturity of the model/project by assessing 11 topics against ten maturity levels. Using either a static table or an interactive Excel tool, I-CMM generates a single maturity score. This is intended to help 'determine the level of maturity of an individual BIM as measured against a set of weighted criteria agreed to be desirable in a Building Information Model' (NIST, 2007, 2015; NIBS, 2007; Suermann et al., 2008).

BIM proficiency matrix

In 2009, Indiana University developed an interactive Excel matrix to assist their own internal team and other facility owners to pre-qualify the supply chain. The matrix includes eight categories measured against four maturity levels that, upon completion, generates a single BIM Maturity Score (Indiana University, 2009). The matrix is completed by candidate project team members who must provide examples of past projects and address each of the eight BIM proficiency categories.

BIM maturity levels

The iBIM model or the 'Wedge' BIM maturity model (Figure 3.1) was developed by Mark Bew, chairman of the HM Government BIM Working Group; and Mervyn Richards OBE, member of the buildingSMART UK managing board (BIM Industry Working Group, 2011; buildingSMART UK, 2015; BIM Task Group, 2015). In its current form, the model reflects both the UK government's BIM strategy and many of the industry's ongoing BIM initiatives. The model identifies different levels of market maturity by grouping standards and working methods into three initial BIM levels. Based on this model, in 2011 the UK government's BIM Task Group mandated that all publicly-procured construction projects are to meet Level 2 requirements by 2016 (Hooper, 2015). While the terminology, standards and methods linked to the 'Wedge' model are mostly UK-specific, it has been used as a guiding framework for BIM policy development by other policymakers in a number of countries.

BIM QuickScan

BIM QuickScan is an online tool developed by the Netherlands Organisation for Applied Scientific Research (TNO). BIM QuickScan is intended to assess the BIM performance of organisations in the Netherlands and generate a performance benchmark. The assessment is conducted against four chapters (categories) and multiple key performance indicators, and is available in two versions: a free *self-scan*, and a more detailed commercial service delivered by a BIM consultant (Sebastian and Berlo, 2010; Van Berlo et al., 2012).

BIM Maturity Matrix

Developed as part of a larger BIM Framework (Succar, 2009), the BIM Maturity Matrix is a static self-assessment tool with ten capability sets, three capability stages and five maturity levels. According to Succar (2010a; 2010b), the matrix is intended for assessing the BIM capability and BIM maturity of organisations; where BIM capability refers to minimum ability, and BIM maturity refers to the quality, repeatability and predictability of these abilities.

Vico BIM score

In 2011, the BIM software tool vendor Vico developed its own scorecard directed towards construction managers. With a focus on clash detection, scheduling and estimating, the declared aim of the tool is to assist organisations to compare their performance against their competitors. Each of these areas is graded based on functionality and capability, best practices and enterprise integration (Vico, 2011).

CPIx-BIM Assessment Form

The CPIx-BIM Assessment Form is a static questionnaire developed by the Construction Project Information Committee in the UK. The questionnaire includes 12 areas grouped under four categories and is 'based on working documentation provided by Skanska' (CPI, 2011).

BIM Excellence

BIM Excellence (BIMe) is a commercial online platform for performance assessment. It is based on the published research of Succar (2010a; 2010b) and Succar et al. (2013). BIMe includes multiple modules for assessing the performance of individuals, organisations, projects and teams. The basic free assessment generates a simple downloadable report, while the more detailed assessments generate competency profiles for comparison against project requirements, pre-qualification criteria and role definitions (BIMe, 2013).

bimSCORE

bimSCORE is a commercial tool based on the VDC Scorecard, the research effort conducted by Kam et al. (2014) at Stanford's Center for Integrated Facility Engineering (CIFE). bimSCORE evaluates BIM practices across ten dimensions, grouped under four areas. The scorecard, which also has a free online version, evaluates the maturity of virtual design and construction practices on construction projects. It does this by comparing the performance of new projects against past projects and industry benchmarks (bimSCORE, 2013; CIFE, 2013).

BIM Planning Guide for Facility Owners

The BIM Planning Guide for Facility Owners is a maturity matrix developed by the Computer Integrated Construction (CIC) Research Program at Pennsylvania State University. The Excel matrix is divided into six categories, five maturity levels and is intended for owners to rate their own organisation (CIC, 2013).

BIM Competency Assessment Tool

The BIM Competency Assessment Tool (BIMCAT) is based on research reported by Giel (2015). BIMCAT is targeted towards facility owners and includes 12 competency categories and 66 factors measured against six maturity levels. The assessment is intended to be self-administered by a manager using an interactive offline questionnaire. Upon completing the assessment, the tool generates a single maturity score as well as a number of radar charts.

For a more comprehensive comparison between these frameworks, please refer to Giel (2013), Azzouz et al. (2015) and NIST (2015).

The way forward

These frameworks for measuring performance are still evolving and are applied differently across the market. In some countries, governments drive the adoption of BIM, while in other countries industry groups take the lead. While the frameworks presented by Succar and Bew/Richards are among the most ambitious, the methods necessary to measure BIM capability and maturity in more depth still require additional development.

If maturity is monitored through a framework such as the iBIM model, it can possibly document a historic process and describe to the outside world which steps the construction sector or company has taken. Industry organisations can position themselves according to what systems they use. Unfortunately, this does not say anything about what they are capable of doing, placing the definition of capability in the limelight. If it is used to describe an organisation's capability to deliver specific outcomes, it will then relate to the ability to deliver projects according to pre-set specification.

If it is viewed in a more dynamic way, it would then inform how an organisation adapts and changes with the use of information-sharing through BIM. This is reflected in Succar's capability stages and maturity levels, which allow the assessment of organisational processes and business models. However, such qualitative indicators may not be sufficient and provide an opportunity to develop quantitative metrics for establishing an organisation's BIM competence in a detailed and measurable way.

Conclusion

While the effect of national BIM mandates is still to be seen, they will certainly challenge clients to be more diligent in identifying their BIM requirements during the procurement process. By clearly identifying their requirements, they would encourage industry actors to adopt technological solutions and develop innovative ways to fulfil these requirements. As innovation diffuses across the market, the need to assess and compare performance levels becomes paramount. This is where authorities and industry actors need to collaborate to adopt, adapt or develop a common BIM performance framework. This is also where they will need to join forces to communicate the benefits of performance assessment across the construction industry.

Bibliography

Andersen, E. S. and Jessen, S. A., 2003. Project maturity in organisations. *International Journal of Project Management*, 21(6), pp. 457–461.

Azzouz, A., Copping, A. and Shepherd, P., 2015. *An Investigation into Building Information Modelling Assessment Methods (BIM-AMs)*, paper presented at the 51st ASC Annual International Conference, Houston, United States, 22–25 April.

Barlish, K. and Sullivan, K., 2012. How to measure the benefits of BIM – a case study approach. *Automation In Construction*, 24, pp. 149–159.

BIMe, 2013. *BIM Excellence.* [Online] Available at: https://bimexcellence.com [Accessed 28 September, 2015].

BIM Industry Working Group, 2011. *A Report for the Government Construction Client Group Building Information Modelling (BIM) Working Party*, strategy paper, Department of Business, Innovation and Skills.

bimSCORE, 2013. *bimSCORE.* Available at: www.bimscore.com/index.php/welcome [Accessed 28 September 2015].

BIM Task Group, 2015. *Meet the Team.* Available at: www.bimtaskgroup.org/meet-the-team/ [Accessed 8 October 2015].

Boverket, 2012. *Harmonisering Av Kommunernas Planarbete I Digitalmiljö – Rapport 2012:2*, Kalmar: Boverket.

buildingSMART UK, 2015. *About Us.* Available at: www.buildingsmart.org.uk/groups2#section-0 [Accessed 8 October 2015].

CIC, 2013. *BIM Planning Guide for Facility Owners.* Version 2.0. Available at: http://bim.psu.edu [Accessed 28 September 2015].

CIFE, 2013. *Center for Integrated Facility Engineering (CIFE)*, Virtual Design and Construction (VDC) Scorecard. Available at: http://cife.pbworks.com/w/browse/#view=ViewFolder¶m=Dimi [Accessed 28 September 2015].

CPI, 2011. *CPIx-BIM Assessment Form.* Available at: www.cpic.org.uk/wp-content/uploads/2013/06/cpix_-_bim_assessment_form_ver_1.0.pdf [Accessed 28 September 2015].

Curtis, D. B., Hefley, W. E. and Miller, S. A., 2002. *The People Capability Maturity Model: Guidelines for Improving the Workforce*, s.l.: Addison-Wesley.

Curtis, B., Hefley, B. and Miller, S., 2009. *People Capability Maturity Model (P-CMM) Version 2.0* (No. CMU/SEI-2009-TR-003). Pittsburgh, PA: Carnegie-Mellon University of Pittsburgh Software Engineering Institute.

Giel, B., 2013. *A Comparison of Existing BIM Assessment Tools.* Available at: http://projects.buildingsmartalliance.org/files/?artifact_id=5140 [Accessed 28 September 2015].

Giel, B., 2015. *Framework for Evaluating the BIM Competencies of Facility Owners*, PhD thesis, Gainesville, FL: University of Florida.

Giel, B. and Issa, R., 2015. Framework for evaluating the BIM competencies of facility owners. *Journal of Management in Engineering* (in press).

Giel, B. and Raja, R.A., 2012. *Quality and Maturity of BIM Implementation within the AECO Industry*, paper presented at the 14th International Conference on Computing in Civil and Building Engineering, Moscow, Russia, 27–29 June.

Grant, K. P. and Pennypacker, J. S., 2006. Project management maturity: an assessment of project management capabilities among and between selected industries. *Engineering Management, IEEE Transactions on*, 53(1), pp. 59–68.

Hooper, M., 2015. *BIM Anatomy II: Standardisation Needs & Support Systems*, doctoral dissertation, Lund: Lund University.

I-CMM., 2009. *Interactive Capability Maturity Model. Part of the United States National Building Information Modeling Standards Version 3 (US- NBIMS-v3).* [Excel document]. Available at: www.buildingsmartalliance.org/client/assets/files/bsa/BIM_CMM_v1.9.xls [Accessed 28 September 2015].

Indiana University, 2009. *IU BIM Proficiency Matrix*. Available at: www.iu.edu/~vpcpf/consultant-contractor/standards/bim-standards.shtml [Accessed 28 September 2015].

IPMA, 2015. Available at: www.ipma.ch [Accessed 10 August 2015].

Jia, G., Chen, Y., Xue, X., Chen, J., Cao, J. and Tang, K., 2011. Program management organization maturity integrated model for mega construction programs in China. *International Journal of Project Management*, 29(7), pp. 834–845.

Jung, Y. and Joo, M., 2011. Building information modelling (BIM) framework for practical implementation. *Automation in Construction*, 20(2), pp. 126–133.

Kam, C., Senaratna, D., McKinney, B. and Xiao, Y., 2014. *The VDC Scorecard: Formulation and Validation*. Stanford: Center for Integrated Facility Engineering.

Kassem, M., Succar, B. and Dawood, N., 2013. *A Proposed Approach to Comparing the BIM Maturity of Countries*, paper presented at CIB W78 30th International Conference, Bejing, China, 9–12 October.

Kolbe, T. H., 2009. Representing and exchanging 3D city models with CityGML, in *3D Geo-Information Sciences*, Berlin: Springer, pp. 15–31.

Lee, A. and Sexton, M.G., 2007. nD modelling: industry uptake considerations. *Construction Innovation*, 7(3), pp. 288–302.

Merschbrock, C. and Figueres-Munoz, A., 2015. Circumventing obstacles in digital construction design: a workaround theory perspective. *Procedia Economics and Finance*, 21, pp. 247–255.

Mihindu, S. and Arayici, Y., 2008. *Digital Construction through BIM Systems Will Drive the Re-engineering of Construction Business Practices*, paper presented at the 12th International Conference: Information Visualisation, London, 9–11 July.

Morin, G., Hassall, S. and Chandler, R., 2014. *Case Study: The Real Life Benefits of Geotechnical Building Information Modelling*, paper presented at the Information Technology in Geo-Engineering, Durham, UK, 21–22 July.

NIBS, 2007. *National Institute for Building Sciences (NIBS) Facility Information council (FIC) – BIM Capability Maturity Model.* Available at: www.facilityinformationcouncil.org/bim/pdfs/BIM_CMM_v1.9.xls [Accessed 11 October 2015].

NIBS, 2015. *United States National Building Information Modelling Standards Version 3*. Available at: www.nationalbimstandard.org [Accessed 30 September 2015].

NIST, 2004. *Cost Analysis of Inadequate Interoperability in the US Capital Facilities Industry: National Institute of Standards and Technology*, Gaithersburg, MD: National Institute of Standards and Technology.

NIST, 2007. *National Building Information Modeling Standard – Version 1.0 – Part 1: Overview, Principles and Methodologies: National Institute of Building Sciences*, Gaithersburg, MD: National Institute of Standards and Technology.

NIST, 2015. *National Building Information Modeling Standard – United States – Version 3.0 Transforming the Building Supply Chain Through Open and Interoperable Information Exchanges*, Gaithersburg, MD: National Institute of Standards and Technology.

Nya Karolinska Solna, 2015. *Nya karolinska solna*. Available at: www.nyakarolinskasolna.se [Accessed 9 September 2015].

Paulk, M. C., Curtis, B., Chrissis, M. B. and Weber, C. V., 1993. Capability maturity model, Version 1.1. *Software, IEEE*, 10(4), pp. 18–27.

PMI, 2015. *Project Management Institute*. Available at: www.pmi.org [Accessed 10 August 2015].

Sebastian, R. and Berlo, L., 2010. Tool for benchmarking BIM performance of design, engineering and construction firms in the Netherlands. *Architectural Engineering and Design Management*, 6(4), pp. 254–263.

Succar, B., 2009. Building information modelling framework: a research and delivery foundation for industry stakeholders. *Automation in Construction*, 18(3), pp. 357–375.

Succar, B., 2010a. Building information modelling maturity matrix, in J. Underwood and U. Isikdag (eds.), *Handbook of Research on Building Information Modeling and Construction Informatics: Concepts and Technologies*, Hershey, PA: IGI Global, pp. 65–103.

Succar, B., 2010b. *The Five Components of BIM Performance Measurement*, paper presented at CIB World Congress 2010, Salford, UK, 10–13 May.

Succar, B., Sher, W. and Williams, A., 2012. Measuring BIM performance: five metrics. *Architectural Engineering and Design Management*, 8(2), pp. 120–142.

Succar, B., Sher, W. and Williams, A., 2013. An integrated approach to BIM competency acquisition, assessment and application, *Automation in Construction*, 35, 174–189.

Suermann, P. C., Issa, R. R. A. and McCuen, T. L., 2008. *Validation of the US National Building Information Modelling Standard Interactive Capability Maturity Model*, paper presented at the 12th International Conference on Computing In Civil and Building Engineering, Beijing, China, 16–18 October.

Travaglini, A., Radujković, M. and Mancini, M., 2014. Building Information Modelling (BIM) and project management: a stakeholders' perspective. *Organization, Technology & Management in Construction: An International Journal*, 6(2), pp. 1001–1008.

Van Berlo, L., Dikkmans, T., Hendriks, H., Spekkink, D. and Pel, W., 2012. *BIM QuickScan: Benchmark of Performance in the Netherlands*, paper presented at the 29th International Conference on Applications of IT in the AEC Industry CIB W078, Beirut, Lebanon, 17–19 October.

Vico, 2011. *Calculating Your BIM Score*. Available at: www.vicosoftware.com/resources/calculating-bim-score/tabid/273811/Default.aspx [Accessed 28 September 2015].

Wong, K. D. and Fan, Q., 2013. Building information modelling (BIM) for sustainable building design. *Facilities*, 31(3/4), pp. 138–157.

Yazici, H. J., 2009. *Does Project Maturity Matter for Organizational Success?* Paper presented at 2009 Industrial Engineering Research Conference, Miami, FL, 30 May–3 June.

4 BIM Benefits Realisation Management

Adriana X. Sanchez, Sherif Mohamed and Keith D. Hampson

Introduction

> Widespread adoption of BIM is largely dependent on how the industry perceives its genuine benefits. Users who are to adopt BIM need to be encouraged by using empirical evidence. Investors also need to justify their investment of time and budget in BIM by discerning clear proof of its benefits. Research has shown that one of the major hurdles for adopting BIM is the justification of the additional cost and benefits.
>
> (Lu et al., 2013)

The value of Building Information Modelling (BIM) is realised through its benefits for different stakeholders. Benefits arise because BIM, like other information technologies, enables people to do things more efficiently by allowing and shaping new ways of working through the redesign of intra- and inter-organisational processes or facilitating new work practices (Peppard et al., 2007). Through this, BIM provides the construction industry with the opportunity to go through a similar transformation to that seen in the automotive industry by reinforcing core construction processes (Sacks et al., 2010).

In order to develop a business case for BIM, a robust process is required to determine if and how benefits are to be achieved, and in which phase of the project life-cycle (Love et al., 2014). This is because expected benefits from any information technology or system implementation are unlikely to emerge automatically. 'Any benefits sought must first be identified along with the changes in ways of working that will bring about and sustain each of the benefits' (Peppard et al., 2007).

This chapter introduces the theory behind Benefit Realisation Management (BRM) and how it can be applied to BIM. It also provides a value realisation framework for the implementation of this approach and a step-by-step guide for its implementation. The framework was developed in consultation with different sectors of the construction industry from infrastructure and buildings, including a number of industry associations in Australia and New Zealand as well as research centres in Australia, Singapore and Sweden.

One objective of this methodology is to be flexible enough to suit a large range of project types and scales while allowing stakeholders to gain a better understanding of what they can achieve by implementing BIM, how they can achieve it and monitor their progress towards those goals. This framework should be seen as complementary to traditional BIM implementation guidelines.

Benefit Realisation Management for BIM

> Benefit Realisation Management is the process of organising and managing, so that potential benefits, arising from investment in change, are actually achieved... a continuous management process running throughout the programme.
>
> (Bradley, 2010)

The BRM approach was originally developed in the 1980s and 1990s to answer the need to understand the return on investment from information technologies and systems, and overcome the limitations of traditional investment appraisal techniques. This aspect of project management has received increasing attention in the past few years (Breese, 2012).

BRM is changing the way performance of major capital investment programmes and projects is thought of by focusing on using benefits to drive, manage and measure this performance. It is based on the idea that vague identification and definition of benefits at the concept stage makes it difficult to achieve and manage said benefits. This can potentially result in the failure to achieve the organisation, programme or project goals (Yates et al., 2009). This process helps answering the questions of *how* benefits will materialise, over what period of time and what the value is; a business imperative for asset owners' executives and managers (Love et al., 2014). This methodology has been adopted by a number of government organisations across the globe to manage and demonstrate the realisation of benefits from business change programmes (Bradley, 2010; Peppard et al., 2007; NSW Government, 2014a). Professional organisations such as the Association for Project Management in the UK and private companies such as Fujitsu Consulting in Canada have also endorsed this approach as a way of obtaining better value from information technology investment (Fujitsu Consulting with John Thorp, 2007; APM, 2012).

A study published in 2015 analysed the effects of implementing BRM methodologies in the management of projects across Brazil, the UK and the US, surveying 331 project management practitioners. The results from this study provide evidence of the contribution of BRM to the creation of business value. Their conclusions suggest that BRM helps stakeholders align around specific goals and set clear success criteria, and contributes to a lower risk of project failure (Martins Serra and Kunc, 2015).

Benefits from BIM can act as the focal point for the development of metrics and action plans that expand across the life-cycle of a built asset. In the past, the implementation of information technology and systems has often been considered only when writing the initial business case, with any ongoing benefits focus rapidly fading away after the start of implementation. Project-focused approaches often also fail to take into account organisational learning and capability development over a significant period of time in which many individual projects may be undertaken (Doherty et al., 2012). BRM can help change existing strategic management methods and practices to ensure that benefits from BIM are realised (Love et al., 2014).

The BRM methodology and theory developed by Ward et al. in 1996 remains one of the most widely used and cited models. This method has been publicly used by 'well over a hundred organizations of all sizes, in both public and private sector across the world' (Peppard et al., 2007). BRM has five general stages: (1) identifying and structuring benefits; (2) planning benefits realisation; (3) executing the benefits realisation plan; (4) evaluating and reviewing the results; and (5) discovering potentials for further benefits (Figure 4.1) (Hesselmann and Mohan, 2014). It is important to note that in order to evaluate the results of the implementation process, access to data that forms a baseline for comparison may also be required to quantify improvements. In some cases, this may mean groups can benefit from identifying the desired benefits and monitoring strategy before the implementation is carried out in order to measure specific indicators and establish this baseline.

The first stage, identification of potential benefits for all stakeholders, provides the foundation for: (1) determining how expected benefits can be achieved; (2) defining the changes needed to realise each benefit; (3) assigning responsibilities for their delivery; and (4) making explicit the resulting benefits (Ward et al., 2007). At this stage, benefits are identified, appropriate measures are derived and the linkages between an investment and the business changes required to realise the anticipated benefits concluded (Hesselmann and Mohan, 2014).

Benefit identification and categorisation

> A benefit is defined as an outcome of change which is perceived as positive by a stakeholder... In relation to a measure, a benefit could be defined as the improvement in a measure arising from a particular programme/project.
>
> (Bradley, 2010)

Benefit identification is a critical process that should engage a broad cross-section of relevant stakeholders (Bradley, 2010). This process identifies and documents potential benefits that can arise from an information technology or system investment. These should include a mixture of tangible

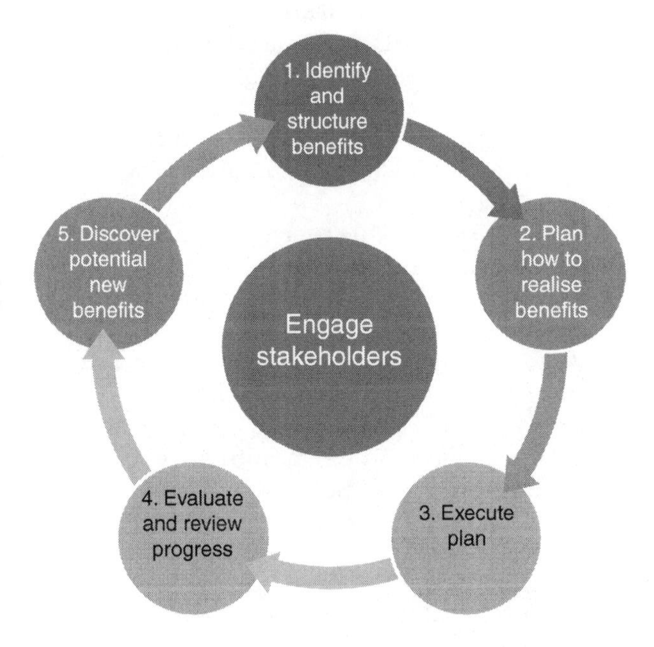

Figure 4.1 BRM general stages according to Hesselmann and Mohan (2014)

and intangible benefits, and associate specific metrics to each benefit that can be used to measure the success of the project (Smith et al., 2008).

Benefit identification is a core feature of BRM, used to create a detailed plan of how those benefits are to be realised throughout the life-cycle of implementation and use of the new technology, process or system. The resulting plan can be used to guide actions throughout implementation, 'and to review progress and achievement both during the project and following its completion' (Peppard et al., 2007).

Classification of benefits can increase understanding of the nature of the benefits, aid analysis and improve communication (Bradley, 2010). Benefits from BIM could for example be categorised in three areas: (1) tangible benefits (quantifiable in monetary terms); (2) semi-tangible benefits (quantifiable, but not in monetary terms); and (3) intangible benefits (non-quantifiable, described qualitatively) (Becerik-Gerber and Rice, 2010).

Although financial quantification of intangible benefits might be attempted by making assumptions and linking to other benefits, these benefits are normally non-quantifiable in monetary terms. They represent a *soft* return on investment and are often measured through qualitative metrics (Becerik and Pollalis, 2006).

Benefits can also be categorised as strategic, tactical and pecuniary; or enhanced productivity, business expansion and risk minimisation (Love et al., 2004). Other classifications include:

- By stakeholder (through, for example, a *Benefit Distribution Matrix* where benefits are on one axis, stakeholders on the other and tick marks are used to signal who benefits (See Table BD.1 of Part II – *Benefits dictionary*).
- By categories that are relatively independent of one another (e.g., cost reduction, revenue generation, risk reduction, productivity, workforce satisfaction and customer service).
- By business impact (helpful when checking strategy alignment or comparing the relative significance of benefits).
- By change type (based on whether the benefit will be achieved by doing new things; stop doing existing things; or doing existing things better).
- By sigma value type (definite financial, expected financial, logical financial, qualitative (or semi-tangible) and intangible) (Bradley, 2010).

The *Benefits dictionary* in Part II of this book has identified a series of benefits that have been acknowledged in the literature to realise the value of BIM in construction projects. This section aims to provide the reader with a starting point from which to identify the benefits they would like to realise and, in time, build their own dictionaries.[1] The dictionary focuses mostly on tangible and semi-tangible benefits and the profiles include definitions for different phases of an asset's life-cycle.

Benefit network mapping

Benefit maps provide a practical approach to the issue of having to use metrics to measure intermediary benefits based on assumed links to other benefits in order to provide a financial value. Measuring productivity through reduced staff salary cost in a project is an example of this. By separating benefits but making explicit their link, the intention can be clearly communicated, double counting of financial benefits can be avoided and the time factor can be included (Bradley, 2010).

A benefits map is also the starting point to identifying metrics for each benefit and is a critical part of BRM (Bradley, 2010). The benefits dependency network (BDN) specifically is a tool developed for BRM that is used to link the overall objectives and benefits to business changes and essential capabilities that enable these changes (Hesselmann and Mohan, 2014). Changes can be separated into business and enabling, where the former are those that are permanent changes to the working practices and enabling changes are one-off changes that are prerequisite for the business changes to occur (Peppard et al., 2007). In this book, changes are limited only to enablers due to the asset life-cycle approach rather than organisational. Enablers are defined here as processes and tools that facilitate the realisation of or enhance the benefit in question.

This type of map provides the means for understanding the route to the ultimate benefit through an analysis of cause and effect linkages. As the understanding of the benefits evolves through implementation and development of new applications, the status of benefits maps will progressively change from a simple wish list to a set of feasible options that can then be prioritised and developed into an action plan. This can then evolve into a way for communicating expectations and eventually become a report of progress towards achieving specific goals (Bradley, 2010). Thus, benefit maps and BDNs specifically, can provide a framework for explicitly linking the overall goals and benefits with the enablers necessary to deliver those benefits (Peppard et al., 2007).

A benefits map is usually created through consultation starting with the objective or end-benefit on the right-hand side and then working backwards to the enabler on the left-hand side. The first step is to determine a set of end-benefits that fully represent the goal of implementation. Once these end-benefits are identified, intermediary benefits can be added and linked. 'Benefits may be added from a pool of benefits, previously identified using another process, or identified using the map logic alone' (Bradley, 2010).

This tool is also central to the plan-do-check-act cycle, where feedback provides the impetus for a process of evaluation and learning (Love et al., 2014). The metrics for each benefit and the benefits themselves are best identified by including a cross-section of key stakeholders in the development of the map. In specific projects, these maps can be used as the framework for project workshops (Bradley, 2010). Once an initial benefit map has been developed, metrics for each of the benefits and responsibilities for all of the benefits and changes are normally assigned and time-scales established (Peppard et al., 2007).

Developing a benefits map for innovative technologies and process is inevitably iterative. This is because benefits are difficult to define and are dependent on the nature of the changes the organisation is willing to make and its ability to develop and deploy new technology and processes (Peppard et al., 2007). This applies to rapidly advancing socio-technical systems such as BIM and, therefore, it is expected that the benefit maps will evolve as the technology changes if this method is applied at the organisational level.

Dictionaries

The *Benefits dictionary* in this book provides profiles of a set of benefits based on an extensive review of literature. In general, profiles should contain a comprehensive description of a single benefit including all its attributes and dependencies (Bradley, 2010). Each profile normally should include: (1) identifier information (title and full description); (2) impact and contribution to end-goals; (3) dependencies to enablers and to other benefits; (4) stakeholder information, including beneficiaries and owners; (5) potential metrics; and (6) assumptions and risk related to the realisation of the benefit. Additionally, it is recommended that users include baseline target, realisation timescales, measurement frequency, reporting mechanism and risks related to their specific projects.

The *Benefits dictionary* provided in this publication is meant to provide a strong starting point for new users by gathering information from a large number of references related to current and soon-expected benefits. Responsibilities and time-scales are not included, as these would depend on project-specific characteristics, but are expected to be assigned when applying the framework.

Metrics dictionaries provide a practical way of avoiding wasted efforts often seen in recording and tracking metrics that are being tracked elsewhere. When this dictionary is developed for a project or an organisation, it should contain all the metrics that are to be measured, each with an expanding history of actual values that can help form benchmarks and targets (Bradley, 2010). The *Metrics dictionary* provided in this book offers a broad range of practical metrics for projects to select and apply based on their goals and needs. These metrics are mostly based on literature but also include KPIs proposed by the authors based on professional experience. The *Enablers dictionary* provides definitions, useful references and examples for all tools and processes mentioned in the *Benefits dictionary* as enabling specific benefits.

BIM value realisation framework

> Benefits, which can be seen as improvements, are increments in the business value from not only a shareholders' perspective but also customers', suppliers', or even societal perspectives… Therefore, the creation of value for business, by the successful execution of business strategy, strongly depends on programmes and projects delivering the expected benefits.
>
> (Martins Serra and Kunc, 2015)

The BIM value realisation framework was developed in Australia in consultation with national and international organisations as part of the research project: Driving Whole-of-life Efficiencies through BIM and Procurement. This project was funded by the Australian Sustainable Built Environment National Research Centre (SBEnrc) and aimed to:

1. define indicators to measure tangible and intangible benefits of BIM across a project's life-cycle in infrastructure and buildings; and
2. pilot test a whole-of-life BIM value realisation framework on leading infrastructure and building case studies.

To do this, the team carried out an extensive literature review, industry consultation and three exemplar case studies in Australia that featured the use of BIM during design, construction and asset management. More information about these case studies can be found in SBEnrc (2015).

It was important that the framework could be easily applied to both infrastructure assets and buildings. The research team studied a number of different approaches available to develop a strategy for measuring the value of information technology in construction. It was also consistently reviewed by a group of industry and academic experts from infrastructure and buildings, and at different levels of the supply chain. An introduction to the framework was also presented at the Australasian Regional Conference organised by the International Roads Federation (IRF) and Roads Australia in May 2015. This was done in order to obtain feedback from a wider audience and address common concerns of BIM guidelines being often directed only to buildings and architectural design.

BRM practices have been shown to be associated with the creation of value (Martins Serra and Kunc, 2015) and been applied to a number of sectors and stakeholder groups (Bradley, 2010; Peppard et al., 2007). However, tailoring this framework for specific organisations and sectors is an essential step towards optimising its value (NSW Government, 2014b).

As shown in Figure 4.2, the framework presented here follows at large the traditional BRM structure and principles but has been tailored to the construction industry based on information provided by a number of sources including academic papers, books and government guidelines, as well as the above-mentioned feedback. The framework implementation is meant to be informed by other sections of this book and acknowledges that 'although benefits are not the only criteria to evaluate project success, they are a measurement of how valuable a project is' (Martins Serra and Kunc, 2015). The methodology presented here also acknowledges that value is realised not only through specifically identified end-benefits but also through unintended benefits. This has been addressed by including 'flow-on' benefits to the benefits profiles. These are benefits that can be obtained once the end-benefit is achieved and are recommended to be identified and monitored in order to account for the full value delivered by implementing BIM.

The framework was partially validated by the Sydney Opera House, which at the time was transitioning into a BIM-based asset management system. Feedback from this effort and the previously outlined consultation was used to finalise the framework into its present form. The Sydney Opera House was chosen because in addition to being in the process of developing an asset management system, this precinct also contains infrastructure and building elements recently designed and constructed using BIM. This organisation has also been actively engaged in BIM research projects since 2004 (CRC for Construction Innovation, 2007) and was in a position to provide feedback not only on the framework itself but on the completeness of the dictionaries provided.

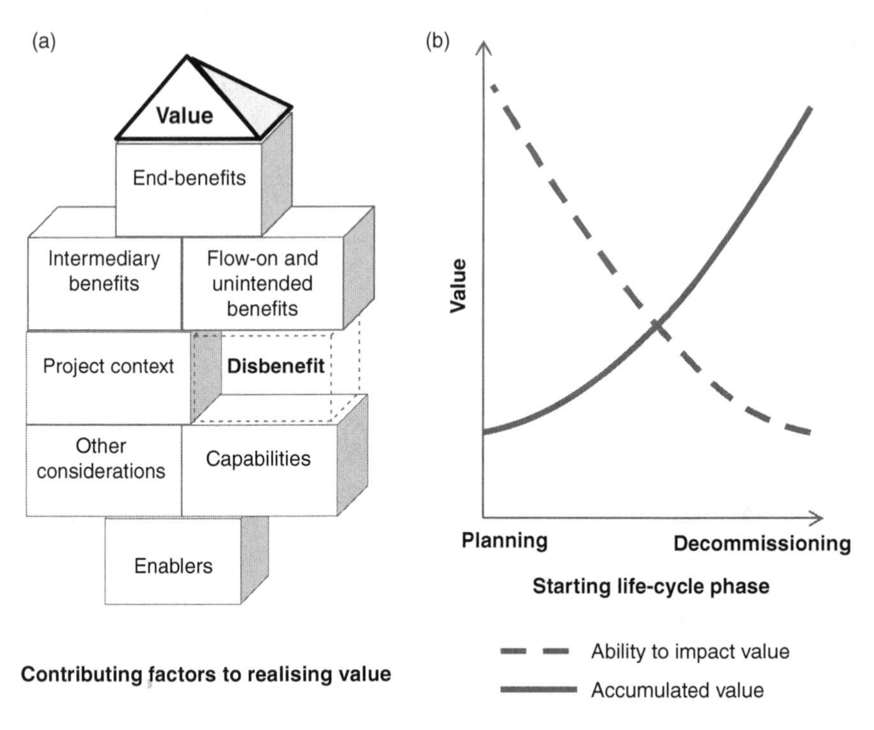

Figure 4.2 Framework principles: (a) contributing factors to realising value; (b) value/life-cycle relationship

Step-by-step guide

This step-by-step guide aims to provide a practical tool for project teams that want to demystify BIM by understanding its potential benefits and implications. The methodology facilitates focusing on the endgame upfront and helps all parties align around overarching objectives rather than focusing on a specific software solution or measurement. This method can be used as a tool to visualise the process involved in the realisation of the value of BIM through specific benefits, including those often seen as intangible. It also provides a set of leading indicators that can be used for proactive value realisation management.

Bradley (2010) explains that when groups start by focusing on a specific software solution and then identify the benefits that can be obtained they are 'putting the cart before the horse'. Nevertheless, the starting point of this methodology is that the user (client or other stakeholder such as designer, contractor, etc.) has decided to implement BIM in a project or is building the business case. Implementing BIM is 'far more complicated than simply implementing an IT project' (Love et al., 2014). BIM is not a single software package that teams can just 'plug in and run' in isolation. It is a new way of

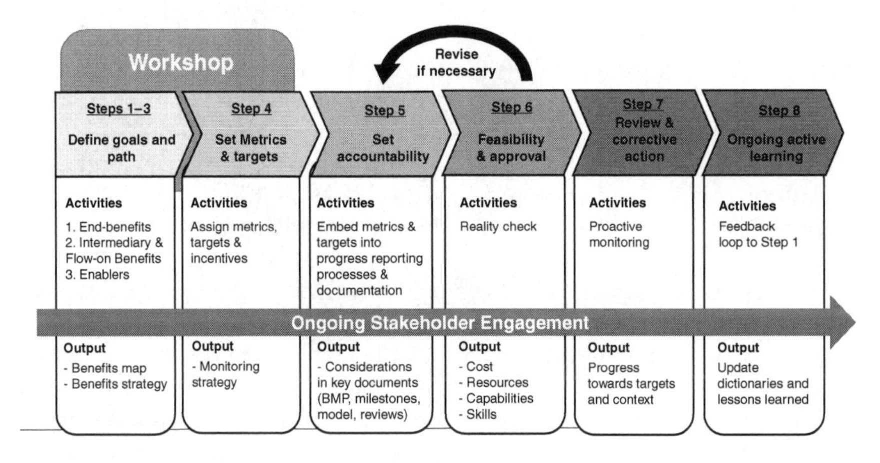

Figure 4.3 Steps of BIM value realisation framework

working that commonly includes the use of a number of tools and software solutions. This can be overwhelming, especially for new users. This methodology aims to provide a strong starting point based on currently achievable benefits from implementing BIM in construction that have been identified by the literature on this topic. It also provides information on the type of tools that would be required in order to realise and monitor this value. It is also based on the understanding that organisations do not need to obtain every technology available associated with BIM to start to realise its value. BIM can be implemented following a progressive and incremental approach by focusing on different benefits at each step. This may be especially relevant to small and medium-sized enterprises (SMEs) and client organisations with limited resources.

The BIM value realisation framework can be applied at any phase of the life-cycle of an asset and is meant to complement BIM implementation guidelines. The MacLeamy Curve (AIA, 2007), however, applies to this process as well. This means that the earlier changes and processes required to implement BIM are introduced, the larger impact they are likely to have on the outcomes of the project and realising the full value of BIM. Therefore, this methodology is recommended the most at the beginning of a project and if applied in different phases, at the start of those project phases.

The methodology is based on eight steps shown in Figure 4.3.

Step 1: define end-benefits

End-benefits

The first step is to define what end-benefits the team will focus on. These are the ultimate objectives – the value that the team definitely wants to have realised from implementing BIM such as lower cost, improved safety and competitive advantage gain. The team should work together to identify and agree on one to three benefits that align with their project and organisational goals. These objectives will provide the main incentive for the strategy developed and will become the central point of the value realisation process. These benefits should therefore be linked to strategic project or organisational objectives (Martins Serra and Kunc, 2015; Bradley, 2010). Defining these end-benefits is recommended to be done through a workshop with key stakeholders.

Benefits are improvement on current standards of practice and project outcomes. In order to identify those that are most important to the team, it may be useful to initiate the conversation by exploring problems and challenges commonly faced in similar projects that can be resolved or prevented. Depending on the context of the implementation strategy, the focus may be to avoid problems or, if current practices are acceptable, to improve upon the norm.

Workshop

Integrated BIM and teams are based on collaborative stakeholder engagement and information integration across the supply chain and asset life-cycle phases. However, who is invited to this exercise will depend on the project type and phase. In fact, this framework may be implemented several times over the life-cycle of the asset and the stakeholders involved may change every time. At concept stage and when considering the use of BIM for the following phases, it is recommended that, as a minimum, the following roles are included: project manager, asset manager, designers and end-users. Including asset managers and end-users will ensure a whole-of-life insight.

Depending on the contract type, the group may also include the contractor, subcontractor coordinator, surveyors and/or suppliers. The selection will be based on answering the questions: who are those that will be affected the most by the changes? And who stands to gain from it? Alternatively, the list can be formed by answering the question: who will need to accept and use the system and/or information created during this or any future phases?

BIM and integrated project delivery implementation guides often recommend carrying out a team integration workshop or series of workshops (AIA, 2007; BSI, 2015; NATSPEC, 2011; ACIF and APCC, 2014). One of these workshops can be used to implement the value realisation framework. Doing so provides an opportunity to align the team around a set of objectives, and gives them ownership over the process by being involved in the map development. It also provides an opportunity to become familiar with the benefits that can be achieved and the path to realisation. For more detailed information about how to plan these workshops, please refer to Bradley (2010, pp. 15–16).

A facilitator will be required to carry out the workshop. This person should be familiar with the methodology and dictionaries. If the project is at an early development stage such as concept design, the programme or project manager can serve as facilitator provided they familiarise themselves with the methodology. If the project is at a more advanced stage, it would be advisable to engage an external facilitator who can remain impartial throughout the process. Finally, if the project is at the asset management phase, the asset or strategic manager may serve as facilitator.

Options for engagement

Defining the end-benefits should normally be done without having specific software solutions in mind (Bradley, 2010). This may be possible if users are first-time BIM-users but unlikely if the workshop group is composed by experienced users who are likely to lean towards those benefits and solutions they are familiar with. To help dissipate this potential bias, it is recommended that the benefits dictionary is made available to the whole group prior to the workshop so they can familiarise themselves with it and expand the range of benefits on sight. The team should be asked to think about three benefits that they consider would be the most inspirational or necessary for this project or asset management. If the team includes users and/or system and asset managers, it may be appropriate to ask for a wish list (e.g., in a world where anything is possible, what would you like your system to be capable of doing? Or what would make your job easier and your time more productive?). These types of exercise are useful for two reasons:

1 It may help push the boundaries of what is possible and foster innovation. Users might be able to make connections more readily between seemingly isolated technologies to propose a new application of existing technologies. For example, the principles of the Nanny Cam might be proposed to integrate security systems and provide remote live feed through a BIM interface in case of emergencies. This may, for example, be proposed to help emergency responders be better prepared and prioritise their action plan.
2 It may also act as a tool to increase stakeholder buy-in, which is an important part of ensuring successful implementation (Arayici et al., 2011; Becerik-Gerber et al., 2012). Cultural bias and resistance to change in the construction industry can lead to rejection of new technologies and tools (Arayici et al., 2011). However, having been involved in the plan development may help users accept implementation.

This can be done in different ways depending on the capabilities of the organisations and workshop invitees. For example, once the roles to be involved in the exercise have been selected, the organiser can add a note to the workshop invitation asking participants to familiarise themselves with the dictionaries and develop their wish list in advance. This may include benefits in the dictionaries or just be the answer to the question 'wouldn't it be great if…?' However, this might make the workshop lengthy depending on the number of stakeholders involved and may prove discouraging for those who are normally shy or have never used BIM and may even be intimidated by the notion.

Another option is to ask invitees to nominate three benefits they would like to achieve via email or other electronic communication system. The organiser can then consolidate this information into a single list to be used as starting point for the workshop, where the organiser can either propose the three most suggested and ask for agreement around those or provide the whole list to the group and ask them to vote on three. To do this the group can use several well-developed tools such as dotmocracy (Diceman, 2014) and/or world café (World Café Community Foundation, 2015). The Australian tool BIM Value (http://bimvaluetool.natspec.org/) developed by SBEnrc and NATSPEC may also be of use by asking participants to go through the steps and send their reports to the organiser.

An additional alternative that could shorten the workshop, if the organiser has time to set up a poll, is to use an online voting or polling tool. These tools are usually easy to use and provide free access. There are a number of online survey and polling tools such as Poll Maker (2015) that could be used to this end. Instructions should be clear about the maximum number of options to choose. Using Poll Maker, for example, the organiser obtains a link to the results and can tick the box to allow users to provide 'other' options to expand the benefits list. Using this option would involve time investment in adding all the benefits from the dictionary as options for the poll and providing the benefits profiles as required reading. If a large number of benefits receive the same number of votes, the organiser can either carry out a second poll based only on the top most suggested benefits or carry out a vote during the workshop.

If this value realisation methodology has been used in the past and the hosting organisation has established an online benefits repository, there may be a possibility to use this system directly to conduct the poll. This would allow the users to have direct access to the different benefit profiles and maintain all records associated with this method in a single source.

Independently of the method chosen, at the end of this step the group should have a set of end-benefits that will serve as final objectives for the project in terms of value to be realised from implementing BIM.

Step 2: define intermediary and flow-on benefits

This step outlines the *story* behind each end-benefit and is carried out in the same workshop environment as the previous step. Here is where the dictionaries can be most useful to the team: by examining the benefit description section of the profiles of the chosen benefits the group should be able to work backwards towards the intermediary benefits. The team can then develop a benefit network map that links all benefits together.

Intermediary benefits are those that are expected to occur between the implementation of early changes and the realisation of the end-benefits (these are sometimes also referred to as enabling benefits) (NSW Government, 2014c).

Depending on the end-benefits chosen by the group, it may be more practical to create separate maps for each end-benefit or a combined map. The aim of this step is to understand how end-benefits are actually to be realised and what intermediary benefits are to be linked to that final goal. For example, an asset owner may have identified lower whole-of-life cost as the final goal. By examining the profile of this benefit, the group would discover the linked benefits according to the literature and build the network map in Figure 4.4. The idea is to keep asking the question 'how do we get to this benefit?' until reaching the point where the only answer left is by using a specific tool or process.

The group must agree on this map because it will be used later to determine the tools needed and the metrics to be used in order to monitor implementation and progress towards the target benefit. Once agreement has been reached, it may be in the best interest of the group to add a priority class to each benefit. This can then be used during the workshop follow-up stage when resources and budgets need to be weighed against the desired

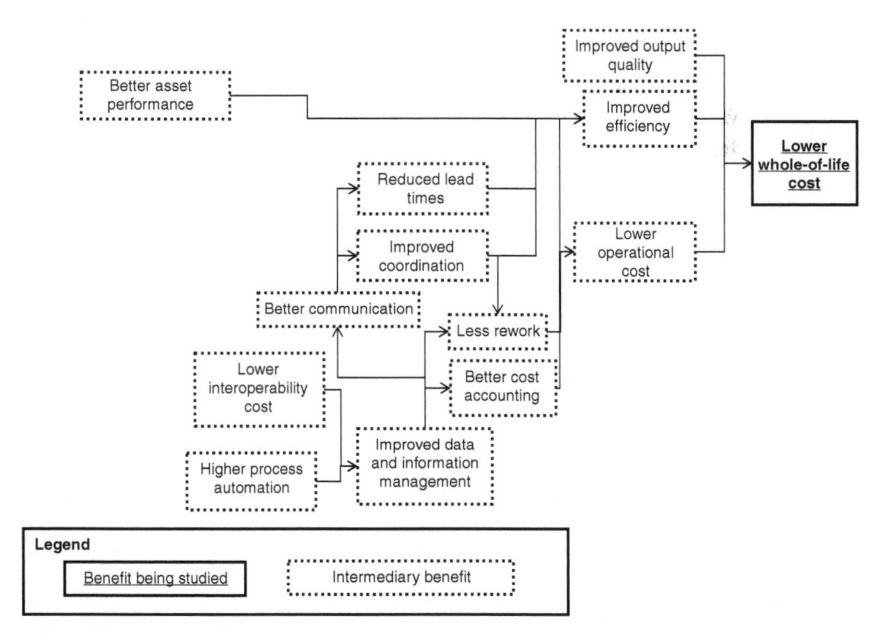

Figure 4.4 Benefit network map example for lower whole-of-life cost

benefits. Bradley (2010, pp. 122–125) proposes a weighting algorithm to develop this priority ranking. However, a simpler approach maybe to use dotmocracy again. This method would consist on providing all participants with three stickers to be allocated as they see fit and then recording this as percentages of the total number of stickers.

The second part of this step involves thinking about what benefits may be derived from achieving the end-benefit. These are called flow-on benefits. Although these benefits are not part of the core objectives, they are likely to be realised and provide extra value as unintended benefits. By being aware of this possibility, the team can choose to monitor these flow-on benefits as well, producing a more complete picture of how value is being delivered with BIM. For example, an end-benefit may be to improve project communication; this is at the core of realising many other benefits from BIM and, if achieved, it is very likely to contribute to achieving a number of other flow-on benefits such as fewer errors and reduced risk.

Step 3: define enablers

Once the benefit maps have been developed and agreed on, priorities can be assigned, and then enablers defined. Enablers are processes and tools related to BIM uses and implementation, without which the first intermediary benefit in the chain could not be achieved or that enhance the likelihood of achieving it. The benefit profiles can also be useful for this step. The enablers section of the profiles can help define the minimum required tools and processes to achieve those benefits selected by the group. The group can use this information to discuss all the enablers and agree on those that will be further investigated.

A risk is associated with each enabler and there are other considerations that need to be observed such as new skills requirements and cost (Martins Serra and Kunc, 2015). 'There are a number of examples where technology-enabled projects produce negative outcomes' (Love et al., 2014), also known as disbenefits. However, if the group is aware of these risks, then a clear strategy can be developed and included in the project risk register. The 'Notes' section of the benefit profile provides some information related to this and the enabler profile also contains information about documented past experiences. However, many of these risks will depend on the asset type, value and phase and would best be determined as part of the risk analysis.

Once enablers have been assigned, the map may look similar to Figure 4.5. While the profiles are meant to provide the group with a starting point based on what has been reported to date, the group should discuss this matter in detail. This is because there may be new tools or processes that had not been recorded at the time of writing or were not part of the literature reviewed.

If the group contains experienced users, it may be appropriate to ask them to propose specific solutions that fit the enabler's profile and functionalities. Depending on the project phase and those present at the workshop, it may

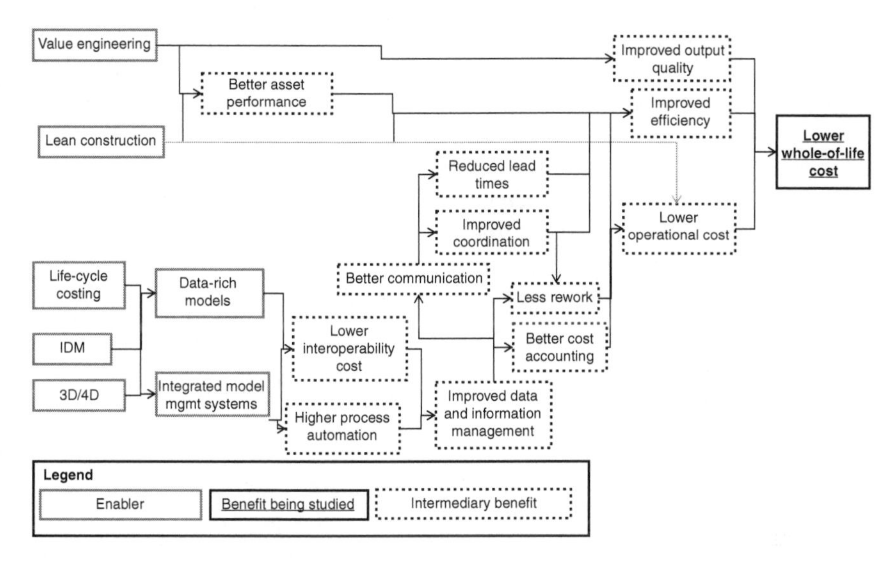

Figure 4.5 Benefit Map example with enablers
Note: IDM – information delivery manuals.

be useful to start a discussion around access to specific solutions and skills required for implementation and use of those enablers. In some cases, some of the team members may have access to some of these solutions and experience using them, while in other cases it may be required to go directly to the next step. In either case, given the rapid change of technology and advances in BIM, it would be in the group's best interest to avoid getting 'locked in' to a specific solution before step 6.

Step 4: assign metrics, targets and incentives

As explained in previous chapters, metrics provide the means to justifying investments made, comparing and ranking benefits, providing targets for success and benchmarking and monitoring progress towards goals. Assigning metrics to benefits is the basic requirement to provide effective accountability (Love et al., 2014). The metrics dictionary offers an introduction to available metrics organised in four categories:

1 People.
2 Processes.
3 Procurement.
4 Sustainability and future-proofing.

It is recommended that the group initially chooses as many in each category that relate to the benefits selected as possible. These can then be examined in terms of practical implementation and the final number reduced to suit the needs of the asset or project. It is important to ensure that the group does not focus solely on measuring one type of benefit from implementation and fails to record others that may provide useful insight into the achievement of the end-benefits.

To start this process, the group can go through all available metrics related to the intermediary and end-benefits. Once they have all been included in the map, the group can start negotiating one by one, choosing those that are to be recorded and monitored throughout the life-cycle of the project. The group may also assign metrics that are only applicable to one phase of the project, as long as this is made clear during the workshop and in the documentation. Please also refer to the 'Development, selection and implementation' section of Chapter 1 to inform this part of the discussion.

Metric profiles provide some examples of case studies where they have been used and levels achieved that may be of value for the second part of this step: setting targets. Although it is not an absolute requirement and will depend on the availability of benchmarks and previous experience with similar projects, this step is recommended for providing a better focus. If targets are associated with each metric and it is deemed appropriate, the group should also consider associating financial incentives to the achievement of benefits and exceeding the targets; achieving targets may be considered a minimum level of success depending on the contract type and phase. Alignment of targets with a reward schema is another key condition to ensure efficient accountability (Love et al., 2014) and has been shown to be successful in other areas such as achieving environmental goals in transport infrastructure construction (Sanchez et al., 2014).

Step 5: embed metrics and targets into progress report documentation and processes

This step is very important. It will ensure accountability and provide a rich source of information based on which the group can make decisions and introduce changes in a timely manner and correct situations that may be hindering the achievement of the goals set in previous steps. The group should make sure that metrics, targets and incentives are embedded in the project documentation including the regular progress report as well as any relevant documentation that may be embedded in the BIM model itself.

It is useful to develop a flowchart for implementation of the different solutions which includes key dates for rollout, responsible personnel, reporting milestones and review processes. This chart can be included in the BIM Management Plan. Targets can be associated with each level of detail and review processes specified for these milestones. 'With suitable governance structures in place managers should be encouraged to continually review and modify the BIM initiative so as to reduce risks and increase benefits' (Love et al., 2014).

The group should also make sure that the reporting process and systems require information about unexpected (or in some cases expected) issues that might be hindering the realisation of a specific objective. For example, one of the end-benefits may be achieving shorter project delivery times. However, a large number of brief changes may have slowed the programme or the diggers may have unexpectedly found an archaeological site, which brought the works to a stop.

This capacity to record context information is important due to the challenges outlined in Chapter 1 associated with measuring benefits from implementing BIM. Additionally, when this information is shared at the progress report meetings, it forces the group to think about the consequences of these issues and develop proactive action plans to correct them and meet the targets.

Alternatively, the group can choose to develop a separate document or repository to record progress towards the desired benefits. If BIM performance metrics are being implemented as well, it may be useful to record both sets of metrics in the same system, providing a more complete view of progress and better understanding of reasons behind missed opportunities.

As mentioned earlier, it is recommended to integrate this reporting process into the repository system used for BIM. In the case of BIM portfolio management, it is recommended to develop a common repository for all projects to provide a benchmarking tool for setting future targets. This repository can also include all the benefits, metrics and enablers dictionaries so they can continue to be updated and developed as technology progresses and experience is gained from implementation.

The BIM Management Plan should also include a stakeholder engagement plan outlining responsibilities and levels of engagement across the life-cycle of the project as they relate to the desired benefits.

Step 6: workshop follow-up – feasibility and approval

Depending on the project phase, it may be required to have a person appointed as responsible for the post-workshop steps. In the case of projects that will go to tender, this may not be necessary as the detail around the specific software solutions to be used can be defined by the consultants and contractors. This holds as long as they provide a format that can be used by all stakeholders that stand to benefit from that specific tool. This is an important point to make, especially if the client is planning on using BIM for asset management at project completion. In cases where the asset management system is in a development phase during the contract negotiations, the client may want to request either basic file outputs such as XML, and open formats such as Industry Foundation Classes (IFC), which should be accepted by most tools or converted easily. Ensuring interoperable formats are used is always recommended, even if files are also required in native (vendor-specific) format. This is due to the rapid rate of technological progress and the fact that some projects can require a number of years for completion. By the end of the construction phase or soon after, the asset manager may choose to change their managing system and having these format requirements could avoid further negotiations or rework.

If the framework is being applied to a project that has already gone through tender, it is important to assign a person within the project team who will be responsible for investigating specific software providers and carrying out a comparative analysis of the different tools and costs. This can be used by the strategic or project manager to negotiate which tools will be used during the project. This person could, for example, be the future BIM manager.

The associated cost will largely depend on the capabilities of the project team and previous experience with the software package chosen as well as licences already purchased. At this stage of cost analysis, the priority weight associated with each benefit may become a useful source of information.

Once budget and cost have been finalised, resources and responsibilities can be confirmed for the different processes that will be required in order to implement the different enabling tools and processes chosen.

Step 7: progress review and correction initiatives

As mentioned in previous steps, advancement towards targets related to benefits should be reported on and reviewed during project progress meetings. This step acknowledges that benefits are not automatically realised and require active monitoring (NSW Government, 2014b). Continual monitoring allows the team to take an evidence-based approach to decision-making that allows actions to be taken in a timely manner. This in turn provides an opportunity to correct any situation that might be hindering the realisation of value.

Additionally, review of 'before and after' measures provides an explicit mechanism for evaluating whether the proposed business benefits have actually been delivered (Ward et al., 1996). In the absence of benchmarks, ongoing monitoring of benefits can show progressive improvement.

Step 8: ongoing active learning

> As a result of the post-project review, it may become apparent that further benefits are now achievable, which were not envisaged at the outset. This stage provides the opportunity to plan for, and realise, these further benefits. In a wider context, this stage also provides the opportunity to review and learn from the overall project process, thus facilitating the transfer of lessons learned to future projects.
>
> (Ward et al., 1996)

Benefits are dynamic and will change as technology develops. Therefore, the dictionaries need to be regularly reviewed and updated (Love et al., 2014). It is recommended that a role is assigned whose responsibilities will include carrying out research on the technology progress and updating the dictionaries based on this research once a year or however often it is deemed suitable. This person may also have the same role assigned to record and follow-up on the specific benefit metrics and targets and compare to the 'user's wish list' as the system and technology evolve.

Through ongoing active learning, this role could also be in charge of reviewing the benefits measured across different projects and the context information in order to diagnose why some projects are successful and others are not (Love et al., 2014).

Next steps

This chapter has introduced a methodology to identify and monitor benefits that will serve to realise and deliver value with BIM across the life-cycle of an asset. However, this is only the beginning of the journey. There are a number of considerations that will have to be addressed following Step 6. Decisions about issues such as standards, protocols, BIM management roles, risk apportioning, skill development plans and system requirements will need to be made. Chapter 5 provides some implementation tips and there are a number of BIM implementation guidelines that cover these and other more technical topics at length, and are freely available. Some examples are: NATSPEC's *Getting Started with BIM* (2014) and *National BIM Guide* (2011) from Australia; the US General Services Administration (GSA) *BIM Guide Series* (2007, 2011), American Institute of Architects' *IPD Guide* (2007) and US Department of Veteran Affairs *BIM Guide* (2010), Singapore's *BIM Guide* (BCA, 2013), the British Standard series (2013, 2014a, 2014b) and International Organization for Standardization (ISO) series (2010, 2012, 2014).

Note

1 Eventually, there could be industry standard dictionaries that help avoid misunderstandings across the industry and a establish benchmarks. However, given the limited uptake and that it would require an organisation taking a global or national coordinating role, implementers of this methodology should just make sure that their project team is aligned around what each benefit means for that project, how it will be measured and who the responsible personnel is.

Bibliography

ACIF and APCC, 2014. *Project Team Integration Workbook*, Canberra: Australian Construction Industry Forum and Australasian Procurement and Construction Council.

AIA, 2007. *Integrated Project Delivery: A Guide*, Chicago, IL: American Institute of Architects.

APM, 2012. *Delivering Benefits from Investment in Change: Beyond 'Business as Usual' to 'Value as Usual'*, Buckinghamshire: Association for Project Management.

Arayici, Y., Coates, P., Koskela, L. and Kagioglou, M., 2011. BIM adoption and implementation for architectural practices. *Structural Survey*, 29(1), pp. 7–25.

BCA, 2013. *Singapore BIM Guide V2.0*, Singapore: Building and Construction Authority.

Becerik, B. and Pollalis, S.N., 2006. *Computer Aided Collaboration in Managing Construction*, Cambridge: Harvard University Graduate School of Design, Design and Technology Report Series 2006-2.

Becerik-Gerber, B. and Rice, S., 2010. The perceived value of building information modeling in the US building industry. *Journal of Information Technology in Construction*, 15(2), pp. 185–201.

Becerik-Gerber, B., Jazizadeh, F., Li, N. and Calis, G., 2012. Application areas and data requirements for BIM-enabled facilities management. *Journal of Construction Engineering and Management*, 138(3), pp. 431–442.

Bradley, G., 2010. *Benefit Realisation Management: A Practical Guide to Achieving Benefits through Change*, 2nd edn, Surrey: Gower.

Breese, R., 2012. Benefits realisation management: panacea or false dawn? *International Journal of Project Management*, 30(3), pp. 341–351.

BSI, 2013. *PAS 1192-2. Specification for Information Management for the Capital/Delivery Phase of Construction Projects Using Building Information Modelling*, London: British Standards Institute.

BSI, 2014a. *BS 1192–4:2014. Collaborative Production of Information Part 4: Fulfilling Employers Information Exchange Requirements Using COBie – Code of Practice*, London: British Standards Institute.

BSI, 2014b. *PAS 1192–3. Specification for Information Management for the Operational Phase of Assets Using Building Information Modelling*, London: British Standards Institute.

BSI, 2015. *Draft BS 8536. Facility Management Briefing for Design and Construction – Code of Practice*, London: British Standards Institute.

CRC for Construction Innovation, 2007. *Adopting BIM for Facilities Management: Solutions for Managing the Sydney Opera House*, Brisbane: Cooperative Research Centre for Construction Innovation.

Diceman, J. 2014. *Dotmocracy.* Available at: http://dotmocracy.org [Accessed 3 February 2015].

Doherty, N.F., Ashurst, C. and Peppard, J., 2012. Factors affecting the successful realisation of benefits from systems development projects: findings from three case studies. *Journal of Information Technology*, 27(1), pp. 1–16.

Fujitsu Consulting with John Thorp, 2007. *The Information Paradox: Realizing the Business Benefits of Information Technology*, s.l.: Fujitsu Consulting.

GSA, 2007. *GSA Building Information Modeling Guide Series 01 – Overview*, Washington, DC: US General Services Administration.

GSA, 2011. *GSA BIM Guide for Facility Management*, Washington, DC: US General Services Administration.

Hesselmann, F. and Mohan, K., 2014. *Where Are We Headed with Benefits Management Research? Current Shortcomings and Avenues for Future Research*, paper presented at Twenty Second European Conference on Information Systems, Tel Aviv, Israel, 5–13 June.

ISO, 2010. *ISO 29481-1:2010. Building Information Modelling – Information Delivery Manual*, Vernier: International Organization for Standardization.

ISO, 2012. *ISO/TS 12911:2012. Framework for Building Information Modelling (BIM) Guidance*, Vernier: International Organization for Standardization.

ISO, 2014. *ISO 55000:2014 Asset Management Series*, Vernier: International Organization for Standardization.

Love, P.E., Irani, Z. and Edwards, D.J., 2004. Industry-centric benchmarking of information technology benefits, costs and risks for small-to-medium sized enterprises in construction. *Automation in Construction*, 13(4), pp. 507–524.

Love, P.E., Matthews, J., Simpson, I., Hill, A. and Olatunji, O.A., 2014. A benefits realization management building information modeling framework for asset owners. *Automation in Construction*, 37, pp. 1–10.

Lu, W., Peng, Y., Shen, Q. and Li, H., 2013. Generic model for measuring benefits of BIM as a learning tool in construction tasks. *Journal of Construction Engineering and Management*, 139(2), pp. 195–203.

Martins Serra, C.E. and Kunc, M., 2015. Benefits realisation management and its influence on project success and on the execution of business strategies. *International Journal of Project Management*, 33(1), pp. 53–66.

NATSPEC, 2011. *NATSPEC National BIM Guide*, Sydney: Construction Information Systems Limited.

NATSPEC, 2014. *Getting Started with BIM*, Sydney: NATSPEC.

NSW Government, 2014a. *Benefit Realisation Management Framework, Part 3, Guide*, Sydney: New South Wales Government.

NSW Government, 2014b. *Benefits Realisation Management Framework. Part 4: Tailoring*, Sydney: NSW Government.

NSW Government, 2014c. *Benefit Realisation Management Framework. Part 5: Glossary*, Sydney: NSW Government.

Peppard, J., Ward, J. and Daniel, E., 2007. Managing the realization of business benefits from IT investments. *MIS Quarterly Executive*, 6(1), pp. 1–11.

Poll Maker, 2015. *Create Your Free and Unlimited Poll.* Available at: www.poll-maker.com [Accessed 3 March 2015].

Sacks, R., Koskela, L., Dave, B.A. and Owen, R., 2010. Interaction of lean and building information modeling in construction. *Journal of Construction Engineering and Management*, 136(9), pp. 968–980.

Sanchez, A.X., Lehtiranta, L.M. and Hampson, K.D., 2014. Use of contract models to improve environmental outcomes in transport infrastructure construction. *Journal of Environmental Planning and Management*, 58(11), pp. 1–21.

SBEnrc, 2015. *Driving Whole-of-Life Efficiencies through BIM and Procurement.* Available at: www.sbenrc.com.au/research-programs/2-34-driving-whole-of-life-efficiencies-through-bim-and-procurement [Accessed 30 September 2015].

Smith, D.C., Dombo, H. and Nkehli, N., 2008. *Benefits Realisation Management in Information Technology Projects*, paper presented at Technology Management for a Sustainable Economy PICMET Conference, Cape Town, South Africa, 27–31 July.

US Department of Veterans Affairs, 2010. *The VA BIM Guide*, s.l.: US Department of Veteran Affairs.

Ward, J., Taylor, P. and Bond, P., 1996. Evaluation and realisation of IS/IT benefits: an empirical study of current practice. *European Journal of Information Systems*, 4(4), pp. 214–225.

Ward, J., De Hertogh, S. and Viaene, S., 2007. *Managing Benefits from IS/IT Investments: An Empirical Investigation into Current Practice*, paper presented at IEEE Computer Science Conference, Waikoloa, HI, 3–6 January, p. 206a.

World Café Community Foundation, 2015. *The World Café Method*. Available at: www.theworldcafe.com/method.html [Accessed 3 February 2015].

Yates, K., Sapountzis, S., Lou, E. and Kagioglou, M., 2009. *BeReal: Tools and Methods for Implementing Benefits Realisation and Management*, Reykjavík: Reykjavík University, pp. 223–232.

5 Implementation tips with hindsight

Chris Linning, Adriana X. Sanchez and Keith D. Hampson

Introduction

As mentioned in Chapter 4, identifying and monitoring benefits that realise the value of Building Information Modelling (BIM) can be used as an overarching framework for the implementation strategy. However, there is a lot more to successfully implementing BIM and maximising its value. This fact can be overwhelming for first-time users and sometimes even discouraging.

Hindsight is a wonderful resource, although its biggest frustration is that people only have access to it after they have laboured for many hours, if not years, trying to achieve it. This chapter provides access to some of the knowledge that was gained by a small team trying to implement the concepts of BIM for asset information management to a well-known global architectural icon: the Sydney Opera House (SOH, Figure 5.1). It will also complement this insight with lessons learned from other projects that have been made available through academic and industry literature.

Unlike most contemporary structures, the SOH is a structure that is expected to be part of the Australian landscape for at least another 200 years. Consequently, the solutions being implemented need to be robust and sufficiently open in design to withstand the ever-evolving developments in software and hardware. This long-term perspective has had a significant impact on the way BIM adoption is approached. It has prompted the team to become well-informed and deeply engaged in the implementation of BIM throughout the planning, design and construction of new sections of the precinct and into asset management through a BIM interface. The BIM interface being implemented aims to be an open solution, robust yet adaptable, able to link to multiple information sources and to become the single source of truth for all information queries. This is sometimes known as BIM Nirvana and this chapter will provide a series of implementation tips that are helping the SOH reach this point. The chapter will cover the following themes: BIM as a single source of truth; standards; culture of engagement; team environment; software choice; and information systems (IS) support group. It is the hope of the authors that new and experienced users will find this chapter useful when developing their BIM implementation strategy.

Figure 5.1 Sydney Opera House by John Hill

Single source of truth

The use of BIM offers a single source of truth. Data and reports produced at any time reflect the most up-to-date state of the model, increasing consistency among datasets and improving version control. BIM also provides a single comprehensive source for tracking the use, performance and maintenance of a built asset for the owner, maintenance team and financial department (NATSPEC, 2014a; Becerik and Pollalis, 2006; Barlish, 2011; Penn State, 2011). However, the successful development of a reliable asset information system is dependent upon the quality and accuracy of the information that is being stored within that system.

The ongoing iterations of design within a single project model delivers potentially a clash-free model available for tender and construction. With integrated project delivery (IPD) protocols in place, the construction model continues to develop, paralleling that of the actual project, eventually to mature into a facility management model at the project completion and handover. Effectively, the matured model now holds or has links to extended accurate information about the building, structure or site and its contents. All of which is to be used as the reference point or library on that same building, structure or site. The maturing model becomes the building information primary model; from the concept and design model to the tender model to the construction handover model, where data-rich assets are placed within the model according to the functional design. All of these developed resources are now held as a single source of truth (SSOT) and can be validated as part of the work-as-executed (WAE) or as-built documentation deliverables such as the operating and maintenance manuals and functional facility management model.

Although a commonly used term in the Australian industry, it is difficult to find a comprehensive definition of SSOT in industry or academic literature. Although a more informal source of information, the SSOT Wikipedia page provides a definition worth exploring.

> In Information Systems design and theory Single Source Of Truth (SSOT) refers to the practice of structuring information models and associated schemata such that every data element is stored exactly once (e.g., in no more than a single row of a single table). Any possible linkages to this data element (possibly in other areas of the relational schema or even in distant federated databases) are by reference only. Because all other locations of the data just refer back to the primary 'source of truth' location, updates to the data element in the primary location propagate to the entire system without the possibility of a duplicate value somewhere being forgotten.
>
> (Wikipedia, 2015)

By holding this concept through all stages of model development and including the ongoing facility management stage, the users of the model are able to interrogate the Building Information System with confidence, knowing that their investigations are based on accuracy and fact. Developing the early stages of concept design to ensure that any reference material is sourced from a reliable, accurate base should be the goal of any project team. In future, that source will be the BIM model and its associated linked databases of asset information.

This incremental asset model management and revision process increasingly develops more and more accurate information about the built asset. For example, the integration of linked geospatially accurate information placeholders allows for increasing model maturity and the development of bespoke model management plans ensures sophisticated handling of model and model revision while not imposing restrictions to model access. The complete life-cycle of any building will be enhanced if all participants in the building's life-cycle embrace the prime meaning of a single source of truth: information is stored only once and the information that is stored is accurate and verified to the best capacity of all parties concerned.

The BIM-SSOT is a very achievable goal with the potential to be an invaluable resource for the built environment industry, but it needs to be engrained in the culture of the group. This means that every stakeholder group needs to embrace this level of accuracy and transparency from the start of the project in order to maximise the value delivered through BIM in the form of better information management and improved efficiency.

Standards

Mark Bew, chair of the HM government BIM Task Group and one of the authors of the BIS BIM report said:

> Standards play an important role in ensuring the wider adoption of BIM technologies, processes and collaboration by ensuring that the same accurate data can be accessed throughout the supply chain.
>
> (BSI, 2013)

Uniformity in layout, consistency in presentation and the basis for comparative evaluation; these terms underpin tangible work results and deliverables. Nevertheless, there must be proven methodologies and academic rigour supporting those terms so users can trust that what they are reading has been developed to meet a country's minimum requirements for construction and long-term usage: our standards. At a macroeconomic level, national standards improve safety, promote international competitiveness and trade, ensure interoperability of technologies and processes, and reduce information asymmetry (Standards Australia, 2013). At an individual level, people depend upon such standards for everyday life; travelling on a train, crossing a bridge, riding an elevator or walking on the floor of a high-rise building; all built to a minimum standard.

> Good standards provide clear requirements that set minimum conformity specifications and strike the right balance between too many and too few varieties; this works in the best interests of both the product supplier and the consumer... standards serve many purposes. They enable trade, improve safety, facilitate efficient use of resources, reduce time, improve quality, permit compatibility and aid integration. Businesses and consumers benefit from them the world over.
>
> (UK National Building Specification, 2014)

At the SOH, drafting standards have a long lineage as a result of the international teams that were involved during its design and construction. Legacy drawings from the UK, Denmark, Austria and France plus other countries complemented those of the team of Australians that prepared the construction documentation. Today, the Opera House still engages international and Australian companies for design, construction and maintenance expertise. This has meant that their teams have had experience with a wide variety of standards and has provided an incentive to remain open to new ways of doing things to improve outcomes.

When the Opera House decided to adopt BIM for construction and asset management, their team had to keep abreast of international and national BIM standards. In 2006, the SOH computer-aided design (CAD) guidelines were released and incorporated into construction contracts as a basis for 2D drafting for new sections being built. The guidelines were developed based on proven industry examples and Australian Standards current at the time (Australian Standard, 1992). In 2007, following the release of the Cooperative Research Centre (CRC) for Construction Innovation FM Exemplar Project report, the Opera House released its first BIM guideline: *BIM Guidelines – Basic Introduction*. This document defined the data standards applicable for the development of an integrated building master model and outlined the BIM tasks ahead: a single federated master model, sub-model organisation, IFC functionality, and the model coding and organisation, among other tasks. At that time, BIM was still a design/construction tool; the concept of an integrated asset management tool was on the SOH horizon but to achieve it would still require time and preparation.

In 2011, to meet SOH requirements for the development of a major design/construction project being undertaken utilising a single design model, SOH released its second BIM guidelines: *BIM Guidelines – Construction Phase (+COBie)*. This document was an adaption of the NATSPEC[1] BIM guidelines, which themselves were an adaptation of the United States Department of Veteran Affairs BIM guidelines. This was done to provide an emphasis on construction and final deliverables. The concept of handing the design model to the construction contractor was a completely new process for the Opera House, as well as for a considerable number of organisations that were approached for assistance.

Standard processes and protocols are key to assigning responsibilities and conducting design reviews and validation (Gu and London, 2010) and the one common theme emerging from most parties was the necessity for a BIM Execution Plan (BEP). This vital document outlines the obligations of the next party (the construction contractor) to manage and maintain the BIM model during the construction phase of the project. The team concluded that this document had to be very prescriptive and detailed in order to ensure that the final deliverable at the completion of construction fit the specific requirements of the facility and asset management teams and the project owner.

Standards set data-exchange formats and vocabulary to increase consistency between projects and software platforms (Gu and London, 2010). At the time of drafting the BEP, however, the project deliverables were three to four years away and the SOH was aware that technology was rapidly evolving. This led to the final requirements of the deliverables being initially left as 'to be advised' in the contract. That particular conversation has recently taken place and at the time of writing, there are still engineering aspects of the project that will have to be handed over as 'bookmarked pdf documents' due to the incompatibility of various software applications for model exchange.

Another document of equal importance and sometimes treated as one with the BEP (NATSPEC, 2011), is the Model Management Plan (MMP) or BIM Management Plan (BMP). This document is specific to each organisation's particular needs for the daily management of the object rich model file. Particular emphasis can be made as to the company's presentation; look and feel, logos, line-weights and hatching. This document also defines the particular topology of the model federation and access. Because of the impact of all these elements, the SOH carried out extensive research to ascertain precedents and address the unique requirements of a performing-arts centre for the integration of facility and asset management tools into the BIM environment. This was a main lesson learned; MMPs' compatibility with the asset management systems should not be overlooked in organisational BIM roll out programmes.

At the time, there were no national standards for these documents and the necessity for compliance to some form of standard resulted in the SOH developing its own guidelines. Those guidelines, however, are expected to be superseded eventually by Australian standards, if not international standards. These guidelines have provided the basis for various parties to develop, manage and refine the processes to ensure a truly collaborative and compliant BIM environment.

By the time this book has been published, the resources, guidelines, discussion papers and standards available to interested parties will likely make the implementation of BIM across the whole-of-life of assets a far easier topic. Hopefully, it will have moved from the realms of research to the realms of reality and practical experience. The potential alignment of international BIM standards will also provide all stakeholders with guidance for true collaboration. Nevertheless, this experience has proven that technology is in a continual flux, always improving, ever-evolving. Whether developing or adopting national or organisational standards and guidelines, the one tip with hindsight is that these documents have to be adaptable to the ever-changing environment. It is vital that the authors of those changes have enough understanding and knowledge of the topic to ensure that a wise path is taken.

Culture of engagement

Organisational culture has been acknowledged in recent years as a contributing factor to competitive advantage, making this a business issue rather than solely a human resources consideration (Brown et al., 2015). A team's culture is based on the underlying assumptions and beliefs of its members about what they share in common, how the world operates and consequently, how they should relate to it (Brewer and Gajendran, 2012). The use of BIM represents a cultural shift in organisations and the level of acceptance of this shift can impact on the efficacy of BIM implementation. In 2013, for example, 23 per cent and 15 per cent of respondents to a survey by the Royal Institution of Chartered Surveyors (RICS) in the UK identified culture and lack of collaboration respectively as barriers to implementing BIM (Muse, 2013).

Integrated BIM and teams are based on collaborative stakeholder engagement and information integration across the supply chain and asset life-cycle phases. The topic of BIM was, however, to most employees at SOH, far from their usual daily tasks of working in one of the busiest performing arts centres of the world. At first, their understanding was that BIM had something to do with the drawing office and a new way of doing traditional drafting. Most had heard of 3D and most had probably seen a 3D model of something being manipulated on a television screen or movie theatre. It existed in the realms of unreality, fantasy and the world of theatre. This meant that stakeholder engagement and active dialogue became key parts of the decision-making and learning processes of the BIM implementation strategy. In order to succeed, everyone had to be taken through the journey to increase ownership and ensure that the final outcome was relevant and useful.

This culture of engagement and open-to-suggestions approach is reflected in the BIM guidelines that state that to achieve this final implementation of the system, 'the portfolio is open to any suggestions and encourages conversations to advance BIM and its adoption' (Sydney Opera House, 2011). End-users of the facility and asset management systems were also included in the development of the list of functionalities. This ensured buy-in from the people who were to use the system and increased the relevance of the changes made.

Further to this, the implementation team has had the ongoing support of senior and executive management, including its CEO, throughout this journey. As shown in Chapter 2, having champions that support and promote new technology implementation programmes at different levels of the decision-making chain can have a significant impact on the success of the implementation efforts and staff motivation. In the case of the SOH, this support and active engagement has provided the leadership and vision necessary for the development of a comprehensive BIM implementation strategy that is expected to provide significant long-term benefits.

The SOH team has also actively engaged with other parts of the industry through collaborative research projects and conferences. This has provided the SOH with the opportunity to learn more about other initiatives that might be relevant and to network with other organisations to stay up to date.

The journey certainly has required engagement with many different parties over the past ten-plus years, to maintain enthusiasm for the eventual delivery of the promised benefits. The culture of engagement of BIM has been consistently embraced by the engineering teams and management alike.

This has led to alignment of process, procedures and systems as well as skills being considered and delivered throughout the implementation journey. Although the tangible results have not been rolled out to all, some major projects are starting to benefit from the BIM programme and end-users are expectant of the new interface functionalities.

This culture of engagement has been vital to the success of the project so far. This will continue into operations through the staff BIM education and training programme, which will yet again ensure the ongoing support and success of the BIM investment made. Although there have been some setbacks along the way, where BIM may have been blamed as the reasons for delays, the final solution is a robust, well-researched, well-delivered resource and, importantly, embraced by all.

Team environment

> Hierarchy is everywhere; command-and-control leadership is still pervasive, yet employees are itching to break free from not being able to help, not being empowered to sort out problems on their own and not being able to proffer ideas that might actually improve the organization itself.
>
> (Pontefract, 2013)

As mentioned in the previous section and Chapter 4, the emphasis on collaborative stakeholder engagement and information integration are the basis for integrated BIM. The use of BIM both supports and requires close team collaboration (Becerik-Gerber et al., 2012). This section deals with the value of working in a team environment. Upon analysing the term *collaborative stakeholder engagement*, the term contains three vital components of a team environment:

- **Collaborative:** produced by or involving two or more parties working together (Oxford University Press, 2015).
- **Stakeholder:** a person with an interest or concern in something (Oxford University Press, 2015).
- **Engagement:** an arrangement to do something (Oxford University Press, 2015).

This term means then two or more parties sharing a common interest or concern and arranging to do something; the team works together in an arrangement to ensure that a common goal is delivered. This takes the culture of engagement down to the individual, hands-on team members. The members of the team are the core resource to ensure the delivery of the idea. The way this is done and its success will be largely defined by issues such as flexibility, empowerment, development, mobility and the type of leadership exerted by management roles (Brown et al., 2015).

Maximising the benefits achieved by implementing BIM in today's globalised industry requires dynamic and high performance multidisciplinary teams. These teams should 'foster and manage genuine interdependence, rotating leadership, high levels of trust in other members of the team, social communication, ability to work on a common group goal', and provide tools to support project management (Becerik-Gerber et al., 2012). The use of BIM therefore requires a new set of skills beyond technical expertise such as ability to place trust in other team members, ability to build and manage interdisciplinary teams and ability to challenge the process (Inguva et al., 2014).

The SOH BIM team,[2] for example, has grown over the past 11 years from one individual with the task of coordinating technical documentation to a group of eight team members. This group is required to cover areas across strategic management, corporate safety, conservation management, facilities and asset strategies, building information management and some minor works projects. This has required it to identify synergies between tasks and roles as well as embracing the possibility that BIM can be a solution to many tasks, far beyond the original concept of 3D modelling. This is even more relevant in global teams, where members may be geographically dispersed and working across multiple organisational boundaries (Becerik-Gerber et al., 2012).

In an interdisciplinary team every member may have different competencies, that is: 'any characteristic required for performing a given task, activity or role successfully' (Murphy, 2014). It is nearly impossible for any one individual to be in possession of all the competencies needed for successful modern-day business acumen. Succar et al. (2013) developed metrics to identify eight domain competencies or skills: managerial, functional, technical, supportive, administration, operations, implementation and research and development. In addition to these, they highlight execution competencies (those that are specific to tools and techniques) and four sets of core competencies that relate to personal abilities.

A team in today's construction and asset management industry will need a combination of skills and competencies; from the technical savvy to the high-level thinkers, innovators and collaborators (Zhao et al., 2015). It is these authors' opinion that a truly well-rounded team will support each other by being aware of what skills they can bring to the table and what skills other team members have that they do not. This philosophy has been paramount to the success within the SOH BIM team. The 'tall poppies'[3] have been welcomed into the group for their particular skills and capacity to be resilient in the face of criticism when bringing new ideas.

This approach recognises that an over-reliance on typical project management strict project controls and hierarchies often stifles innovation which can lead to difficulties to align BIM implementation with traditional practices (Murphy, 2014). The development of groundbreaking ideas usually will upset the normality of a working environment. It is, however, exactly those groundbreaking ideas that the SOH has welcomed, discussed and embraced as the team has moved towards its goal of a fully functioning BIM interface for asset and facility management.

This approach may break the working paradigms and cause disruptive change, and it will mean a change in the way we do business and the way services are provided, but if others wish to participate in this groundbreaking journey into a new way to work, being part of the team is the first step on that journey. It is the opinion of these authors that discussion, constructive criticism, new ideas and innovation should always be welcomed; most of all when implementing rapidly advancing socio-technical systems such as BIM. Obviously, ground rules need to be negotiated or set, and maintained with in-built reality checks but the understanding of encouragement and support from fellow members should always be present as part of the team environment.

Software choice

> How do you choose the right software solution?
> Do you define exactly what you want?
> How to weigh version and subscription issues against software capabilities?

Choosing the right software will always be a challenge. Although all BIM software platforms are conceptually similar, some are easier to use; some can create and use more complex geometries; and some are better suited for large-scale implementations while others are better for small-scale. Additionally, it is often unknown what new software is in the pipeline and when currently used software could be 'no longer supported'. Different software platforms and versions also have different functionalities and interoperability capabilities with different add-ons or tools as well as different prices. New versions and software packages keep appearing in the market at a higher speed every year. It is a minefield. This point may be well illustrated through the SOH building information management journey. SOH originally used Bentley's Microstation but in the early 2000s a new building information manager was appointed whose experience and preference favoured AutoCAD. At the time, making a total transition between software platforms was deemed so difficult that both the Bentley and Autodesk products were maintained as viable software options.

In early 2002, the Opera House embarked on its Venue Improvement Plan; a major capital works programme to public (front-of-house) and private (back-of-house) areas of the building and site. SOH's architects and consulting engineers were able to work in different software platforms but they were all required to have the ability to exchange design files by using DXF formats. This could have been called 'OpenCAD' if such a term had existed then. This ability to read DXF files in both software platforms allowed the SOH to maintain the Bentley/Autodesk environment.

In 2007, the Venue Improvement Plan had been delivered on many projects affecting both front-of-house and back-of-house areas of the building. Most of the concept and design work was a combination of 2D and 3D work with many of the visualisations being derived from the 3D model files. Handover to construction continued in the traditional 2D printed drawing files; hundreds per project.

The same year, the CRC for Construction Innovation Exemplar Project had made the SOH management aware of the potential of 3D modelling in a BIM environment; not only for concept design, but for construction and facility management. This was the result of a collaborative initiative that had started in 2005 (Figure 5.2). The software tools of choice for BIM, however, although cutting-edge at the time, presented themselves as partial solutions and not total packages. The full BIM building life-cycle, from concept to ongoing occupation, became a target for the future. Their challenge was to ensure that they 'got their house in order' while not locking into a software solution that may later become a one-way path with no turning back. They needed to be sure that whatever software they embraced could allow flexibility as the journey continued.

At the time, they continued using both Autodesk and Bentley Microstation solutions because the sheer volume of users provided them with sufficient confidence that these two providers would be around for a while. Microstation was used by the structural engineering consultants in the development of a major design initiative to upgrade one of the major theatre spaces. The many concepts and designs were re-purposed to a variety of analysis programmes and the SOH witnessed some of their first ventures in the use of point-cloud scanning to capture as-installed features of the building.

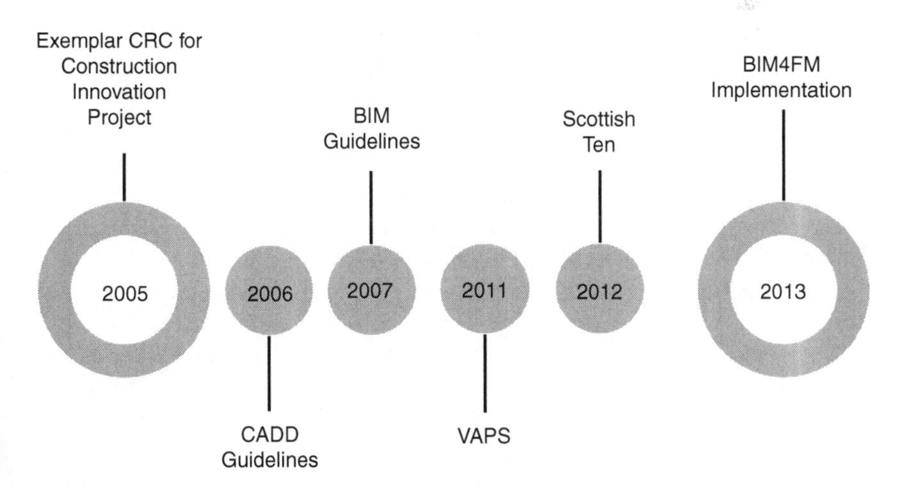

Figure 5.2 SOH BIM journey timeline

The Venue Improvement Plan identified construction projects where the consulting engineering firms continued to use 3D model design and presentation of concepts. The use of 3D became the norm for presentation of ideas to all and this was the opportunity for the SOH Building Development and Maintenance group to request (but not mandate yet) the use of 3D modelling in a multidisciplined collaborative manner. It was an opportunity to test how it was to be done, the degrees of software compatibility, what formats were required to exchange files, how to carry out their file management and naming, etc. In June 2007, the first *Sydney Opera House BIM Guidelines – Basic* had been written and incorporated into SOH construction project tender documentation sets. This document introduced, to all parties working with the SOH, the idea that a single BIM model would one day be developed as a single source of truth for all asset and engineering information sets relative to SOH.

Autodesk's AutoCAD and Bentley's Microstation continued to be the normal software selected by most parties in the development of concept, design and tender set development. In early 2010, the Vehicle and Pedestrian Safety (VAPS) project was announced. This AU$152 million (US$112 million) project was to provide a large underground loading dock beneath the southern end of the existing building. Access to the underground loading dock would be via a tunnel opening on the south-west site boundary and from there via a covered ramp to the underground dock; 16 metres below sea level. Two tunnels would run from the underground loading dock to two new and three refurbished lifts serving back-of-house areas in both the concert hall and opera theatre structures. This project would remove in excess of 500 large vehicle movements per week from the forecourt and broad-walk spaces of the site, returning these spaces to the 8.2 million pedestrians who visit the building on an annual basis.

Formal design commenced in 2010 using Autodesk Revit software. All existing forecourt services and the building structure that would be affected by the works were modelled using Revit. Virtual design reviews, clash detection and phasing were undertaken using Navisworks. Progression of the model through the design process was coordinated by the lead architects. Structural elements were detailed in Bentley software and transferred to the Revit using IFC, DXF, DGN and DWG formats. By the time tender documentation was finalised, a 96 per cent clash-free design had been completed.

The *Sydney Opera House BIM Guidelines – Construction Phase (+COBie)* had been written by this stage and was part of the tender documentation set. This provided guidance regarding expected information deliverables. The BEP accompanying the tender model contained specific requirements for model management through the various stages of the project. The VAPS project plan was planned for a minimum of three years from project commencement to handover and occupation. As mentioned earlier, in consideration of the ever-changing and evolving world of CAD and BIM software, the final deliverables were identified in the BEP as 'to be determined'. What corporate takeovers may have evolved? Would Revit be superseded by some new start-up? Who could predict how software would change over the next three years? That was the challenge at the time. Three requirements were identified within the tender documentation to accommodate such unknowns:

- Preference is given to object-oriented software applications that comply with current industry interoperability standards and are able to be used in a collaborative environment. All software platforms used shall be compliant with:
  - the most current version of Industry Foundation Classes (IFC) file format;
  - commercially available collaboration software that provides interoperability between the different software applications.
- Traditional 2D documentation shall be prepared with approved IFC compliant BIM authoring software and plans, elevations, sections, schedules, and details shall be derived and fully coordinated with the coordinated building model. All other documents are to be submitted in compliance with contract requirements.
- Software other than those from a pre-approved list (Table 5.1) may only be used subject to the above compliance requirements and the principal's approval (Sydney Opera House, 2011).

In late 2011, SOH was approached to apply to be one of the five international sites of 'The Scottish Ten' project. This is an ambitious five-year project using cutting-edge technology to create exceptionally accurate digital models of Scotland's five UNESCO designated World Heritage Sites and five international sites in order to better conserve and manage them. The project team consisted of:

- Scotland's heritage agency, Historic Scotland, and its partner Glasgow School of Art, under their collaborative venture the Centre for Digital Documentation and Visualisation; and
- the CyArk Foundation, a US non-profit organisation.

In October 2012, SOH was advised of its successful selection and preparations began for the Scottish team to travel to the site to undertake the documentation work. The team would process the data and the deliverables would be gifted to the host. In April 2013, the SOH was scanned and in December 2013, 2.8 terabytes of data were delivered to the SOH for ongoing conservation management and interpretation. This included a collection of almost 1,000 raw scan data, point cloud images, 56,000 panoramic digital photographs, survey traverses and compiled proprietary 3D model files.

SOH was now in a position to fully develop an accurate 3D model of the building. Utilising a coordinated approach, SOH commenced a 3D jigsaw

Table 5.1 Sydney Opera House accepted software packages organised by purpose/type. No endorsements are to be inferred by the inclusion or exclusion of software in this list

Type	*Software*
Planning/preliminary cost estimates	CodeBook, Tokmo, DProfiler, Onuma Planning Software (OPS), CostX
Authoring – design (architectural, structural)	Revit Architecture, Bentley BIM, ArchiCAD, Tekla, Vectorworks, Sketchup
Authoring – MEPF (engineering and construction)	ArchiCAD MEP, Revit MEP, AutoCAD MEP, CAD-Duct, CAD-Pipe, AutoSprink, PipeDesigner 3D
Authoring – civil	Bentley Inroads & Geopack, Autodesk Civil 3D
Coordination (clash detection)	NavisWorks Manager, Bentley Navigator, Solibri Model Checker, Horizontal Glue, EPM Model Server, BIMServer
4D Scheduling	Synchro, Vico, Navisworks Simulate, Primavera, MS Project, Bentley Navigator, CostX
5D cost estimating	Innovaya, Vico, Tokmo, CostX
Specifications	eSpecs, NATSPEC, Auspec
Model checking validation	Solibri or equal
IFC File Optimiser	
COBie	Tokmo COBie Exchange, Onuma Systems
Energy analysis	EcoDesigner, Ecotect, eQuest, Green Building Studio, EnergyPlus, Trane/Trace, DOE2, Sefaira
Point cloud manipulation	Autodesk Recap, Leica CloudWorx & TruView, Bentley Cyclone
Render/visualisation	3D Max, VLC (video viewer)
Manufacturer's tool	Autodesk Inventor
Document management Systems	Zip, HP Trim, Aconex, MS Access
Operating systems	Citrix, MS Office, Internet Explorer, Google Chrome

puzzle combining the models to a controlled x, y, z coordinate system. The programme would combine its library of 3D construction project model files, survey works, point cloud technology derived models and the yet to be extracted 3D model files from the Scottish Ten project. All the lower levels of the VAPS project (basement levels one to four) to the uppermost reaches of the shell structures could now be combined into a single model with an accuracy of up to ten millimetres.[4]

In 2013, SOH embarked on a capital works masterplan programme, looking into the future needs of the building to meet its operational and performance needs for the next ten, 20 and 50 years. They employed an experienced Revit technician to realise their dream of having an accurate 3D model of the whole building; from sub-basement to the geometric intricacies of the spherical shell structures that form the famous profile of this iconic building. As the building is primarily a concrete structure, the typical aspects of this building tended to merge traditional engineering disciplines. The sub-structure, super-structure and architectural disciplines became effectively one. Exposed concrete surfaces of structural elements form walls and ceilings. Modelling an existing building presents far more challenges than working on a greenfield building site. Many of the Opera House walls are arcs, wall thicknesses change as contiguous walls transit from level to level through the building. Non-vertical walls and variable height floor slabs and many other challenges had to be faced.

A majority of consultants and contractors to the SOH used Revit as a tool for concept and design. The VAPS project had now committed Revit to be their tool for construction. As mentioned earlier, however, they had yet to define the software platform for handover and asset management. AutoCAD, Revit and Sketch-up were used both internally and externally. Other departments within the Opera House used other 2D and 3D software for theatre staging, lighting and acoustic projects. The SOH software list was growing.

In late 2013, SOH commenced a worldwide expression of interest (EOI) opportunity for companies to deliver a facility management interface to their BIM resources. This was referred to as the BIM4FM Interface. The first stage of this EOI was to develop a resource pool of proven practical software solution implementers. By early 2014 they had selected a company[5] from the resource pool to develop a technical specification for their BIM4FM Interface. In late 2014, the SOH tendered for the implementation of the BIM4FM Interface based on the developed technical specification. This work highlighted the next challenge in the ever-evolving BIM for asset management journey; not only did they require a bespoke BIM4FM Interface solution, but they additionally required a secure environment for the interface to be hosted. As this chapter is being written, the SOH is investigating the best software solutions to achieve the interface between BIM models, asset management databases, live building management control systems and the facility management interface.

As mentioned in a previous section, BIM has been seen as an opportunity to merge a number of disparate sources of information currently available in different databases. The prime requirement is, therefore, to ensure a *single source of truth* for all parties involved. The primary driver is to be able to retrieve asset management and technical documents and other information from existing systems within an intuitive, visual search solution. The model will be hosted on an internal server; all personnel accessing the model will log in to a hosted application session on the SOH server and all file activity will be contained within the hosted session. This means that no files will be copied to local client machines at any time. The SOH will then have achieved the final stage of the current BIM journey: a privately hosted, federated model environment.

This case illustrates some of the challenges presented to clients aiming to select the software platforms to be used in the planning, design, construction and management of assets. In order to maximise the benefits from implementing BIM across the life-cycle of the asset, the choice of software must take into account factors such as interoperability, desired functions, end-user, life-cycle phase, skills requirements, and budget. The sheer number of options alone presents a challenge for managers aiming to maximise BIM capabilities.

The objectives of implementing BIM may be one of the most important criteria when choosing a platform. Discussing the requirements established by these goals against software functionalities and consulting 'other organisations that have already implemented BIM may give insight into the right package to choose' (NATSPEC, 2014b).

Cost and budget are other important factors to consider. The initial cost and ongoing licensing/training costs of BIM software is often greater than a traditional 2D CAD package. However, Autodesk has announced that, as of 2016, they will start a Subscription Licensing Program and Bentley, Graphisoft and Nemetschek are also launching flexible licensing programmes. This means a move towards software rental rather than purchase may eventually become the norm for software providers. It may also prove to be a cheaper option for organisations that need to use different software for different projects such as consultants and contractors (Hughes, 2015).

Other issues that may come into consideration are:

- extent of training and support availability;
- availability of experienced users in the organisation, project or marketplace;
- availability of object libraries and other tools that can be used with each platform;
- interoperability with other platforms-ability to import, export and link;
- software currently used by other stakeholders;
- mobile and remote access capabilities;
- hardware requirements; and
- ongoing legacy maintenance of file formats (NATSPEC, 2014b; Scott et al., 2014).

IS support group

CAD was originally developed internally by manufacturers and ran on mainframe computers. Two decades later, the world of CAD changed with the introduction of the cheaper and more convenient personal computer, leading to the new world of parametric technology and stand-alone PCs (Tornincasa and Di Monaco, 2010). The implementer and the information technology (IT) or information systems (IS) team were both discovering the new world of PCs; people were in charge of their own destiny and their expectations were only limited by the amount of RAM and storage space on their machines. Then CAD systems progressively joined the world of computer networks, which required a higher understanding of network licensing, libraries and projects. In the early days of this new way of working, most CAD managers managed their own networks given that few in IS clearly understood the particular needs of a CAD environment. They became quasi-members of the IS/IT fraternity but in their own specialist world. The CAD journey then took a major stride forward as managers started to embrace BIM.

As they say in Australia: this really put the cat among the pigeons;[6] technology requirements multiplied overnight. The hardware and software demands of BIM exceed the traditional gamut of most IT groups' expectations. BIM teams expect optimum performance at all times, which has led some businesses to introduce radical changes to work in a different environment to what had been the norm for CAD requirements. This is a major paradigm change where IT/IS teams are needed even more than before but these IT/IS needs must be coordinated. It must also be accompanied by support for items such as network speeds, network space allocation, group policies, external access controls and specific IT monitoring management. These types of tasks fall into the grey space between both the traditional modelling functions and traditional information services functions.

The final goal is to have a single source of truth for a whole-of-life asset management model that is accessible to all stakeholders from anywhere and at any time. This model not only provides concept and working project options but is the prime source and resource for the interface between stakeholders and information. This certainly challenges all parties to ensure a fully supported and functioning system that is constantly available, providing optimal response and management. The model manager of the future will require a thorough understanding of the BIM software of their choice, plus the capacity to understand and negotiate within the IS/IT environment. They will need to be a new resource and a new manager, highly skilled in both BIM and IT, a champion for BIM implementation and management and a champion for negotiating the crossover into information systems and technology. This role will be vital for the successful future of BIM in design, construction and asset management.

In addition to the manager, each team member will have individual experiences and expertise that can be extremely valuable to other team members when navigating through the rapidly changing world of BIM. This is why a BIM/IS support team should be established; this is a group of team members who are designated as points of contact for consultation on specific issues. These are the *super-users*: highly capable and BIM-experienced team members who form a special task force of sorts to start the project and help less-experienced users. This requires a true culture of collaboration and engagement where the super-users have two roles: BIM IS support and in their usual project roles (Merschbrock and Munkvold, 2014).

Conclusions

The implementation of the concepts and process of BIM into any existing or new architecture, engineering, construction and owner-operated (AECO) organisation will begin a massive process of change. It is a major paradigm shift to the way work has traditionally been undertaken. The implementation will have an impact on the people, process and systems across the organisation and its supply chain network. Implementers can also expect resistance; people will feel challenged and it will be necessary for all team members to feel involved and to understand why the change is necessary.

Many governments are of the view that with a higher implementation of BIM sky is the limit.[7] We as a society need to get ready for the new transition ahead and the continuation of an era of fast technological advancement. The next few years will drive the future of this evolving change in the way we do business and the exciting and challenging times ahead in a more interconnected digital built environment. The current generation is starting to embrace this change and it is the hope of the authors that this chapter has provided the readers with some insight from the experience of those who have already embarked on this journey.

Notes

1 NATSPEC is an Australian not-for-profit organisation aiming to improve the construction quality and productivity of the built environment through leadership of information, which in more recent years has positioned itself as a provider of information that aims to help implement BIM in the construction industry.
2 Formally known as Building Strategy & Planning.
3 Australian slang for the tendency to criticise highly successful people (i.e., tall poppies), and 'cut them down'.
4 Historic note – drawing accuracy: The legacy 2D CAD files of the building had been developed in the mid-1980s utilising large format digitisers to transfer line-work from original architectural drawings and prints at scales of forty foot to the inch. The potential inaccuracies of the manually digitised point meant that every point 1mm out of placement could result in 500mm inaccuracy within the model. Survey work, prior to 2006, undertaken at the SOH were mostly focused on a particular isolated area with no connection to a master network or system. In 2006, 165 brass survey plaques were installed throughout the building, providing the basis of a controlled survey network. Where possible, all surveys were now undertaken with reference to the controlled survey network (CSN). The accurate coordinated re-drafting of the then 17 levels of the building, to correct for any inaccuracies, was a major time-consuming task which was constantly postponed until 2013, following the Scottish Ten project deliverables.
5 BIM Academy, Northumbria University, Newcastle upon Tyne, UK.
6 British and Australian expression meaning an action that causes trouble and makes other people angry or worried.

7 See UK case study in Chapter 2, also reference to video produced by a Dutch initiative from Uneto-Vni, Department of Public Works – Ministry of Infrastructure and Environment, Building Information Council. English translation was done by OpenBIMlab. BIM – Sky is the Limit (2011) available online at www.youtube.com/watch?v=cTX9mQbOjuY.

Bibliography

Australian Standard, 1992. *AS 1100.101–1992. Technical Drawings*, Sydney: Australian Standard.

Barlish, K., 2011. *How to Measure the Benefits of BIM: A Case Study Approach*, Phoenix, AZ: Arizona State University.

Becerik, B. and Pollalis, S. N., 2006. *Computer Aided Collaboration in Managing Construction*, Cambridge: Harvard University Graduate School of Design, Design and Technology Report Series 2006-2.

Becerik-Gerber, A. M., Ku, K. and Jazizadeh, F., 2012. BIM-enabled virtual and collaborative construction engineering and management. *Journal of Professional Issues in Engineering Education and Practice*, 138(3), pp. 234–245.

Brewer, G. and Gajendran, T., 2012. Attitudes, behaviors and the transmission of cultural traits: impacts on ICT/BIM use in a project team. *Construction Innovation*, 12(2), pp. 198–215.

Brown, D., Chheng, S., Melian, V., Parker, K. and Solow, M., 2015. Culture and engagement: the naked organisation, in J. Bersin, D. Agarwal, B. Pelster and J. Schwartz (eds.), *Global Human Capital Trends 2015: Leading in the New World of Work*, Westlake, TX: Deloitte University Press, pp. 35–41.

BSI, 2013. *New Standard for BIM to Help Meet 2016 Government Savings Target*. Available at: www.bsigroup.com/en-GB/about-bsi/media-centre/press-releases/2013/3/new-standard-for-bim-to-help-meet-2016-government-savings-target/#.U4Ql3_mSySo [Accessed 27 May 2014].

Gu, N. and London, K., 2010. Understanding and facilitating BIM adoption in the AEC industry. *Automation in Construction*, 19, pp. 988–999.

Hughes, D., 2015. *Three Winning Advantages of BIM Software Rental Licensing*. Available at: https://sourceable.net/three-winning-advantages-of-bim-software-rental-licensing/# [Accessed 24 June 2015].

Inguva, G., Clevenger, C. M. and Ozbek, M. E., 2014. Differences in skills reported by construction professionals who use BIM/VDC, in D. Castro-Lacouture, J. Irizarry and B. Ashuri (eds.), *Proceedings of Construction Research Congress, Construction in a Global Network*, paper presented at Construction Research Congress, Construction in a Global Network, ASCE, Atlanta, GA, 19–21 May, pp. 61–69.

Merschbrock, C. and Munkvold, B. E., 2014. *Succeeding with Building Information Modelling: A Case Study of BIM Diffusion in a Healthcare Construction Project*, paper presented at the 47th Hawaii International Conference on System Science, Hawaii Island, USA, 6–9 January.

Murphy, M. E., 2014. Implementing innovation: a stakeholder competency-based approach for BIM. *Construction Innovation*, 14(4), pp. 433–452.

Muse, A., 2013. *BIM: Cultural Shift*. Available at: www.building.co.uk/bim-cultural-shift/5054645.article [Accessed 5 June 2015].

NATSPEC, 2011. *NATSPEC National BIM Guide*, Sydney: Construction Information Systems Limited.

NATSPEC, 2014a. *Introduction to BIM*. Available at: http://bim.natspec.org/index.php/resources/introduction-to-bim [Accessed 25 August 2014].

NATSPEC, 2014b. *Getting Started With BIM*, Sydney: NATSPEC.

Oxford University Press, 2015. *Oxford Dictionary: Online Edition*. Oxford: Oxford University Press.

Penn State, 2011. *BIM Execution Plan*. Available at: http://bim.psu.edu/Uses/Resources/default.aspx [Accessed 5 September 2014].

Pontefract, D., 2013. How'd we get so rigid? in *Flat Army: Creating a Connected and Engaged Organization*, Somerset, NJ: John Wiley & Sons, pp. 27–45.

Scott, T., Montgomery-Hribar, J., Barda, P., Marshall, C., Kane, C., Eynon, D., Schuck, R., Burt, N., Canham, C., Mitchell, J., Collard, S. and Jurgens, D., 2014. *A Framework for the Adoption of Project Team Integration and Building Information Modelling*, Canberra: Australian Construction Industry Forum and Australasian Procurement and Construction Council.

Standards Australia, 2013. *The Economic Benefits of Standardisation*, Sydney: Standards Australia.

Succar, B., Sher, W. and Williams, A., 2013. An integrated approach to BIM competency assessment, acquisition and application. *Automation in Construction*, 35, pp. 174–189.

Sydney Opera House, 2011. *BIM Guideline: Construction Phase (+COBIE)*, Sydney: Sydney Opera House.

Tornincasa, S. and Di Monaco, F., 2010. *The Future and Evolution of CAD*, paper presented at 14th International Research/Expert Conference – Trends in the Development of Machinery and Associated Technology, Budapest, Hungary, 11–18 September.

UK National Building Specification, 2014. *NBS National BIM Report 2014*, Newcastle upon Tyne: RIBA Enterprises.

Wikipedia, 2015. *Single Source of Truth*. Available at: http://en.wikipedia.org/wiki/Single_Source_of_Truth [Accessed 13 April 2015].

Zhao, D., McCoy, A.P., Bulbul, T., Fiori, C. and Nikkhoo, P., 2015. Building collaborative construction skills through BIM-integrated learning environments. *International Journal of Construction Education and Research*, 11(2), pp. 97–120.

Part II

Dictionaries

Benefits dictionary

Adriana X. Sanchez and Will Joske

This *Benefits dictionary* was developed through an extensive review of literature related to: Building Information Modelling (BIM), virtual design and construction (VDC), Building Information Modelling and Management (BIM(M)), computer-aided visualisation and design, computer advanced visualisation tools (CAVT), Bridge Information Modelling (BrIM), digital engineering, computer-aided collaboration for managing construction, and related terms. These are all considered to be enveloped within the BIM concept described in Chapter 1.

The benefits listed in this dictionary can already be realised, although some have only been measured in pilot or R&D projects with limited current adoption. This dictionary therefore includes all benefits that can be realised with the mainstream and state-of-the-art tools currently available. Similar and directly related benefits were grouped under overarching benefits, which are enabled by the same processes and tools and have similar flow-on benefits leading to other overarching benefits. The benefits listed in this dictionary exist across different levels of goals; that is, some are end-benefits (high-level goals) and others are intermediary benefits (benefits that lead to further benefits). For example, 'fewer errors' flow on to improved documentation quality and processes, which can then flow on to improved output quality and competitive advantage gains.

The benefit profiles were informed by the review of more than 150 references available at the end of the dictionary in the Bibliography section. Each profile is formed by seven sections:

1. **Benefit description:** General introduction to the benefit that could be applied to different phases of the project life-cycle. This section also provides specific descriptions of the benefit for some phases where specific tools or interpretations of the benefit only apply to that stage. This description includes information about how the use of specific enablers can facilitate or enhance the profiled benefit when used with BIM.

Table BD.1 Benefit distribution matrix by beneficiary

Benefit	*Client/ owner*	*Designer*	*Contractor*	*Subcontractor*	*Fabricator/ manufacturer*	*Surveyor*	*Asset manager*	*Supplier*	*End-user*
Asset management labour utilisation savings	●						●		
Better change management	●	●	●				●		
Better cost accounting	●		●		●		●	●	
Better data/information capturing		●	●	●	●		●		
Better environmental performance	●	●	●				●		
Better programming/scheduling	●		●	●			●		●
Better scenario and alternatives analysis	●	●	●				●		
Better space management			●				●		
Better use of supply chain knowledge	●	●	●	●	●	●	●	●	●
Competitive advantage gain		●	●	●	●	●	● *		
Faster regulation and requirement compliance		●	●	●	●				
Fewer errors	●	●	●		●			●	
Higher customer satisfaction	●	●	●	●	●	●	●	●	
Higher process automation	●	●	●	●	●	●	●		
Improved communications	●	●	●	●	●	●	●	●	●
Improved coordination	●	●	●	●	●	●	●		
Improved Data and Information Management	●	●	●	●			●		●
Improved Documentation Quality and Processes	●	●	●	●	●	●	●	●	●
Improved efficiency		●	●	●	●		●		
Improved information exchange	●	●	●	●	●	●	●	●	●
Improved learning curve		●	●	●	●		●		
Improved output quality	●	●	●				●		
Improved productivity	●	●	●	●	●	●	●	●	
Improved safety			●	●	●		●		●
Less rework	●	●	●	●	●		●		
Lower cost	●	●	●				●		●
More accurate quantity take-off			●	●	●		●		
More effective emergency management			●	●	●		●		●
Optimisation of construction sequence			●	●	●				
Reduced execution time and lead times	●	●	●	●	●	●	●	●	●
Reduced risk	●		●				●		●

Note: * specific type of assets (e.g., entertainment, housing developments, etc.)

2. **Beneficiaries:** Specific stakeholders who stand to gain from the realisation of this benefit. Designers can be engineering or architectural, contractors refer mostly to head or leading contractors, and asset managers include facilities and operations managers.
3. **Enablers:** Links to the *Enablers dictionary*, outlining those tools and processes that facilitate the realisation of and enhance the benefit being profiled. These can often be implemented in non-BIM projects but when used as part of the BIM implementation strategy help maximise the value of BIM through the profiled benefit.
4. **Flow-on benefits:** Provides first a set of overarching benefits that can be derived from achieving the profiled benefit and then the intermediary benefits between that being profiled and the flow-on ones.
5. **Metrics:** Links to the *Metrics dictionary* suggesting indicators that can be associated to the profiled benefit.
6. **Examples:** Information about specific instances where that benefit has been experienced and measured.
7. **Notes:** Additional information related to the realisation of the benefit being profiled.

The aim of this dictionary is to serve as a basis for stakeholders to develop their own benefit realisation management strategy based on those benefits that they identify as being most relevant to their organisational goals. Table BD.1 provides a benefit distribution matrix by beneficiary based on the literature reviewed.

It should be noted that, in general, the accuracy and appropriateness of a model and the information that can be taken from it is only as good as the data used to build it and the people who built it. In turn, without a properly defined plan on how the model will be created and what the BIM uses and deliverables are, there is no certainty for specific benefits.

> Data is at the heart of the BIM process. A key benefit of BIM is consistency of data throughout a project, and the avoidance of data loss as the building moves from design to construction to management. However, it is vital that the data fed into the model is itself accurate and consistent, if the full potential of BIM is to be realized.
>
> (Building: Product Research, 2014)

Asset management labour utilisation savings

Benefit description

This benefit refers to cost savings during the operational phase of the asset's life-cycle due to the more efficient and productive use of asset (and facility) management staff labour. For example, well-structured data associated with data-rich model elements that are accessible through BIM-based computerised maintenance management system (CMMS) or computer-aided asset management (CAAM) systems make information about the asset and its parts more easily accessible. This results in more efficient asset management, which enables more efficient utilisation of staff time and shorter work order times.

When these systems are used with radio-frequency identification (RFID) and handheld devices and applications, information can be updated more easily and faster. This and data-rich objects reduce the time needed for data entry, validation, searching and surveying. Staff will also require less time and effort to determine current conditions due to easy access to life-cycle data and historic records as well as the use of record modelling. For example, there is no need to break walls or ceilings to determine where pipes are or do *re-walks*. This information can be found through the use of RFID, augmented reality or simply accessing a model component linked to the BIM-based asset management system. The use of BIM during this phase thus reduces wasted time by eliminating additional trips to a specific location to carry out unscheduled work orders.

Data and information is more readily available when needed; up-to-date data such as maintenance schedules, warranties, cost data, upgrades, replacements, damages/deterioration, maintenance records, manufacturer's data and equipment functionality can be accessed at any time. This includes, for example, quickly finding the contact person responsible for a determined task or the organisation responsible for repairs of specific equipment. This information can be accessed, for example, by choosing the component in the BIM model on a mobile platform (handheld devices).

Additionally, there is a digital document trail and more complete inventories that allow more efficient audits and tracking during operations, as well as faster adaptation of standards. The use of BIM-based management systems and CMMS facilitate the production of performance reports and equipment inventories for plans, specifications, rent bill management and submittals. These systems can also help new staff learn how to operate the asset and equipment more quickly through more intuitive visualisation and easier access to information about previous activities.

Beneficiaries

Beneficiaries include clients/owners, and asset managers.

Enablers

• BIM-based asset management systems	• Design authoring (3D visualisation)
• Data-rich, geometrically accurate model components	• Handheld devices
• Record model	• RFID
	• Well-structured data

Flow-on benefits

Asset management labour utilisation savings can help reduce operational costs.

Metrics

- Labour intensity (asset management).

Examples

In the UK, the Department of Trade and Industry (DTI, now Department for Business, Enterprise and Regulatory Reform and Department for Innovation, Universities and Skills) sponsored a project where one of the targets was to eliminate delays and cost due to repopulating the facilities management system after handover. For a typical hospital under normal conditions, this task would have amounted in 2004 to 6–12 months and more than £200,000 (US$310,000). The handover packages were created and loaded to a commercial asset management system (including maintenance schedules and types and operations instructions) automatically 'in a couple of minutes' (Fallon and Palmer, 2007).

A study in Taiwan developed and applied a BIM-based facilities management system. This system used 3D visualisation to assist users to easily and quickly find the location of elements of the facility as well as increase their work efficiency (Su et al., 2011).

In the Shanghai Disaster Tolerance Centre project, asset managers were included during the design phase. This building was designed in 2010 with a construction area of 28,124 square metres. By including asset managers during this early stage, designers were able to reduce maintenance time by defining and optimising the travelling path of asset managers as well as ensure easy access to object location information (Wang et al., 2013).

Notes

- Although most of the industry and academia seem to agree that the largest benefits from BIM can be achieved in asset management, this is still relatively uncommon. Especially in existing buildings, developing the model, updating it and handling uncertain data can be challenging (Volk et al., 2014).

- It is also important to be realistic about how much data can be embedded in the BIM model itself at this time. A more pragmatic approach can be to hold most of the data in the CMMS, CAAM or CAFM database, and link 3D components to the database. Only a single piece of data is needed to do this. This approach can be useful when some of the stakeholders involved in producing the asset information model (or record model) do not have the capacity or ability to embed all the data required into the 3D components. In this case, the BIM model can act as visual interface to the information content in the asset management databases.
- It is also noteworthy that the BIM model itself must be maintained from the moment it is first delivered so as to mirror changes to the hard and soft assets in the facility.
- Re-walks refer to having to physically go to areas of the asset to check existing conditions.

Better change management

Benefit description

Changes are a critical source of conflicts, claims and disputes. This benefit refers to the more efficient and effective management of changes to the design, construction methods and schedules, and material and equipment supply and services. This benefit also includes the reduction of the number of changes and conflicts associated with them. Unbudgeted changes and change orders can be significantly reduced through better informed decision-making. When they occur, change order processing can be done faster with lower response latency (reduced time to clarify a problem). Owner- or client-initiated changes can be better managed due to higher accuracy of bills of quantities and faster production of cost estimates based on changes. Digital fabrication also facilitates adapting late changes to the design. Changes needed due to clashes can be identified easier, faster and more accurately, and be made in a more timely manner.

Increased transparency and fewer changes help reduce the number of conflicts related to change orders. When conflict occurs, being able to track and document changes more effectively can, for example, facilitate the retrieval of information that will be needed to quickly resolve the claims and understand the resulting damages. Automated updates across different modelled systems additionally enable quicker reaction to design problems or site issues and better conflict resolution.

Design

The use of parametric objects that can be easily amended or replaced (facilitated by object libraries) and integrated model management systems allow changes to design to be made more quickly, easily and accurately. Because of the automated coordination of all 2D representations derived from the 3D model, these changes can be done without the need of manually introducing the same change in each element and drawing, and with less conflict. The 3D visualisation of the design allows participating stakeholders to identify consequences of design changes immediately by reviewing the model directly in the authoring platform or associated reviewing tools.

Construction

There is more information required at earlier phases and automated clash detection can help identify issues that would normally only be found during construction. Any changes required during construction can be reviewed prior to the start of construction, fabrication and installation, and this can reduce the number of changes made during construction when the cost will likely be higher.

Operations

Having accurate and complete asset management information in a single system facilitates seamless changes between asset managers and suppliers if required. For example, if a specific service supplier would have to be substituted, the new supplier would have access to accurate historic and current data to facilitate the transition.

Beneficiaries

Beneficiaries include clients/owners, designers, contractors and asset managers.

Enablers

Table BD.2 Better change management enablers

	Planning	*Design*	*Construction*	*Operations*	*Whole-of-life*
Automated clash detection	X	X	X	X	X
BIM-based asset management systems				X	
Cost planning (5D modelling)	X	X	X	X	X
Design authoring (3D visualisation)	X	X	X	X	X
Digital fabrication			X		
Integrated model management systems	X	X	X	X	X
Object libraries	X	X			
Phase planning (4D modelling)			X	X	

Flow-on benefits

Better change management can help improve claim management, reduce cost through better asset management and fewer construction changes, and reduce design cost.

Metrics

• Cost of change	• Request for information – changes
• Variations and change orders	• Conflict – changes
• Time for change	• Latency – change orders

Examples

The use of BIM has been reported to be responsible for a reduction of between 10 per cent and 40 per cent of unbudgeted changes and 60 per cent of requests for information (Leite et al., 2011; Azhar, 2011; Lu et al., 2013; Gilligan and Kunz, 2007). In a study published in 2013, for example, where three sets of comparable projects were analysed against similar projects built without BIM, all BIM-assisted projects achieved reductions in the number of change orders (37–48 per cent reduction) and requests for information (RFIs) (34–68 per cent). The total cost of change orders as a percentage of the initial contract value was also lower in the BIM-enabled projects than in the traditional ones. One of the projects still had a large number of RFIs. However, 15 per cent were resolved during the pre-construction phase, assisted by the model. This study also estimated that between 9.3 per cent and 74 per cent of the total cost of change orders in the traditional projects might have been prevented by using BIM (Giel and Issa, 2013).

In a study published in 2007, a medical centre and a pilot plant that used BIM for their design and construction were reported to have significantly fewer change orders than expected for such complex projects. In one of these projects, there was only one contractor-initiated change order for the scope of work modelled in 3D while in the other project, there were no change orders 'related to field conflicts after the construction of MEP systems for the first six quadrants' (Staub-French and Khanzode, 2007).

General Motors (GM) Worldwide Facilities Group used lean construction and BIM in four design-bid-build projects between 2004 and 2006. Change orders had traditionally accounted for 8–10 per cent of the project cost and were reduced to less than 0.5 per cent due to building component interference and rework (Fallon and Palmer, 2007).

Note

- It has been reported that approximately 15 per cent of the final contract value can be generated by work related to change orders (Roper and McLin, 2005).

Better cost accounting

Benefit description

This benefit refers to more accurate cost estimates, which are closer to the final cost and bid results, and require less effort and time. The greater level of detail and stakeholder engagement at earlier phases of projects helps provide a higher level of certainty in early cost estimates. There are fewer unforeseen costs by more accurately defining the scope of works of bid packages. This in turn can reduce overpay contingency for unforeseen change orders and material allowance. There is better control over whole-of-life economic and environmental cost; cost estimates can include the cost of construction, operations, maintenance and decommissioning of the asset in a single database.

Cost estimates can be directly extracted from a bill of quantities and precise quantity take-offs generated from 3D models. Effects of changes to the design and materials can be easily and quickly integrated into the cost budgeting of the project and asset management strategy. The use of 5D BIM, combining schedules and costs, enables cost estimators to track budgets throughout the life-cycle of the asset more accurately and timely.

Design

Cost estimation during design and feedback to designers can be accelerated leading to faster and better project budget calculations for different design options.

Construction

Detailed and accurate cost estimates can be updated daily from accurate quantity surveys and data associated with the model elements.

Operations

Asset managers have the ability to track asset components more accurately, identify inefficiencies and respond quickly to requests based on accurate inventories and maintenance cost data. This can reduce cost overruns during operations.

Beneficiaries

Beneficiaries include clients/owners, contractors, suppliers, manufacturers and asset managers.

Enablers

Table BD.3 Better cost accounting enablers

	Planning	*Design*	*Construction*	*Operations*	*Whole-of-life*
BIM-based asset management system				X	
Cost estimation (quantity take-off)	X	X	X	X	X
Cost planning (5D modelling)	X	X	X	X	X
Data-rich, geometrically accurate model components	X	X	X	X	X
Early and effective stakeholder engagement	X	X	X	X	
Field and management tracking			X	X	

Flow-on benefits

Better cost accounting can help reduce contracting risk and increase certainty of cost during earlier stages. It can also help lower the operational cost by facilitating easier and more accurate estimation of effects of changes on the performance of the built asset. Whole-of-life cost can also be reduced by facilitating easier and more accurate estimation of effects of changes on the different life-cycle phases. Better cost accounting also provides a greater level of transparency about costs associated with variations during construction, reducing the potential for conflict.

Metrics

- Labour intensity – cost estimates.
- Cost predictability.

Examples

The US-based Centre for Integrated Facility Engineering (CIFE) carried out a study of 32 projects completed before 2007. They concluded that the accuracy of BIM-based cost estimates were within 3 per cent of the final values and could be produced in 44–80 per cent less time. In a separate case study of a municipal fire station in the city of Salisbury, USA, Gillian and

Kunz found that the project was within 0.6 per cent of the original budget estimates (Azhar et al., 2012; Leite et al., 2011; Gilligan and Kunz, 2007).

In a case study of a pharmaceutical pilot plant in California 'cost control was a key concern for the owner'. In this project, the use of BIM helped reduce cost growth to an average of 1 per cent for the mechanical, electrical and plumbing (MEP) subcontractors. This 1 per cent was mostly due to owner-initiated design changes and lower than the industry average for projects of this complexity of 2–10 per cent (Staub-French and Khanzode, 2007).

In a US hospital case study, the use of BIM allowed for quick, automatic in certain instances, quantity take-off updates in order to effectively assess potential cost (Manning and Messner, 2008).

Notes

- Although having a greater level of detail at earlier phases greatly enhances this benefit, particular attention needs to be paid to finding the right balance between enough information to carry out the pertinent calculations and too much information so that it restricts design and supply later in the project (Philip, 2015).
- The lessons learned from a road expansion procured as a design-bid-build in Finland found that when using BIM cost estimation tools, the cost is calculated by specialists from corresponding technical areas. This is why, in order to achieve better cost accounting, it is important for the group to agree on who will carry out each calculation during options analysis so the same cost is not counted twice (Gerbov, 2014).
- If cost accounting is a known objective for a project, then a higher standard of modelling would be expected and this should be clarified during early stages.

Better data/information capturing

Benefit description

This benefit refers to the easier and faster capture of more accurate and comprehensive datasets and information that is long-lasting and easier to retrieve. When using BIM, 3D objects can either co-exist in a single database that captures all the knowledge, information and data known for a specific asset or be easily accessed through a single interface while residing in separate databases through the use of unique identifiers. This information can, for example, include the geospatial location of discrete elements of the asset that can later be referenced in any phase and project. Online collaboration and project management also allows capturing any design, construction or asset management decision and the reasons behind it. Once this data and information is captured, it can be automatically transmitted to any other relevant system or document, and stored for later use. In cases where future software data platforms have not been defined yet, data can be placed in an open format for later use while ensuring interoperability. This data can also be tracked by using integrated programme management tools that create a separate and linked database that can track issues around brief management.

Combining online integrated systems with handheld devices provides the opportunity for more timely and easier collection and updating information and data as events occur. Mobile devices, for example, provide a flexible platform for on-site data-capturing in real-time. Some software solutions also allow photographs of components, processes or tasks to be incorporated into the model, reducing preparation time and increasing accuracy of the information reflected in the model.

Operations

Existing conditions can be more accurately modelled through, for example, the use of laser scanning coupled with BIM. Having this more accurate data about existing conditions allows quantity verification and detailed layouts creation that is faster and more accurate. This approach can also be used to capture pre- and post-disaster conditions. Equipment performance data can additionally be automatically captured into the model and management systems through the use of management interfaces for BIM asset management platforms. Although BIM for asset management and post-disaster assessment is not a common use at the time of writing, this has been implemented in a number of projects and is expected to become more common over the coming years.

Beneficiaries

Beneficiaries include designers, contractors, subcontractors, fabricators/ manufacturers and asset managers.

Enablers

Table BD.4 Better data/information capturing enablers

	Planning	*Design*	*Construction*	*Operations*	*Whole-of-life*
3D laser scanning	X	X	X	X	X
BIM-based asset management system				X	
Handheld devices			X	X	
Integrated model and programme management tools	X	X	X	X	X
Online collaboration and project management	X	X	X	X	X
Photogrammetry		X	X	X	

Flow-on benefits

Better data/information capturing can have the following flow-on benefits:

- Asset management labour utilisation savings due to less time invested in revisiting objects for verification purposes, more efficient management processes, more accurate inventories and less uncertainty when identifying components and existing conditions.
- Improved documentation quality and processes due to more accurate representation of reality.
- Higher customer satisfaction through faster response to client requests.
- Lower operational cost achieved by having more accurate inventories, which can lead to less uncertainty in budgeting and planning, fewer cost overruns, more efficient asset operations and more timely access to up-to-date information needed for decision-making.
- Better programming/scheduling due to more timely access to more accurate on-site conditions data, and better tracking of historic changes.
- Improved project coordination can be achieved by having more timely access to more accurate site condition data.
- More effective emergency management is achieved through access to more accurate and visual representation of pre- and post-disaster conditions.
- Reduced project execution time can be achieved due to easier access to real-time data and less time needed for information acquisition.
- Less rework due to more appropriate work orders, more accurate information, and easier, faster and more visual access to defects data.
- More accurate quantity take-off is enabled by having access to more up-to-date and accurate inventories.

Metrics

- Overall time required to capture and document data.
- Number of RFIs related to ambiguous information.
- Volume of rework related to data and information capturing.

Examples

The US General Services Administration (GSA) has advocated that '3D laser scanning and advanced scanning technologies allow for better documentation of as-built drawings and existing conditions'. They also argue that 3D visualisation allows clients to see 'historic preservation and site context with respect to new project' (GSA, 2007). In 2014, a research group also proposed that construction progress could be monitored through the use of surveillance cameras or laser scanners to record the state of the building at certain time steps (Tuttas et al., 2014). Methodologies to automatically generate textured as-built models have also been proposed (Lagüela et al., 2013).

Laser scanning was used in a UK rail and tunnel electrification project including eight tunnels ranging in length from 700m to 7km. This technology allowed fast, accurate and safe measurements to be taken and used to create the BIM model. In this case, the team used a software platform that provided template tunnel cross-sections that allowed the data to be processed more quickly, so real-time 3D visualisation was possible. A scalable terrain model was also created at a rate of 400km/week instead of 40km/week with traditional methods. Tunnel modelling time was also improved by using laser point cloud technology allowing the team to model 1.74km of tunnel in one to two days as opposed to one week with standard CAD tools (Bentley, 2014).

In 2014, the BIM Academy was using laser scanning as part of restoration works of the 900-year-old Chapter House in the Durham Cathedral in England. In this project, capturing the data as quickly as possible was important due to the conservation and protected status of this well-visited monument (Northumbria University, 2014; Barker, 2015). The resulting data-rich, easily accessible model was developed to allow asset managers to:

- plan and showcase adaption scenarios of the building;
- attach and visually represent condition survey data to the model;
- take accurate measurements and volumes for restoration works;
- create accurate plans, section and elevations; and
- utilise mobile technology to explore the model while on site (Northumbria University, 2014; Barker, 2015).

Once the model was functional asset managers were able to create surveys and reports easily as well as update the model on site using handheld devices (Northumbria University, 2014; Barker, 2015).

Notes

- 3D laser scanning can be very useful to create a 3D model of built assets. However, this has traditionally required time-consuming and difficult data quality assurance to make sure that the model is truly representative, which can also be expensive (Tang et al., 2011). Nevertheless, studies such as Tang et al. (2011) have tested faster ways to identify quality issues. There are also certain tools that allow directly incorporating laser scans into the FM database, highlight sections of the scans and assign them to already registered assets. See, for example, Ecodomus (2015).
- Recent research has also started to investigate the use of non-destructive testing to create more comprehensive models of existing buildings. Examples include material or texture-based recognition, ground penetrating radars, radiography, magnetic particle inspection, sonars

or electromagnetic waves and tags installed during retrofits (Volk et al., 2014).

Better environmental performance

Benefit description

This benefit refers to improving the efficiency and effectiveness of procedures that help optimise the use of resources by assets and processes, as well as reduce their environmental cost. The use of BIM can help increase control over environmental data by increasing the ability to capture, monitor, report and test environmental performance measures. This can help make environmental performance more predictable and better-managed.

Earlier involvement of and cooperation between different stakeholders can enhance the sustainability of the final asset through active feedback, design reviews with sustainability evaluation criteria and greater understanding of interdependencies. More environmentally sustainable outcomes can be achieved through easier, faster, more reliable and more rigorous analyses of environmental and economic impacts of design options over the whole life-cycle of the asset during earlier phases. Resource use can be better tracked, understood and managed.

Better equipment and more durable materials can be selected based on feedback from stakeholders across the supply chain more effectively. More information about the performance of the end-product can be made available at earlier phases when changes have the largest effect on the final environmental performance of the asset. Sustainability considerations and constraints can also be applied more consistently throughout the life-cycle of the asset for decision-making in a more affordable way. The use of BIM for energy analysis also reduces the duplication of work normally required for recreating the building design to conduct the energy analysis.

Design

Rule-checking and information-rich objects enable continuous validation of design against environmental requirements. Tight *test-review-amend* workflow cycles can be carried out to improve the design in terms of environmental performance that would not be achievable using more traditional workflows.

Construction

Changes made during the design phase, access to more accurate quantity take-off, better understanding of the design and layout, higher precision of fabrication processes and fewer errors all help reduce waste during the construction phase.

Operations and decommissioning

Dangerous and toxic materials can be easily identified. This is particularly important during decommissioning when this information serves as a basis for handling procedures and recycling strategies. The use of BIM for asset management can also facilitate implementation of sustainability measures by making information more accessible to end-users.

Beneficiaries

Beneficiaries include clients/owners, designers, contractors, asset managers and end-users.

Enablers

Table BD.5 Better environmental performance enablers

	Planning	*Design*	*Construction*	*Operations*	*Decommissioning*	*Whole-of-life*
BIM-based asset management systems				X	X	
Design authoring (3D visualisation)	X	X	X	X	X	X
Energy simulation and analysis tools	X	X	X	X		X
Engineering analysis	X	X	X		X	X
Data-rich, geometrically accurate model components		X	X	X	X	X
Interoperable formats	X	X	X	X	X	X
Sustainability evaluation	X	X		X		X

Flow-on benefits

Achieving better environmental performance can have the following flow-on benefits:

- Lower whole-of-life cost through lower whole-of-life environmental cost, better understanding of life-cycle cost and lower resource consumption.

- Higher output quality in the form of more sustainable assets.
- Competitive advantage gain and higher customer satisfaction due to higher environmental certification.
- Lower operational cost due to lower use of resources (e.g., energy and water).

Metrics

- Sustainability and environmental performance scores.
- Resource use and management.
- Carbon footprint.

Examples

A study published in 2015 used BIM-enhanced comparative life-cycle assessment to more accurately compare a number of energy-efficient alternatives for a 2,100m^2 two-floor primary-school building. This study allowed comparing the overall environmental impact of using different materials versus renewable energy technology. They were also able to study the relationships between building configuration, global warming potential and human health (Ajayi et al., 2015). Another study published in 2011 reported the use of BIM to model the annual energy and fuel consumption, and annual CO_2 emissions of a single family house in Florida, US. This group concluded that using BIM with computer-aided building simulation can improve the accurate study of energy performance of buildings and the decision-making process (Raheem et al., 2011).

Also in the US, the GSA has stated that the use of BIM is allowing them to automatically calculate relevant space data such as net area and efficiency ratio. This agency is using BIM-based energy models 'to conduct more efficient, accurate and reliable energy simulations to predict building performance during facility operations' (GSA, 2007).

In 2013, a Spanish research group also proposed using BIM in conjunction with laser scanning, infrared thermography and photography on existing buildings for more efficient rehabilitation projects. The aim of this method was to identify more effectively the most severe faults for energy rehabilitation projects (Lagüela et al., 2013).

Notes

- It has been estimated that an increase of 2 per cent in upfront costs to support sustainable design results in life-cycle savings of approximately 20 per cent of total construction cost (Azhar et al., 2011).
- It has been noted that, as of January 2015, no single BIM tool or platform had 'incorporated enough intelligence for whole building LCA studies'. Future integration of BIM software platforms with LCA tools would

assist in improving the early assessment of building life-cycle impact (Ajayi et al., 2015). A number of tools have, however, been proposed and beta-tested in the past few years for eBIM (Energy Building Information Modelling). These tools integrated building management systems with wireless sensing and metering tools (Ahmed et al., 2010; Rowe, 2013).

- Almost two-thirds of contractors surveyed globally by McGraw Hill in 2014 reported using BIM-based coordination systems to improve energy performance (McGraw Hill Construction, 2014a).

Better programming/scheduling

Benefit description

This benefit refers to more accurate, better informed and more efficiently managed programmes and schedules, which are also more easily communicated. Programming and scheduling can be improved, for example, by using an accurate model of the design that includes all materials required for each element and task. This allows just-in-time arrival of staff, equipment and materials, and enables advanced purchase of materials and better inventory management based on less schedule uncertainty.

Programme/schedule alternatives can be quickly generated and compared to achieve a more efficient programme. These can be studied based on task interdependencies and requirements such as completion of preceding tasks, space, information, safety reviews, resources crews, materials, and equipment. The use of BIM can allow using libraries of construction method 'recipes'. Thus, programme/schedule changes can be made and evaluated in less time and construction processes can be rehearsed and iteratively optimised.

Work flows and standardised procedures can be better communicated to non-technical people. BIM also provides a medium through which staff can report work done in a more timely manner, facilitating monitoring and active programme/schedule management. Site organisation and space usage can therefore be updated more quickly and the time needed for space planning reduced. Additionally, systems that actively integrate input from subcontractors help produce more effective schedules.

Status can be easily visualised, including information about crew's location, state of completion and materials availability. This type of schedule information can be valuable in developing aggressive strategies. The master schedule can also be made more reliable and executable, increasing schedule conformance. This improves decision-making processes based on the programme/schedule.

Planning

Work scope and interference between trades can be identified earlier during planning phases. This provides the opportunity to optimise the programme or schedule based on constraints being made explicit. Planned activities and resources can be analysed against different spatial and temporal data.

Construction and planning can be synchronised, reducing unnecessary time requirements and allowing simultaneous work by different disciplines to be coordinated more effectively. Concurrent operations and construction activities can also be streamlined.

The use of BIM work programme planning tools enables better control over the programme and schedule, by allowing the main office to monitor activities occurring at the job-site. This can improve programming across different projects allowing stopping or diverting flows before they happen on site, more effective assembling of teams and improved resource allocation.

Design

The design can be more accurately assessed in terms of space needed for construction and operational programmed activities.

Construction

Programme progress can be better visualised and activities can be updated in real time through the use of phase planning (4D modelling). Logistics can be optimised by including traffic layouts and potential job-site hazards. This allows better planning of site activities and more predictable planning. Construction activities can be simulated helping to optimise activities programming and scheduling. Site use layout can be efficiently generated for temporary activities, assembly areas and material deliveries for all life-cycle phases. Temporary structures can also be evaluated more effectively. This can allow faster identification of space and time conflicts and accurately evaluate the site layout based on safety concerns. Interactive project scheduling tools can help reduce human errors during the construction process and enhance understanding of processes involved.

Operations

Operational programming can also be improved because the physical location of equipment, systems and materials can be clearly understood by maintenance and asset managers.

Beneficiaries

Beneficiaries include client-owners, contractors, subcontractors and asset/operation managers.

Enablers

Table BD.6 Better programming/scheduling enablers

	Planning	*Design*	*Construction*	*Operations*	*Decommissioning*
3D control and planning		X	X	X	X
Asset (preventative) maintenance scheduling				X	
Constructability analysis	X	X	X		X
Construction system design (virtual mock-up)	X	X	X	X	X
Data-rich geometrically accurate model components	X	X	X	X	X
Integrated model and programme management systems	X	X	X	X	X
Phase planning (4D modelling)	X	X	X	X	X
Site coordination models			X		X
Site utilisation planning			X		X

Flow-on benefits

Improved programming and scheduling can help realise the following benefits:

- Improved safety by developing more realistic safety plans.
- Optimisation of construction sequence through site layout optimisation, more use of virtual lifts and reduced equipment stand-by time.
- Reduced project delivery time due to better planning leading to schedule compression.
- Improved communication provided by more effective communication of construction sequence and layouts, and fewer or no language barriers, which can lead to fewer internal communication errors.

- Improved efficiency can be achieved through higher management control and smaller and more accurate inventories of components and finished pieces, which in turn lead to fewer delays due to material shortages and non-delivery of components.
- Fewer errors due to higher degree of integration between model and geographic coordinates.
- Improved productivity in the form of less time spent on field planning.
- Less rework due to better communication of control points.
- Lower operational cost provided by more proactive maintenance programming, which leads to improved asset performance. This helps reduce the number of repairs needed as well as maintenance and corrective emergency maintenance cost. The physical location of equipment and systems is better understood requiring better use of asset management staff time.

Metrics

Planning and design

- Conflict
- Time predictability – tasks
- Speed of production
- Volume of rework due to programming errors
- Cost predictability
- Programme capacity
- Resource use and management

Construction, operations and decommissioning

- Conflict
- Time predictability – tasks
- Speed of production
- Volume of rework due to programming errors
- Worksite clashes
- Staff stand-by and idle time
- Cost predictability
- Field material delivery
- Programme capacity
- Resource use and management
- Equipment stand-by time
- Cost per unit – materials

Examples

Better programming due to the use of BIM has been reported to produce up to 30 per cent savings in electrical material by eliminating the need for material restocking and leftovers (Thomas et al., 2004). Using virtual lifts and cranes in asset construction and maintenance programming has been additionally shown to result in smoother field operations and reduced equipment stand-by time. One project reported by Thomas et al. (2004), for example, attributed three months' savings to the use of virtual lifts of a reactor.

The use of BIM for construction planning of a 16-storey building in Lithuania contributed to the construction site and project manager saving approximately 10 per cent of time that would have normally been used to check for conflicts. The use of BIM was particularly helpful during the planning and managing of construction processes and materials delivery to a space-constrained site with 'demanding neighbourhoods of office workers' (Migilinskas et al., 2013).

In the US, the GSA has reported that using BIM in pilot projects has contributed to optimising the construction schedule leading to, for example, a reduction of 19 per cent of the duration of construction of the Los Angeles Federal Office Building (GSA, 2007). The communication of construction programmes has also been reported to have improved by using BIM in the US. For example, during the construction of the Worcester Trail Courthouse, 17 versions of the 3D model showing phase development of the construction process were created (in two-week intervals). These were posted on the contractor's website as a communication aid during meetings with subcontractors and helped improve communication and coordination across trades (Salazar et al., 2006).

In an Australian 11.4km rail project carried out by Parsons-Brinckerhoff, the use of BIM during design and construction helped the team to finish the first phase of construction four weeks earlier than originally scheduled. It was also expected that if the same rate of schedule acceleration was achieved for the second phase, the project construction could be shortened by up to a year (Bentley, 2012a).

BIM was used for the reconstruction of the Soldier Field project where structural engineers used 3D modelling and handover 'throughout the steel supply chain to enable the stadium's construction to be completed 17 per cent faster than industry best performance'. BIM was also used in this project for quantity take-offs, identifying interferences, validating steel geometry and digital surveying equipment to position steel (Fallon and Palmer, 2007).

Notes

- It has been noted that assigning a person, such as a BIM modeller, responsible to create and update the model for this purpose can help maximise this benefit (Salazar et al., 2006).
- A report from McGraw Hill on the value of BIM for infrastructure reported that 44 per cent of the owners surveyed assigned high value to the use of BIM for the planning phase due to the ability to visually convey complex engineering solutions. This was highlighted to be 'increasingly valuable for review and approval processes with non-technical stakeholder groups' (McGraw Hill Construction, 2012).

Better scenario and alternatives analysis

Benefit description

This benefit refers to higher accuracy and efficiency of scenario and alternatives production and analysis tasks. Simulation processes are more productive, less prone to error, and faster than the traditional method of compiling analysis models from scratch. By using object libraries and virtual mock-up, a larger number of alternatives can be analysed in a specific period of time, so the added value to the client is higher for the same or similar cost. Simulations can also include more accurate product information provided by manufacturers and suppliers, increasing the accuracy of the option analysis. Automated rule-checking can be used for automated analysis of different scenarios based on pre-specified criteria.

All phases of the asset life-cycle can be simulated at a greater level of detail: information can be extracted directly from the model, analysed under different conditions and options, and optimised to reduce whole-of-life cost (including decommissioning). BIM can be used to simulate a number of aspects of the designed asset once it is built and operational. These aspects may include structural, thermal, acoustic, lighting and fire performance. For example, energy performance of the completed asset can be simulated and compared for different options. The consequences of the most energy-efficient option on other aspects such as acoustics or thermal comfort can then be quickly analysed. The use of BIM can therefore help improve the efficiency of simulations used for budgeting of new assets by integrating elements such as major maintenance, security and dislocation during construction activity. The interrelations of these aspects with the design and construction options can also be understood more easily leading to better decision-making.

Planning and design

At concept stage, financial officers can efficiently analyse different types of assets through an increased level of visualisation. Faster and easier scenario and alternatives analysis also enables testing solutions to complex problems in a more efficient and effective way. For example, dynamic plans of occupancy offer multiple options and solutions to space conflicts. BIM can also be used to simulate the assets' operational phase during planning to evaluate its expected performance and maintainability. This allows specific materials or design options to be chosen in order to improve maintenance and repair.

The use of BIM facilitates and increases the speed at which performance aspects can be recalculated based on changes to the model. This allows different design alternatives and configurations to be evaluated, analysed and optimised in less time and with less effort. It allows rigorous analyses of different options.

Designers can take advantage of the parametric relationships, behavioural intelligence, and automated generation and layout of detailed components to produce multiple design alternatives. Different design alternatives and configurations can be easily modelled and changed in real-time during design review based on end-users and/or owner feedback. The use of virtual mock-up and 3D accurate visualisation eliminates the need of physical models to understand different design solutions.

Construction

Different construction methods can be simulated in order to improve delivery time, cost, processes, constructability and risk during the design and construction phases to avoid on-site problems. This provides users with spatial and temporal insights of the construction process. Production options can also be simulated for multiple 4D models from different perspectives, allowing project stakeholders to compare construction alternatives.

Operations

In the operational phase, different maintenance and repair approaches can also be simulated and evaluated based on cost and efficiency information included in the model.

Beneficiaries

Beneficiaries include clients/owners, designers, contractors, subcontractors, asset managers and end-users (tenants and emergency response units).

Enablers

Table BD.7 Better scenario and alternatives analysis enablers

	Planning	*Design*	*Construction*	*Operations*	*Decommissioning*	*Whole-of-life*
Automated rule-checking	X	X	X	X	X	X
Construction system design (virtual mock-up)	X	X	X	X	X	X
Data-rich, geometrically accurate model components	X	X	X	X	X	X
Design reviews	X	X	X	X	X	X
Energy simulation tools		X		X		X
Engineering analysis		X	X	X	X	
Interoperable formats	X	X	X	X	X	X
Object libraries	X	X	X	X	X	X
Phase planning (4D modelling)	X	X	X	X	X	X

Flow-on benefits

Achieving better scenario and alternatives analysis can have the following flow-on benefits:

- Improved output quality due to an optimised design and improved constructability and maintainability.
- Better environmental performance achieved through more sustainable designs that have been optimised based on environmental criteria leading to, for example, higher resource use efficiency and reduced waste.
- Lower operational cost achieved through optimised design and systems, with higher resource use efficiency and improved asset performance, maintainability and life-cycle cost.
- Better programming/scheduling based on better informed decision-making.
- Lower whole-of-life cost based on higher return on investment, improved audits and analysis tools, and reduced waste.
- Competitive advantage gain through improved specialised services.
- Improved productivity in the form of higher construction productivity.
- Improved safety through increased safety awareness and enhanced health, safety and welfare project performance.
- Improved communication by decreasing language barriers.
- Lower design cost achieved through reduced cycle times for design analyses.
- Higher customer satisfaction due to improved internal comfort and amenities for building occupants due to better sustainability and building performance alternatives analysis.
- Reduced risk in the form of less uncertainty.
- More effective emergency management ensured by designs that have been optimised for emergency situations.

Metrics

Planning, design and whole-of-life.

• Cost predictability • Quality	• Labour intensity for scenario and alternatives analysis

Construction, operations and decommissioning

• Time predictability • Cost predictability • Labour intensity for scenario and alternatives analysis	• Conflict • Quality • Resource use and management

Examples

A case study carried out by Kam et al. observed that a design team was able to evaluate the survival probability of 60 different plant types depending on different landscape options over 23 acres. This analysis included variables such as the species, the orientation, and shade from building/terrace. BIM allowed the team to objectively predict annual solar exposure of each terrace, and calibrate the results with the plant species characteristics to optimise the landscape design (Kam et al., 2013).

In another recent case study, BIM was integrated with construction process simulations for project scheduling support. The project was a 14,966m^2 building for which the team was able to create 1,000 simulations with different project durations in order to optimise the resource allocation strategy (Wang et al., 2014a).

In a road and bridge infrastructure project in Norway, the project team was able to develop 17 road alternatives and eight bridge designs using GIS, 2D CAD, 3D existing models and raster data. These models were also automatically modified by making changes directly to the original data sources (Autodesk, 2013).

The US National Institute of Standards and Technology (NIST) has also reported significant benefits in relation to scenario and alternative analysis. In one project where BIM was used for energy and cost analysis, it allowed 'rapid, reliable analysis of design options and encouraged new forms of collaboration between team members'. Using interoperable formats also helped reduce rework 'and provided more detailed and actionable feedback from those analyses' (Fallon and Palmer, 2007).

Notes

- Some of the software used for scenario analysis of resource use and environmental conditions do not use BIM files directly. However, the information can be extracted from the model and transferred into these software platforms, and back into the model with relative ease.
- The US GSA has stated that the use of BIM-based energy models allows project teams to carry out 'more efficient, accurate and reliable energy simulations to predict building performance during facility operations' (GSA, 2007).

Better space management

Benefit description

This benefit refers to achieving more efficient and effective management of space in order to reduce occupancy cost and space management conflicts. The use of BIM modelling tools allows team members to better simulate the space environment and make decisions based on these options. Space use can be optimised through, for example, the use of smart algorithms.

Planning

The use of geospatial information and site analysis with BIM facilitates determining whether potential sites meet required criteria according to project requirements, and technical and financial factors.

Construction and operations

The use of 4D modelling enables early detection of potential site logistics and accessibility constraints as well as provide spatial and temporal insight into the job-site space use. Site utilisation planning allows quick identification of potential and critical space conflicts. It also allows easy updating of site organisation and space usage as specific tasks progress, while minimising the time spent in site utilisation planning. This is also true during operations of assets with unusual space management requirements, such as entertainment venues and production plants.

In the case of assets where the layout changes often, such as entertainment facilities, BIM asset management systems can help staff identify work areas and pathways more effectively. BIM also facilitates space planning. Space management and tracking ensures the appropriate and easier allocation of spatial resources throughout the operations of the asset. These tools also enable proficiently tracking the use of current and future space.

Beneficiaries

Beneficiaries include contractors and asset managers.

Enablers

Table BD.8 Better space management enablers

	Planning	*Construction*	*Operations*	*Decommissioning*
BIM asset management systems			X	X
Design authoring (3D visualisation)	X	X	X	X
GIS-BIM	X	X	X	X
Phase planning (4D modelling)	X	X	X	X
Space management and tracking			X	
Site analysis		X		X

Flow-on benefits

Better space management, through more accurate site analysis, can help lower operational cost by improving the effectiveness of transition planning and asset performance. It can also improve programming and scheduling by helping reduce the number of space conflicts.

Metrics

- Conflict (space management).

Examples

In the US, the GSA has stated that the use of BIM is allowing them to automatically calculate relevant space data such as net area and efficiency ratio. This agency is using BIM during preliminary and final concept design, to 'validate spatial program requirements more accurately and quickly than using traditional 2D approaches' (GSA, 2007).

In an AU$2.1 billion (US$1.6 billion) Australian rail project, the use of integrated BIM modelling tools helped the team to analyse a number of options regarding track alignment and structure. This effort resulted in amendments that saved up to AU$1 million (US$770,000) (Bentley, 2012a).

Note

- The US GSA has required since 2007 that all major projects submit a spatial program BIM to the Office of Chief Architect (OCA, part of the Office of Design and Construction) prior to final concept presentation (GSA, 2007).

Better use of supply chain knowledge

Benefit description

> To increase the productivity and efficiency, knowledge management (KM) plays an important role in construction lifecycle processes by managing the corporate knowledge, which provides organisations with competitive advantage... knowledge has become the most important resource and asset for companies today... The role of knowledge is to facilitate the processes of transforming data into information through data interpretation, deriving new information from existing through elaboration, and acquiring new knowledge through learning.
>
> (Liu et al., 2013)

This benefit refers to the more effective and efficient use and management of knowledge acquired by different stakeholders in order to improve outcomes. Improved communication, for example, provides the opportunity to engage all relevant stakeholders at earlier stages of the asset's life-cycle where the greatest value can be derived from their input. This exchange is also made more efficient and effective. The use of BIM repositories and management tools also enables collaborative knowledge management and sharing throughout the asset's life-cycle. Users can integrate and reuse asset information and domain knowledge throughout the life-cycle of an asset and in other projects. BIM-based knowledge management systems can further facilitate this by providing automated alerts so when best practices and lessons learned in a specific area are added to the repository, interested parties receive a message when starting a similar project.

Knowledge can be more easily updated and shared across project phases and organisations and interactively built. Decisions and the reasons behind them can be better captured through the use of information-rich objects, articulating and improving the permanence of tacit knowledge. Experiences and processes can be recorded and retrieved when new situations arise in the same or different projects. For example, the reasons for using a specific process, tool or material can be associated with the object so other projects can access this information even if the person who made the analysis and decision is no longer in the organisation.

Knowledge is also made more accessible; iterative communication exchange processes between different disciplines being faster and recordable. Supply chain value knowledge can be accessed and used at the right time in order to achieve overall project objectives in a timely, accurate, error-free and cost-effective manner. Visualisation tools additionally allow non-technical stakeholders to better understand the details of the work, provide more timely advice for decision-making and have more influence on the decision-making process. All of this contributes towards stakeholders focusing their efforts in more value adding tasks instead of information and knowledge retrieval efforts.

Design

In the design phase, there is better supply chain knowledge integration through, for example, BIM design reviews engaging more disciplines and increasing access to the model. These provide the opportunity, for example, for constructability analysis to be carried out more frequently. Site-based skills can be applied to the design and construction programme to solve clashes and constructability issues. End-users and operators can also be involved in the review process during the design phase in order to integrate usability and maintenance criteria that will optimise the final asset. Rework can therefore be reduced by fully utilising expert knowledge to facilitate optimised designs.

Beneficiaries

Beneficiaries include client/owner, designers, contractors, subcontractors, surveyors, fabricators/manufacturers, suppliers, asset managers and end-users.

Enablers

Table BD.9 Better use of supply chain knowledge enablers

	Planning	*Design*	*Construction*	*Operations*	*Decommissioning*	*Whole-of-life*
Asset knowledge management				X	X	
Constructability analysis	X	X	X		X	
Cost estimation (quantity take-off)	X	X	X	X	X	X
Design reviews	X	X	X	X	X	X
Early engagement of stakeholders	X	X	X	X	X	X
Online collaboration and project management	X	X	X	X	X	X

Flow-on benefits

Better change management can have the following flow-on benefits:

- Lower construction cost due to constructability issues being identified during the design phase.
- Fewer errors due to access to more supply chain knowledge.
- Improved productivity through better knowledge management.
- More effective emergency management through emergency planning considerations being introduced during the design phase through more active involvement and feedback of asset managers.
- Asset management labour utilisations savings due to more efficient knowledge retrieval and decision-making process.
- Better cost accounting due to more efficient use of estimator's time in value adding tasks.
- Reduced risk through more knowledge-based decision-making.
- Less rework through the use of expert knowledge.
- Better environmental performance through the integration of expert knowledge during design.
- Competitive advantage gain through better internal knowledge management.

Metrics

- Knowledge management metrics and stakeholder involvement.

Examples

In a case study in the UK, the Nationwide Building Society used an online collaboration and project management (OCPM)-based knowledge management tool to capture and distribute existing knowledge. This tool was found to contribute towards time and cost savings in practical issues. The knowledge store contains information about issues such as 'how difficult-shaped branches were handled, imaginative use of materials, and effective building techniques'. This tool was also used by competing suppliers to learn from each other and share experiences. The use of the tool across projects has also allowed contractors in different projects to share their knowledge regardless of their affiliation, and for questions to be answered in very short periods of time. It has also been used to resolve building-related legal issues by having access to the knowledge of legal experts (Becerik and Pollalis, 2006).

In a 20-storey building constructed in Lithuania (2002–2004), BIM was used for bill of quantities production and design reviews among other tasks. This project reported reduced time wasted in dispute resolution related to volumes of work and a specialist being able to pay more attention to the quality of the works as well as discussing construction method improvements (Migilinskas et al., 2013).

In the construction of a hospital, phase planning (4D modelling) was used to plan the construction process, including equipment location and type. This information was presented to the hospital staff, who were then able to quickly point out that some of this equipment would block the emergency helicopter flight path and therefore could not be used. This allowed the team to resolve the issue prior to construction (Kivits and Furneaux, 2013).

Notes

- Although BIM-based knowledge management systems are not currently common in the industry, Charlesraj (2014) and Lin (2014) proposed a process for developing this type of tools.
- Nationwide Building Society was, in 2006, the 'UK's fourth largest mortgage lender and eighth largest retail banker' (Becerik and Pollalis, 2006).

Competitive advantage gain

Benefit description

This benefit refers to being able to improve current services or provide new ones as well as improving profitability in order for organisations to be in a superior business position. The use of BIM can, for example, provide access to new sources of revenue and business opportunities by offering new specialised services as well as providing a new way of marketing the firm.

The use of 3D visualisation can help facilitate the client's decision to engage a firm by allowing quick comparisons of various alternatives and improved understanding of products or services offered. Primary and support activities can also be carried out faster and with less effort. This can allow firms to offer more competitive tender bids, improving performance cost and providing higher engineering standards. The use of models and animations can also facilitate showcasing ability to carry out the work. Information is more easily shared when common data protocols are in use, allowing firms to add more value to the outputs delivered to clients who can also then work more efficiently.

Improved documentation quality and processes can additionally help to better accommodate owner's requirements, helping foster a stronger relationship and promoting repeat business. Enhanced communication and output quality can improve relationships with the client and customers. This in turn can facilitate subsequent work with the same client and help increase access to market.

The use of BIM-enabled knowledge management systems allows organisations to retain and access their intellectual capital, improving their competitive advantage. Online collaboration and project management can also allow easier access to international markets and staff, advantageous source of equipment, supplies and manpower development and better services to clients.

Finally, lower cost and improved efficiency can help increase profitability and provide earlier access to market.

Operations

BIM online collaborative tools can also allow building new relationships with end-users and increase visibility of sales. Better environmental performance can also help achieve higher environmental certifications, which may attract specific end-users and buyers. Additionally, reducing project execution time of certain types of construction can facilitate faster launch to market and earlier return on investment. This can be important, for example, for residential developments.

In the case of assets that provide services to the public that can be improved by better information management, using BIM asset management systems can help improve customer satisfaction and the kind of services provided. This, for example, applies to entertainment districts and venues.

Beneficiaries

Beneficiaries include designers, contractors, subcontractors, fabricators, surveyors and asset managers.

Enablers

Table BD.10 Competitive advantage gain enablers

	Design	*Construction*	*Operations*
BIM asset management system			X
Asset knowledge management	X	X	X
Construction system design (virtual mock-up)	X	X	X
Common data protocol	X	X	X
Design authoring (3D visualisation)	X	X	X
Engineering analysis	X	X	X
Online collaboration and project management	X	X	X
Phase planning (4D modelling)	X	X	X
Sustainability evaluation	X	X	X

Flow-on benefits

Competitive advantage gain is considered an end-benefit with no further flow-on benefits.

Metrics

- Access to market
- Globalisation
- Profit

Examples

In an Australian case study about the construction of a hospital, one of the consultants said they had gained a competitive advantage by providing BIM management expertise that gave them access to new markets. Internal automation processes also helped increase their competitiveness. Automating the model federation process reduced the number of resources required for this task and this made smaller projects more economically viable. In the same study, it was pointed out that certain tasks such as set-up require less time and resources, which can help contractors and subcontractors provide more competitive prices and schedules (Sanchez et al., 2015).

Notes

- This benefit applies to a specific type of asset managers (e.g., entertainment, housing developments, etc.).
- In a survey carried out by CIFE in 2007, 'nearly two thirds of respondents saw significant value of new projects won based in part on competitive advantage gained through VDC use' (Gilligan and Kunz, 2007).

- The enablers mentioned in this profile are based on cases reported in the literature where these tools and processes have provided a competitive advantage through the processes explained in the description of the benefit. However, enablers used to gain a competitive advantage will be specific to the area in which organisations operate.

Faster regulation and requirement compliance

Benefit description

This benefit refers to achieving compliance with regulation and client requirements in less time than when using traditional methods. For example, achieving compliance with client requirements can be done faster because the final product can be examined in detail by the client during the design phase through 3D visualisation and virtual walk-throughs. This ensures that expectations and scope of works are clearly defined and project requirements are more likely to be met and approved. Shop drawing submissions can be made more complete and approved faster due to faster information exchange and better data management. Additionally, subcontractors using and developing the BIM model can be better coordinated and carry out clash detection prior to manufacture and installation of asset elements.

Turnaround of construction permitting and approvals by planning officials can also be made faster; officials and project team need to spend less time meeting with code commissioners, visiting the site and fixing code violations during punch list or closeout phases.

Design

Automated rule-checking of the design against building and safety codes can accelerate the speed of regulation compliance. Design progress reporting can provide direct and continued feedback on code and requirement compliance based on automated rule-checking. This process can also be improved through visual compliance checking. Library objects can additionally be designed to incorporate clearance and/or service areas that can be used to check whether they have been included, such as disabled access features.

Beneficiaries

Beneficiaries include designers, contractors, subcontractors and fabricators.

Enablers

• Automated rule-checking	• Code validation
• Construction system design (virtual mock-up)	• Object libraries

Flow-on benefits

Increasing the speed of regulation and requirement compliance can help improve the efficiency of the design process and lower design cost.

Metrics

- Overall time for compliance.

Examples

Model-checking software that includes the ability to check against building codes and best practices can be very useful in expediting compliance processes. For example, one such software tool was being used by Fiatech in the US in their autocodes projects to flag safety issues during construction. In this project, the software checked the model against relevant regulation and flagged objects that did not meet it. This allowed them to identify 137 items and review them within a 3D modelling environment. This software was, for example, used to check that all fire extinguishers were positioned correctly (Sawyer, 2014). The same software was also used in the design validation process of the refurbishment and extension of the San Gerardo Hospital in Monza, Italy. Here, rule-checking was used to quickly analyse compliance of the design with building codes such as those related to accessibility of people with disabilities (Read, 2014).

A case study published in 2010 applied a BIM-based integrated approach to automated code 'compliance checking for building envelope design based on simulation results and building codes'. The tool covered codes such as those of American Society of Heating, Refrigerating, and Air-Conditioning Engineers (ASHRAE), Model National Energy Code of Canada for Buildings and Houses (MNECB), and the National Building Code of Canada (NBCC), among other. This process automatically generated a complete report of compliance with specific design regulations and advice based on simulations and decision tables (Tan et al., 2010).

Note

- Yang and Xu (2004) developed a pilot implementation of an online code-checking system. This study included a number of application scenarios that demonstrated the capabilities of such algorithms to support building conformance checking.

Fewer errors

Benefit description

This benefit refers to the reduction of total number of errors and omissions throughout the life-cycle of an asset, as well as easier identification of these mistakes when they are least costly to resolve. Information can be created once and reused throughout the life-cycle of the asset. This leads to reducing the possibilities of introduction of errors during data re-entry. Additionally, by using a central model as a single source of truth where changes are propagated to all elements, there are fewer possibilities of errors arising from version mismatch and information asymmetry. Having more control over the data used by all stakeholders can further reduce the likelihood of errors and omissions.

All the details about geometry, connections between parts and geographic location of end-products can be more accurately known, reducing the likelihood of errors. Improved visual representation and rule-checking can also make mistakes more visible, increasing the likelihood of these errors being corrected at earlier phases. Fault identification is improved throughout the life-cycle of the asset and having a formalised set of standard protocols and procedures can further reduce the number of errors. Quantity and location of furniture, fixtures and equipment (FF&E) items can also be scheduled from the model and compared to briefing requirements reducing the number of FF&E procurement errors.

Design

Automated clash detection and rule-checking allows detecting more errors during design, reducing the total number of errors and omissions in the final design output. The ability to quickly and easily compare the as-designed modelled content to the brief requirements can also reduce the number of errors, including FF&E and room schedule information.

Construction

Digital fabrication reduces variation and errors during construction by increasing the level of pre-fabrication and precast. Better communication and coordination also allows errors and conflicts to be identified more quickly and before construction starts.

Streamlining logistic processes and using field and management tracking tools can additionally contribute to reducing communication errors and lags. Field and management tracking processes can also improve work management so it is carried out correctly the first time. Finally, the use of digital layouts can reduce layout errors by linking model elements to specific geographic coordinates.

Beneficiaries

Beneficiaries include clients/owners, designers, contractors, fabricators/manufacturers and suppliers.

Enablers

Table BD.11 Fewer errors enablers

	Planning	*Design*	*Construction*	*Operations*	*Decommissioning*	*Whole-of-life*
Automated rule-checking	X	X	X	X	X	X
Automated clash detection	X	X	X	X	X	X
Digital fabrication – pre-fabrication and precast			X			
Field and management tracking			X	X	X	
Interoperable formats	X	X	X	X	X	X
Integrated model and programme management systems	X	X	X	X	X	X
Online collaboration and project management	X	X	X	X	X	X
Phase planning (4D modelling)			X	X	X	
Streamlined logistics			X	X	X	

Flow-on benefits

Having fewer errors can help reduce the cost of design and engineering, and reduce the volume of rework due to errors.

Metrics

Planning, design and whole-of-life

- Accuracy and number of errors/omissions
- Conflict due to errors/omissions
- Request for information – errors and omissions
- Variation and change orders – errors and omissions
- Cost of change
- Cost avoidance – errors and omissions

Construction

- Accuracy and number of errors/omissions
- Conflict due to errors/omissions
- Request for information – errors and omissions
- Schedule conformance
- Variation and change orders – errors and omissions
- Cost of change
- Cost avoidance – errors and omissions

Operations and decommissioning

- Accuracy and number of errors/omissions
- Conflict due to errors/omissions
- Request for information – errors and omissions
- Schedule conformance
- Cost of change
- Cost avoidance – errors and omissions

Examples

A study related to the benefits of adopting 3D modelling for precast engineering assessed data of more than 32,500 pieces from numerous companies in 2005. This study revealed a total reduction of errors of 0.46 per cent of total project cost achieved across assembly design, drafting, piece-detailing and design coordination. This estimate was calculated based on company historic cost of engineering and manufacturing errors, which is normally around 1 per cent of erected sale price. Even at this early BIM adoption stage, the study estimated that approximately 60 per cent of errors can be eliminated by using BIM (Sacks et al., 2005).

Another case study carried out in the US showed how design errors can be identified easier and at earlier phases when using BIM. In this building project, a design error was identified related to the chiller size and the space allocated. This conflict was identified three months before installation and the piping was rerouted to a new location with minimal rework (Staub-French and Khanzode, 2007).

In a pilot project of a road and bridge project, clash detection was used to identify two major clashes that would otherwise not have been found until construction started. One of the issues was related to pipelines that were to be built with the road and bridge. The project team was able to contact the company building the pipelines and the avoided cost was estimated to have completely covered the cost of implementing BIM (Gerbov, 2014).

Note

- In traditional projects, design changes and omission errors have been reported to account for up to 79 per cent of the total rework costs experienced, of which omission errors alone can account for up to 38 per cent (Love et al., 2009).

Higher customer satisfaction

Benefit description

This benefit refers to higher satisfaction of clients with project outputs. The use of BIM enables better customer service because proposals are better understood by clients due to improved visualisation leading to better-informed decisions and more effective consultation. Better quality of services can be delivered based on better, more visual communication of intended project outcomes. Project outcomes are therefore more likely to meet user and client expectations and there is greater flexibility to meet client demands. This in turn can reduce the number of disputes and unscheduled work-orders, which can be resolved faster. The use of integrated BIM has been shown to increase responsiveness to client's requests for changes and reduced lead times, all leading to higher customer satisfaction.

The use of object libraries, information-rich models, common data protocols and common data environments allows more value to be delivered in less time and cost through, for example, enhanced information flows. Information can be accessed in multiple formats allowing different users to access the same data more quickly and in a way that suits their system needs and skills level.

Operations

BIM visualisation tools can also be used to improve public relationships and better communicate work orders to tenants and maintenance staff. Additionally, better scenario analysis allows improved asset operational performance, including optimised thermal performance and daylight use. This can lead to higher productivity of end-users and higher customer satisfaction.

Beneficiaries

Beneficiaries include client/owners, designers, contractors, subcontractors, fabricators/manufacturers, surveyors, suppliers and asset managers.

Enablers

Table BD.12 Higher customer satisfaction enablers

	Planning	*Design*	*Construction*	*Operations*
Common data protocol and environments	X	X	X	X
Construction system design (virtual mock-up)	X	X	X	X
Data-rich geometrically accurate model components	X	X	X	X
Design authoring (3D visualisation)	X	X	X	X
Object libraries	X	X	X	X
Phase planning (4D modelling)	X	X	X	X

Flow-on benefits

Higher customer satisfaction enables competitive advantage gain by gaining access to markets and improving customer relationships.

Metrics

- Satisfaction.

Examples

The use of BIM was reported to be a central process in increasing client satisfaction in the design of a train maintenance centre in Australia. This project is part of a rail network upgrade worth AU$4.4 billion (US$3.2 billion) that will increase the fleet capacity by 30 per cent. The use of BIM in this project was mainly driven by the contractor as a response to a general client strategy: aiming to increase productivity and value for money. The 3D visualisations helped improve communications with the client leading to better feedback, more clarity on the scope of the project and better aligned expectations. The model was found to have provided a common ground for clear communication between the client and the design team. This experience was so satisfying for the client representative interviewed in the case study that, although initially hesitant, they later expressed interest in continuing to use BIM processes in future projects (Utiome et al., 2015).

Higher process automation

Benefit description

This benefit refers to the reduction of manual labour required to carry out tasks through the use of computer software processes that automate these tasks. The use of BIM and its enabling tools allows the automatic generation of documents in response to various criteria throughout the life-cycle of an asset. For example, drawings are produced directly from the model, including schematic, design development, construction and shop drawings. Reports and aspects of the model can be automatically updated, propagating any model changes to the reports and eliminating the need to manually update each drawing or document for each design change. This process includes auto-populating schedules, automatically updating drawing sets based on changes in the model. Accurate, complete and unambiguous information can be directly transferred between project phases and stakeholders.

Quantities for bills of materials can be directly extracted from the information in a BIM model, enabling estimators to quickly analyse the cost of different options at planning, design, construction and operations phases. Here, the level of accuracy will depend on the level of detail of the model at each stage. Evaluation of conformance to client value and building and safety codes can be automated through rule-checking as well as automated on-site verification, and guidance and tracking of construction activities.

Design

Automated evaluation of building design to enhance fault-finding processes allow pre-programming design BIM protocols to detect internal conflicts with the design intent. Model viewing systems can detect and highlight conflicts between the models and other information imported into the viewer.

Tagging with family, schedule and quantity generation can be automated. Relevant space data calculations (e.g., net area and efficiency ratio), and low-level corrections when design changes are made can also be automated.

Automated clash, interference or collision detection processes identify elements of the design that occupy the same physical space in the building and are used to locate errors in the model.

Construction

Automated clash detection is also useful during pre-construction and construction stages, allowing teams to identify conflicts before work commences and coordinating to avoid issues that will hinder assembly and installation tasks. This process can refer not only to geometric clash detection but also to crew scheduling, space logistics and construction sequence.

Fabrication

Automated assembly, computer-controlled manufacturing, digital fabrication and automated production processes use digitised information and product data to facilitate downstream processes, manufacturing of materials and assembly of structural systems. Data provided by the model can be used for laser layout projection systems and for computer-controlled machines such as cranes, robotic applicators for sandblasting and acid etching, and other piece-handling equipment. Examples include: sheet metal fabrication, structural steel fabrication, pipe-cutting, reinforcing bar-benders, machine-welding, milling and/or laser-cutting for production of Styrofoam mould parts, wire mesh-bending, prototyping for design intent reviews and precast concrete. This higher degree of automation can prompt a change from on-site processes towards factory pre-fabrication and other automated high-quality processes.

Operations

Scheduled work orders for maintenance staff can be automatically generated from the model, or BIM enabled CAFM/CMMS system, providing access to all the information contained in associated objects. BIM also supports automated analysis of the performance of building products and designs rather than manual, iterative and time-consuming traditional approaches. Life-cycle cost analysis can be automatically transferred between models, databases and applications to enhance these analyses through data-mining.

Beneficiaries

Beneficiaries include clients/owners, designers, surveyors, contractors, subcontractors fabricators/manufacturers and asset managers.

Enablers

Table BD.13 Higher process automation enablers

	Planning	*Design*	*Construction*	*Operations*	*Decommissioning*	*Whole-of-life*
Automated rule-checking	X	X	X	X	X	X
Automated clash detection		X	X			
Data-rich geometrically accurate model components	X	X	X	X	X	X
Digital fabrication			X			
Field and management tracking			X	X	X	
Integrated model management systems	X	X	X	X	X	X
Interoperable formats	X	X	X	X	X	X

Flow-on benefits

Higher process automation can have the following flow-on benefits:

- Lower cost (construction) may be achieved by reducing waste and conflict during construction as a result of higher process automation in areas such as cost estimation, clash and interference detection, field management, communication of information and management of changes.
- Lower cost (design) may be achieved through, for example, faster performance simulations, use of off-the-shelf building products from object libraries and manufacturers, and faster generation of multiple views.

- Lower cost (whole-of-life) may be achieved through better data/information capturing (automated transfer of accurate, unambiguous and complete data and information), easier adaptation of late changes and a proactive approach to value engineering allowing design integrity to remain.
- Faster regulation and requirements compliance in the form of faster planning approval process due to faster regulation conformance checking as well as enhanced capacity to produce drawings.
- Improved project coordination due to accurate off-site manufacture, helping reduce the number of site conflicts. Higher automation can also enable faster and more accurate (better) validation of spatial programme requirements, reduced coordination problems and reduced language barrier.
- Better cost accounting through more accurate cost estimates.
- Improved data and information management due to higher level of consistency between datasets, improved standardisation and better data and information quality.
- Improved output quality due to enhanced geometric integrity of design, more accurate data provided to manufacturers, improved design, more pre-assembly, and higher level of pre-fabrication.
- Improved productivity in the form of (1) enhanced capacity to produce drawings; (2) faster production of construction documents; (3) more pre-assembly; (4) improved fabrication productivity; (5) higher level of pre-fabrication; (6) reduced field survey time; (7) reduced data re-entry; and (8) parallel off-site work.
- Less risk through less uncertainty, fewer conflicts and more reliable performance simulations.
- Less rework due to fewer inconsistencies in production of drawings, fewer errors in drawings, improved fault identification and reduced data re-entry.
- Improved safety can be achieved due to having more pre-fabrication and pre-assembly, which increases construction safety.
- Reduced project execution time and lead times due to schedule acceleration and parallel off-site work.
- Fewer errors due to improved linking of layout to geographic coordinates and reduced language barrier.
- Better environmental performance in the form of reduced dependency on paper and more interoperable life-cycle data.

Metrics

• Labour intensity of tasks that are now automated	• Accuracy and number of errors/omissions
• Speed of production (documents)	• Clashes
• Volume of rework	• Off-site manufacturing
• Cost of change	• Cost savings/avoidance

Examples

In the Sequus Pharmaceuticals Pilot Plant Facility project (in the US) many of the different pipe runs were fabricated directly from the 3D models, resulting in time and cost savings and fewer errors. It was estimated that fixing errors in those pipes had they not been pre-fabricated could have cost approximately US$700 per error. In addition, the model was used by the supplier to fabricate the pipe for chillers in the shop at about one-third of the cost. The project manager for Rountree Plumbing stated that 'virtually everything pre-fabricated from the 3D model was installed as planned' (Staub-French and Khanzode, 2007).

In a case study reported by Kam et al., automated clash detection was estimated to have saved US$1.3 million by identifying 16 issues that would have been undetected in a traditional design process and unresolved prior to on-site installation and construction (Kam et al., 2013). In other reports, savings of up to 10 per cent of contract value have also been estimated due to clash detection (Azhar, 2011).

In 2010, Swinerton Builders (US-based construction firm with 125 years in the market and US$1.17 billion in revenue in 2012) estimated that, based on data from ten major projects, the average cost of a change order was US$17,000. In a building project carried out with BIM, they estimated that the use of clash detection contributed to avoiding 450 change orders, accounting for more than US$6.7 million in savings (Brown, 2013; Allison, 2010).

Notes

- Interoperability between software used for design and construction, and that used by fabricators has been mentioned as a challenge that hinders realising the full value of BIM through higher process automation (Sanchez et al., 2015).
- Almost 90 per cent of respondents in a study by Gillian and Kunz (2007) reported using 3D automated clash detection as a business purpose of VDC.
- Although the use of pre-fabrication from BIM models has been reported to be less frequent globally, more than half of South Korean and French firms surveyed in 2014 by McGraw Hill indicated using pre-fabrication in order to create tighter building envelopes (McGraw Hill Construction, 2014a).
- Building detailing (standard construction details) can be done based directly on the model or completely separate. Certain software used – for example, for water processing projects – can alternatively link diagram schematic drawings directly to the 3D model objects.
- Live BIM from Model the Planet in the US claims that this tool can reduce work orders processing costs by up to 75 per cent through

improved data management and automatic information flows (Model the Planet Corp., 2015).

Improved communications

Benefit description

This benefit refers to facilitating communication between stakeholders that is more accurate, effective, transparent and timely. The use of 3D visualisation tools reduces the likelihood of misinterpretation by allowing more clarity around a task's scope and facilitating communicating different solutions and issues from the outset and throughout the asset's life-cycle. Outputs can also be transferred to other visualisation and review platforms that may be more easily accessed by untrained stakeholders.

Through visual communication and the use of enablers such as collaborative/common data environments, BIM can facilitate two-way multi-stakeholder communication through more efficient data and information exchange and decreased language barriers. It can simplify and improve the quality of communication through, for example, computer-aided visualisation of processes and design options. This means that information can be expressed more intuitively, issues immediately grasped and interrelations can be clearly visualised; enabling view-based frameworks of collaboration. Collaborative data environments used with BIM further allow web-based communication without the need for face-to-face meetings as well as clear definition of tasks and elements, well-defined and clear boundaries, and faster information exchange.

Design

Design intent can be clearly communicated to technical and non-technical staff in a more intuitive way.

Construction

Construction system design (virtual mock-up) can be used to design and analyse the construction of a complex asset system in order to improve planning. Walk-throughs and construction sequence animations allow construction projects to be seen in near real-life fidelity and become alive before they are built. The use of 4D modelling tools in construction allows the visualisation and communication of processes and product status.

Effective communication between on-site and off-site stakeholders can be facilitated by the use of collaborative/common data environment, BIM management and collaboration tools and smart handheld devices. These tools can provide a clear visual context required for effective communication of aspects related to the collection of data using model views by surveying and construction teams as well as fast information exchange between crews.

Beneficiaries

All stakeholders benefit through better understanding of the design intent and project development. More specifically, clients/owners benefit through better understanding of the asset being delivered, more certainty of information, and better communication to service and goods suppliers; designers and contractors can explain issues and tasks in a more intuitive way to other stakeholders; fabricators/manufacturers and subcontractors can visualise and understand the intricacy of framing and connection details in a 3D structural model; and maintenance staff and tenants can have better understanding of schedule work orders and the asset.

Enablers

Table BD.14 Improved communications enablers

	Planning	*Design*	*Construction*	*Operations*	*Decommissioning*
Common data environments	X	X	X	X	X
Construction system design (virtual mock-up)	X	X	X	X	X
Design reviews	X	X	X		X
Design authoring (3D visualisation)	X	X	X	X	X
Effective stakeholder engagement	X	X	X	X	X
Handheld devices	X	X	X	X	X
Online collaboration and project management	X	X	X	X	X
Phase planning (4D modelling)	X	X	X	X	X
Walk-through and animations	X	X	X	X	X

Flow-on benefits

Improved communications can help achieve the following benefits:

- Improved project coordination, less errors, optimisation of the construction sequence, better scenario and alternatives analysis, better programming and scheduling, and reduced risk. This can be done due to all stakeholders (non-technical included) having a greater understanding of scope, objectives, design intent, needs, geometrical relations, construction sequence, space requirements, flow of operations and project status. Additionally, there is more awareness of progress, and fabricators and subcontractors visualise and understand the complexity of framing and connection details in a 3D structural model. Better-quality conflict analysis and faster and more accurate detection of more logical errors also help achieve the above benefits.
- Higher customer satisfaction due to better customer service (proposals can be better understood), more certainty of information and better tools to enhance social awareness and align expectations.
- Better use of supply chain knowledge and improved productivity due to stakeholders being engaged more effectively and provide better quality feedback as well as extended collaboration between disciplines and phases leading to better analysis.
- Competitive advantage gain through better marketing.
- Reduced lead times through web-based communications that do not need to be coordinated (asynchronous communication) and reduction in latency.
- Improved efficiency in the form of more efficient design, fabrication and construction process, increased cost-efficiencies and profitability, and improved project process outcomes such as fewer field coordination problems.
- Reduced risk due to problems being solved more collaboratively, increased trust and better appreciation of other participants' due diligence and concerns as well as faster conflict resolution and easier task monitoring.
- Improved output quality due to stakeholders being engaged more effectively and providing better quality feedback, better handling of design options, improved project definition; all of these leading to better decision-making.
- Improved coordination due to improved conflicts and constructability analysis, site layout planning and task monitoring.
- Improved learning curve due to new staff training based on highly visual processes.

Metrics

- Meeting effectiveness
- Meeting efficiency
- Meeting agenda appropriateness
- Overall time required to resolve issues
- Latency
- Conflict (related to misunderstandings)
- Number of RFIs
- Visualisation

Examples

In an Australian case study of a new hospital, end-users had access to accurate models of the rooms they would use when the hospital was completed and could *walk* the room using an avatar of their height and build. This allowed demonstration of line of sight and movement patterns within the model for an area that was disputed based on a misunderstanding of the 2D drawings. This led to almost immediate sign-off of individual spaces by end-users (Sanchez et al., 2015).

In the same project, a situation was found around the location and size of slab penetration and set down. The team was able to create views during meetings that allowed isolating disciplines one at a time so that issues could be discussed and solved individually. In another example, models were used to produce first-person viewpoints from the position of CCTV cameras. This allowed the client to have a better understanding of the scope of coverage that the cameras could provide based on their specifications. This exercise provided the client with complete confidence about the CCTV camera layout required for sign-off (Sanchez et al., 2015).

In a case study of a pilot plant facility, the project observed a reduction of RFIs of 60 per cent with respect to that expected for a project of such complexity. In this case, it was thought that the improved visualisation provided by 3D and 4D tools allowed project teams to better understand spatial needs of each discipline over time and improve overall communication (Staub-French and Khanzode, 2007).

The use of BIM in pilot projects by the GSA helped improve the means for communication with tenant agencies and during pre-bidding conferences (GSA, 2007). In a courthouse case study for example, BIM was reported to enhance 'the communication process among the different construction trades and facilitates coordination' (Salazar et al., 2006).

In the upgrade of the Great Eastern Highway in Australia, 3D visualisation tools were used to provide information to public users about the current and planned works in their area of residence. During construction, the site team had access to the model through handheld devices that allowed them to more accurately define where to dig (Wang, 2015).

Note

- Having the right tools alone may not provide improved communication if the team does not embrace the new more collaborative way of working and digital 3D visualisation-based processes.

Improved coordination

Benefit description

This benefit refers to more effective and efficient coordination of documents, processes and tasks. Better coordination between documents can, for example, be achieved because all drawings and reports are derived from the same model, eliminating coordination issues between documents prepared or modified separately. More efficient coordination between disciplines can be achieved because the use of BIM promotes earlier engagement of stakeholders and more transparency. This increases motivation to collaborate. Additionally, changes made to different elements and documents can be better coordinated across all relevant disciplines based on a single model. BIM can enhance project collaboration and control among stakeholders by enabling early identification of work scope and interferences between trades and disciplines.

The use of integrated systems and 3D coordination tools enables better coordination of computer-controlled manufacturing data for accurate off-site manufacturing. BIM also enhances business-to-business integration and coordination of activities. Design reviews also increase coordination and communication between different parties, enabling better decision-making.

Design

Elements of the design can be better coordinated, within and across disciplines, through design reviews and clash detection software. This reduces the possibility of misaligned connections, incorrect architectural features and geometry conflicts. This reduces the need for cross-coordination between drawings and designers. Sharing of work-in-progress models in design-authoring software allows a constant and proactive approach to model coordination.

Construction

Site coordination can be made more effective through the use of 4D site coordination models. These allow planning site logistics, developing traffic layouts and identifying potential hazards at the job-site. Schedule changes due to site conditions can also be better coordinated, informing scheduling decisions in advance at the least cost. Construction models can be coordinated prior to fabrication, installation and construction on-site.

Operations

As-built BIM models can serve as reference for coordination of future works and maintenance tasks.

Beneficiaries

Beneficiaries include clients/owners, designers, contractors, subcontractors, fabricators/manufacturers, surveyors and asset managers.

Enablers

Table BD.15 Improved coordination

	Planning	*Design*	*Construction*	*Operations*	*Decommissioning*	*Whole-of-life*
Automated clash detection		X	X	X	X	
Design authoring (3D visualisation)	X	X	X	X	X	X
Design reviews	X	X	X	X	X	
Early and effective stakeholder engagement	X	X	X	X	X	X
Integrated model management systems	X	X	X	X	X	X
Phase planning (4D modelling)			X	X	X	
Site-coordination			X		X	

Flow-on benefits

Improved coordination can have the following flow-on benefits:

- Improved output quality in the form of enhanced design.
- Lower construction cost due to resource savings, lower cost growth, and reduced waste and more pre-fabrication.
- Improved efficiency in the form of improved cost-efficiency and reduced waste.
- Improved productivity due to improved on-site productivity and more pre-fabrication.
- Reduced project execution time due to decreased construction time and reduced time required for conflict resolution.
- Fewer errors and omissions during project delivery and implementation.

Metrics

Planning, design, operations and whole-of-life

• Volume of rework • Number of RFIs related to coordination issues	• Number of clashes • Number of variations and change orders related to coordination issues

Construction and decommissioning

- Conflicts – field
- Volume of rework
- Overall time – field coordination related activities
- Number of RFIs related to field coordination issues
- Number of variations and change orders related to field coordination issues
- Off-site manufacturing – pre-fabrication
- Number of field clashes
- Equipment stand-by time
- Staff and work-site crew idle time

Examples

Data analysis carried out by Khanzode (2010) suggests that the use of BIM for coordination purposes creates more opportunities for pre-fabrication for mechanical systems. In a case study in North Carolina in the US, the team using BIM for coordination and pre-fabrication reported realising benefits such as: virtually zero field conflicts between various systems; less than 0.2 per cent rework; productivity improvement of more than 30 per cent for the mechanical contractor; less than two hours per month spent on field coordination issues by the superintendent for the general contractor; only two field issues-related requests for information; and zero change orders related to field conflict issues (Khanzode, 2010).

On a different project that used BIM, after 23,225m^2 were constructed the superintendent said that he had never experienced this level of accuracy of field installation before in his 35 years of experience. He also estimated that he spent significantly less time resolving field issues compared to past projects; he estimated spending 2–3 hours/day in previous projects dealing with these issues while only spending a total of 10–15 hours over an eight-month period after the MEP installation began as part of the BIM project. This reduction in conflict and rework was attributed to improved coordination prior to construction (Staub-French and Khanzode, 2007).

The SEK10.5 billion (US$1.3 billion) Swedish Hallandsås tunnel project consisted of two 8.7km parallel railway tunnels due to be completed in mid-2015. This project faced a number of challenges such as excavation works through hard rock, soft rock and clay, high water pressure, and significant restrictions regarding leakage to ground water due to sensitive land area. These led to the manufacturing of 40,000 segments to provide a water tight tunnel lining. BIM has been used to: (1) optimise production throughout the different construction stages; (2) coordinate activities between disciplines; (3) increase quality throughout the life-cycle (better design and methodology as well as better facility documentation for operations); (4) control cost; and (5) reduce risk (Sanchez et al., 2014b). The BIM-based coordination process allowed the identification of 200 non-constructible conflicts and 3,000 unique collisions during the design phase. This reduced the production cost due to rework by 50 per cent (corresponding to about 7 per cent of the total contract sum) (Bentley, 2012b).

BIM was also used for machine control/guidance, logistics and survey layouts (Sanchez et al., 2014).

The US GSA has also reported that the use of 3D coordination reduces the number of RFIs, errors and omissions (GSA, 2007). The creators of the software Trimble have additionally stated that the use of clash detection can reduce coordination time by 40 per cent based on data gathered during projects they have worked on (Henrich, 2010).

Note

- In Australia and New Zealand, BIM coordination and pre-fabrication were reported to be more common among large firms than small firms among those surveyed by McGraw Hill in 2014 (McGraw Hill Construction, 2014b).

Improved data and information management

Benefit description

This benefit refers to ensuring data and information is more accurate, interoperable and long-lasting, and that it can be more easily and efficiently found, queried and used. In BIM, data and information is associated with graphical representations. This is created once, then automatically updated when modified or repurposed, and can be continued to be reused over the whole-of-life of the asset. Associated information can include: fire ratings and safety information, submittal documents, links to external documents, procurement details, material specifications, cost of materials, schedules, as-built and as-installed conditions (e.g., precise location of ductwork, piping and electrical systems), geospatial data, historic records including RFIs and change orders, and service providers and suppliers. Documents and databases can be easily interrelated, reducing abstraction and integrating information and data from all disciplines. By using common data protocols and environments, information can be easily integrated into different processes and therefore its exchange is made more efficient with no communication lags and errors.

Data can be made interoperable so it can be collected from disparate software platforms and centralised in the model or reused for further analyses and processes. If the model is maintained and appropriate protocols are put in place, BIM allows having a single source of truth from which data and reports produced always reflect the current state of the model. This increases consistency among datasets and improves version control. BIM also provides a single comprehensive source of truth for tracking the use, performance and maintenance of assets for the owner, maintenance team and financial department. Datasets are more complete and long-lasting with higher ownership and security of information. This allows for more complete, simple and faster audits that can be performed at any time, thus promoting better data quality as a result of enabling more verification. It also improves the integrity of the data and provides the ability to easily refer back to it. Information can be better managed through the use of extensive file management systems with specific user rights across the life-cycle of the asset.

BIM also allows making information more easily and intuitively accessible through, for example, the creation of user-friendly interfaces that use augmented reality and handheld devices; displaying assets behind walls, under floors or above ceiling while in place. BIM models can be used as an interface to improve data and information identification and classification by using unique identifiers to link components with other management systems. It also allows the creation of zones that can identify areas serviced by common components. Storage and transmission of information and data is made easier. Data associated with leading indicators can be easily transferred across life-cycle phases to take proactive, preventive action versus reactive, corrective action. Finally, BIM has the potential to be used for portfolio information and data management.

Design

Information embedded in modelled objects allows consistent display of information (i.e., tagging) and automatically coordinated schedules (i.e., component schedules, quantity and material take-offs). This reduces the repetition of information and data across disaggregated databases.

Construction

BIM tools facilitate two-way communication between field and design offices, improving data- and information-capturing and data review and feedback processes. Field personnel can access, read and update data contained in the BIM model or linked databases.

Operations

BIM improves inventory data management by supporting more complete, accurate and reliable information.

Beneficiaries

Beneficiaries include clients/owners, designers, contractors, subcontractors, asset managers and end-users (emergency response units).

Enablers

Table BD.16 Improved data and information management enablers

	Planning	*Design*	*Construction*	*Operations*	*Decommissioning*	*Whole-of-life*
Augmented reality (AR)		X	X	X	X	
Common data protocol and environments	X	X	X	X	X	X
Data-rich geometrically accurate model components	X	X	X	X	X	X
Design authoring	X	X	X	X	X	X
Field and management tracking			X	X	X	
GIS-BIM	X	X	X	X	X	X
Handheld devices	X	X	X	X	X	
Interoperable formats	X	X	X	X	X	X
Open data exchange standards	X	X	X	X	X	X
Photogrammetry		X	X	X	X	
RFID			X	X	X	
Record modelling				X	X	
Well-structured data	X	X	X	X	X	X

Flow-on benefits

Improved data and information management can help achieve the following flow-on benefits:

- Reduced risk of errors, omissions or misinterpretation of data during and after handover. Risk is also reduced by having clearer responsibility of authorship and more transparency.
- Fewer errors due to greater consistency, easier capturing of accurate data, and increased opportunity for measurement and verification of systems.
- Less rework due to less data re-entry and optimised system and component performance based on owner project requirements.
- Better cost accounting and more accurate quantity take-off due to more geometrically accurate components.

- Improved safety due to more accurate information for health and safety plans, and improved prevention of site hazards and job-site safety.
- More effective emergency management due to: (1) more accurate information about piping, electrical systems, and other relevant characteristics of the built asset; (2) faster access to required information in case of emergency; and (3) faster access to information required for emergency response planning such as security analyses, and fire and life safety calculations.
- Improved productivity due to optimised use of time and resources, increased insight into the process state and measures of improvement, and faster hand-off.
- Improved output quality due to better handling of options, better quality control and assurance and simpler audit processes.
- Better environmental performance due to improved accessibility to energy and sustainability analysis information.
- Lower cost (operations) due to more accurate inventories and fewer costs due to unforeseen repairs.
- Lower cost (whole-of-life) due to reduced document preparation time in all phases, fewer unforeseen repairs and optimised use of time and resources.

Metrics

• Model (or drawing) coordination consistency	• Accuracy and number of errors/omissions
• Volume of rework due to data entry	• Quality of data and documentation

Examples

Pennsylvania State University's Office of Physical Plant included in their BIM Project Execution Plan and Owners BIM Requirements to ensure all data relevant to asset management is easily and readily transferred into their central database before handover. They expect that this improved data and information management system will allow their Work Control Centre staff to only need to field-verify 'asset attributes, rather than spending resources on data acquisition and verification' (Kasprzak and Dubler, 2012). The requirements include that operation and maintenance manuals are incorporated as well as asset specific notes.

The US Coast Guard is using BIM for portfolio information management by owning 100 per cent of their real property in both individual and portfolio BIM models. This organisation has created *BIM-blobs* for all their assets and is progressively adding data as required (Suermann, 2009).

Improved documentation quality and processes

Benefit description

This benefit refers to more accurate and complete documentation, which can be produced faster and with less effort. By using information-rich models that contain data from different stakeholders and life-cycle phases, documentation created from the model contains more complete information about the asset, its systems and specifications of its parts. In particular, as-built documentation has more value-adding information and allows owners to find synergies between BIM uses more easily.

More information can also be stored in a single digital file and less physical space for storage of documentation is required when compared to projects where physical 2D documentation is required. The quality of documents is also improved by reducing the number of errors and requiring objects to be modelled only once; documentation can be produced in a more consistent and accurate manner. Production of specifications can be facilitated at early design phases through the use of BIM repositories or object libraries.

Documents can be generated faster and in a more structured, coordinated and standardised way. This in turn creates more consistent, detailed and accurate documents that can be easily modified to include future data as needed.

Beneficiaries

Beneficiaries include clients/owners, designers, contractors, subcontractors, surveyors, suppliers, fabricators/manufacturers, asset managers and end-users (emergency response units).

Enablers

Table BD.17 Improved documentation quality and processes

	Planning	*Design*	*Construction*	*Operations*	*Decommissioning*	*Whole-of-life*
Data-rich accurate models	X	X	X	X	X	X
Object libraries	X	X	X	X		
Interoperable formats	X	X	X	X	X	X
Open data exchange standards	X	X	X	X	X	X
Record modelling				X	X	

Flow-on benefits

Improved documentation quality and processes can have the following flow-on benefits:

- Improved data and information management because it is better documented.
- Asset management labour utilisation savings due to less time needed in surveys and documentation checking.
- Faster regulation and requirements compliance through easier tracking of changes versus specified codes of practice.
- Competitive advantage gain through better accommodating owner's requirements to help foster a stronger relationship and promote repeat business.

Metrics

- Document consistency (model coordination consistency).
- Quality (documents).

Examples

The GSA has cited as a driver (for implementing BIM more widely throughout their asset management programme) that they can have space measurements to, at least, 90 per cent accuracy within minutes (Suermann, 2009). A study published in 2013 on the construction of a 20-storey, 76.85m tall building and adjacent five-storey and three-storey buildings using BIM in Lithuania achieved 20 per cent time savings for plan and view drawings production. This was in part due to less time required for redrawing and mistake correction when changes occurred (Migilinskas et al., 2013).

Note

- It should be noted that less effort to produce documentation is related to the ability to produce documents and reports directly from the model as well as being able to use documentation created during earlier phases for downstream purposes (e.g., documentation being interoperable between design and asset management systems so it does not have to be recreated after handover).

Improved efficiency

Benefit description

Many of the benefits included in this dictionary refer to more efficient ways of carrying out specific tasks by reducing the resources and time required to complete them. This specific benefit profile aims to acknowledge this fact in a more obvious manner. Using BIM can provide the tools to carry out primary and support activities more effectively. There are also less resources and time required for construction management and information exchange tasks.

Certain processes can be expedited through the use of BIM with handheld devices and RFID technologies such as material tracking and related activities. RFID-BIM technologies in particular can foster innovative efficiency practices through, for example, more effective delivery of lean construction practices. The use of BIM can help increase efficiency by eliminating waste of resources in the form of overproduction, idle time and waiting, unnecessary transportation, inappropriate processing, unnecessary inventory, unnecessary movement and defects.

Space management and tracking increase the efficiency of transition planning and management while more effective coordination and collaboration can increase cost efficiencies and profitability. Faster information exchange can improve efficiency through smoother information flows, more pre-fabrication, more effective processes and more effective construction and asset management.

Beneficiaries

Beneficiaries include designers, contractors, subcontractors, fabricators/manufacturers and asset managers.

Enablers

Table BD.18 Improved efficiency enablers

	Planning	*Design*	*Construction*	*Operations*	*Decommissioning*	*Whole-of-life*
Common data environments and protocols	X	X	X	X	X	X
Handheld devices		X	X	X	X	
Integrated model management systems	X	X	X	X	X	X
Interoperable formats	X	X	X	X	X	X
Online collaboration and project management	X	X	X	X	X	X
RFID			X	X	X	
Space management and tracking				X		

Flow-on benefits

Improved efficiency can help gain competitive advantage by achieving higher profitability.

Metrics

- Labour intensity
- Time per unit

Examples

A case study of a hospital in Australia found a 2,000 per cent efficiency increase for clash detection tasks which were done 20 times faster requiring two hours instead of 40 (Sanchez et al., 2015). In a bridge project in India costing approximately INR1.9 billion (US$29 million), BIM (or BrIM) was used to design the box girder. This would normally require an experienced bridge engineer 128 hours (15–20 days), using modelling tools; however it only required 16 hours (2–3 days) providing a significant increase in efficiency (Bentley, 2013).

Note

- Readers should note that many of the benefits included in this dictionary refer to more efficient ways of carrying out specific tasks by reducing the resources and time required to complete them. Those enablers included in this specific profile are just a few examples that have been specifically mentioned in the literature.

Improved information exchange

Benefit description

This benefit refers to easier, faster and more cost-effective information exchange within and between organisations and individuals. Provided a project team is properly managed and is integrated in its approach, the use of BIM can greatly enhance the availability of current project information. This in turn reduces the risk of uncoordinated or duplicated information, and minimises delays in coordination, and management of change orders, delays and unforeseen costs.

Integrated collaborative systems, for example, allow reduction of cycle times between reviews, workflow turnaround and latency. These systems allow automated information exchange; once a change is made in the model, this can propagate to any related reports and notifications are sent to the relevant parties. These can then be downloaded by, for example, shop fabricators as they become available and use them to create isometric drawings with corresponding materials lists, which can then be transmitted automatically back to the project team. Some systems also allow bi-directional editing which facilitates this exchange. It should be noted, however, that current common practices in most countries manage information exchange with most stakeholders by creating repetitive exchanges (i.e., work in progress is shared weekly or fortnightly).

Online collaboration and project management tools also reduce the number of information bottlenecks, allowing faster reporting and feedback. Field and management tracking tools can reduce operations and maintenance hand-off, on-boarding and uptime processes.

Beneficiaries

Beneficiaries include clients/owners, designers, contractors, subcontractors, surveyors, fabricators/manufacturers, suppliers, asset managers and end-users (tenants and emergency responders).

Enablers

Table BD.19 Improved information exchange enablers

	Planning	*Design*	*Construction*	*Operations*	*Decommissioning*	*Whole-of-life*
Field and management tracking			X	X	X	
Integrated model and programme management systems	X	X	X	X	X	X
Online collaboration and project management	X	X	X	X	X	X

Flow-on benefits

Improved information exchange can have the following flow-on benefits:

- Better use of supply chain knowledge through information being easily shared internally and across organisation boundaries.
- Reduced project execution time due to faster shop fabrication, more pre-fabrication and more effective construction management.
- Improved efficiency due to smoother information flow, more pre-fabrication, more effective construction processes and management, and more effective asset management.

- Improved communications and information management due to more efficient exchange and update cycles.

Metrics

• Latency	• Meeting effectiveness
• Request for information	• Meeting efficiency

Examples

In a case study of a hospital in Australia, the consultants were able to create and send accurate schedule information within minutes from receiving the request. This information was used by the contractor for quality assurance and accountability regarding subcontractor submissions (Sanchez et al., 2015).

Improved learning curve

Benefit description

This benefit refers to faster learning of specific tasks. Improved communication, for example, makes it easier to train new staff based on the highly visual process. 3D visualisation of objects and processes can be used as a learning tool for training staff for occupational health and safety, operation of construction machinery and logistics planning. Training can also be made more effective by focusing on key issues identified through automated rule-checking for health and safety considerations. Engineering analysis tools are also often easier to learn than other tools and are less disruptive to established workflows.

Learning from past experiences can also be made more efficient. Re-engineering and workflow management through online collaboration and project management can, for example, allow organisations to learn from their performance and past activities. BIM knowledge management tools can provide access to lessons learned in a specific area where interested parties can, for example, receive a message when starting a similar project to learn about these processes.

Construction

During construction, BIM can be used for constructability analysis and process identification, which contain learning processes that can be facilitated by visualising the construction sequence and other animations.

Operations

During operations, BIM can help staff to understand operation tasks faster through improved communication and documentation.

Beneficiaries

Beneficiaries include designers, contractors, subcontractors, fabricators/manufacturers and asset managers.

Enablers

Table BD.20 Improved learning curve enablers

	Planning	*Design*	*Construction*	*Operations*	*Decommissioning*	*Whole-of-life*
Augmented reality (AR)		X	X	X	X	
BIM-based asset management system				X		
Asset knowledge management tools	X	X	X	X	X	X
Constructability analysis			X		X	
Design authoring (3D visualisation)	X	X	X	X	X	X
Engineering analysis			X		X	
Online collaboration and project management	X	X	X	X	X	X
Phase planning (4D modelling)			X	X	X	

Flow-on benefits

Improving the learning curve helps realise asset management labour utilisation savings and improve productivity and safety.

Metrics

- Learning curve.
- Overall time required for training.

Examples

A 2003 study where AR was used to instruct field workers in how to carry out certain assembly tasks found that ‘overlaying 3D instructions on the actual work pieces reduced the error rate for an assembly task by 82 per cent, particularly diminishing cumulative errors – errors due to previous assembly mistakes'. The measured mental effort was also reduced (Tang et al., 2003).

Note

- 'The learning effects might be phenomenal in many repetitive construction field operations ranging from smaller ones, such as fixing, molding, and concreting, to larger ones, including the construction of a standard floor. Unfortunately, these learning effects contributed by BIM have rarely been identified and measured' (Lu et al., 2013).

Improved output quality

Benefit description

This benefit refers to project outputs being more accurate and built assets having fewer defects, as well as having optimised features based on better informed decision-making. In terms of individual task outcomes, these are more repeatable and information exchange can be formalised, mapped and broken down into parts based on specific functionalities. This makes it possible to verify the results from information exchange and validation tools.

Outputs include reports that can be produced faster and more frequently to improve different processes across the life-cycle of the asset. BIM also enables better-informed decision-making based on having access to more and better data as well as input from a variety of stakeholders. This can improve the quality of outputs. The use of BIM visualisation tools also allows users to better understand the inter-dependencies between different measures and strategies and their effects on the built asset and whole-of-life cost.

Design

Designs can be improved by facilitating rigorous analysis of proposed designs based on information control, review, revision and validation. The use of standardised object libraries, parametric modelling, clash detection (including 3D coordination) and rule-checking enables fast production of simulation and automated analyses. This can lead to optimised and innovative designs with fewer errors. Better coordination between design, structural and other systems can also help improve the quality of design outputs. Improved communications and design reviews also provide more opportunities to receive feedback from clients and other disciplines, and to better understand construction challenges and expectations.

BIM-based engineering analysis and alternatives analysis provides the basis for establishing realistic energy, cost and environmental targets at early stages of development. For example, the design can be improved by trying different design concepts and materials, and performing quick simulations to test the effects on heating, lighting, comfort or life-cycle cost.

Design outputs are also improved, in that they can reflect the design intent more accurately and provide additional views to aid understanding by non-technical stakeholders. 2D drawings, derived from the 3D model, are more accurate, consistent and have fewer errors.

Construction

Field and management tracking helps avoid contractor call-backs and warranty claims due to construction defects. 3D coordination can also be used based on subcontractor models being created to improve construction.

Operations

As-built documentation delivered to asset managers can contain a greater level of detail and accuracy for use during operation and maintenance phases. Quality of the built asset can also be improved by using pre-fabrication, pre-assembly and off-site fabrication. This improves the production quality by encouraging more standardisation of parts, reducing the risk of failure, making documentation outputs flexible and exploiting automation.

Beneficiaries

Beneficiaries include client/owners, designers, contractors and asset managers.

Enablers

Table BD.21 Improved output quality enablers

	Planning	*Design*	*Construction*	*Operations*	*Decommissioning*	*Whole-of-life*
Automated rule-checking	X	X	X	X	X	X
Automated clash detection	X	X	X	X	X	X
Construction system design (virtual mock-up)	X	X	X	X	X	X
Data-rich, geometrically accurate model components	X	X	X	X	X	X

Table BD.21 (cont.)

	Planning	*Design*	*Construction*	*Operations*	*Decommissioning*	*Whole-of-life*
Design reviews	X	X	X	X	X	X
Digital fabrication			X			
Engineering analysis			X		X	
Field and management tracking			X	X	X	
Integrated model management systems	X	X	X	X	X	X
Object libraries	X	X	X	X		

Flow-on benefits

Flow-on benefits include lower whole-of-life cost and higher customer satisfaction.

Metrics

Planning, design, construction and decommissioning

- Satisfaction
- Quality
- Model (drawings) consistency
- Volume of rework
- Cost per defects-warranty
- Accuracy and number of errors/omissions

Operations and whole-of-life

- Accuracy and number of errors/omissions
- Satisfaction
- Quality
- Model (drawings) consistency
- Volume of rework
- Sustainability and environmental performance Scores
- Resource use and management
- Carbon footprint
- Cost per defects-warranty
- Fire safety

Examples

The use of BIM with alternative analysis and construction simulations was reported to have helped the project team of the New Generation Rollingstock Depot in Australia to improve output quality. In this case, the team was able to use BIM tools to carry out thorough constructability and maintenance analysis. This led to the team developing the working mantra of 'if it can be designed, it can be built' and carrying out significant front-end planning that would ensure better quality outputs throughout the life-cycle of the asset (Utiome et al., 2015).

Note

- 44 per cent of infrastructure owners surveyed internationally by McGraw Hill in 2012 reported overall better project outcomes as one of the top benefits from using BIM (McGraw Hill Construction, 2012).

Improved productivity

Benefit description

This benefit refers to the reduction of man-hours required to carry out a task or generate an output. With increasing experience in the use of BIM, a team can carry out a wider range of activities within a specific period of time providing more added-value to clients. This is attributable to, for example, less time being required due to rework, conflict resolution, duplication of efforts and resolution of RFIs and change orders. Staff can also use their time for more productive activities instead of repetitive and surveying tasks because information can be better managed and more easily retrieved.

Higher stakeholder engagement and improved communication can help decrease the time required for decision-making and answering RFIs (latency) and increase productivity. This is done by, for example, improving the understanding of timelines and quality requirements, which can help increase schedule conformance and even shorten it.

Design

By using object libraries and creating design outputs directly from the model, engineering and designing firms can increase the production, detailing and amount of documentation with fewer or the same number of resources. Laser scan surveys can also allow office personnel to take new on-site measures without leaving the office by accessing the point cloud data.

Construction

Modelling the construction process can allow better planning of site activities and optimisation of the construction sequence, methods and processes to maximise the productivity of the construction phase. Assets can therefore be built faster and with fewer defects and less waste.

More of the asset can also be pre-fabricated off-site to a greater precision and be incorporated into the asset being constructed with less effort.

Computer-controlled assembly equipment can additionally shorten construction programmes and costs to obtain a higher-quality asset within less time and using fewer resources. Cast-in-place reinforced concrete structures can be produced with fewer resources and more accuracy.

Operations

In addition to higher productivity of asset management staff due to the processes outlined in the general description of this benefit, the productivity of asset users can also be improved. Improving the quality of the design through better scenario and alternatives analysis and improved output quality can lead to higher asset user productivity. For example, sustainable design strategies and good use of daylight can raise the productivity rate of employees by 1 per cent or more. This can be included as a criterion during design development and modelled for different design options.

Beneficiaries

Beneficiaries include client/owners, designers, contractors, subcontractors, fabricators/manufacturers, surveyors, suppliers and asset managers.

Enablers

Table BD.22 Improved productivity enablers

	Planning	*Design*	*Construction*	*Operations*	*Decommissioning*
3D control and planning (digital layout)			X	X	X
Asset (preventative) maintenance scheduling				X	
BIM-based asset management systems				X	X
Common data protocol and environments	X	X	X	X	X
Construction system design (virtual mock-up)	X	X	X	X	X

Table BD.22 (cont.)

	Planning	*Design*	*Construction*	*Operations*	*Decommissioning*
Cost estimation (quantity take-off)	X	X	X	X	X
Digital fabrication			X		
Effective stakeholder engagement	X	X	X	X	X
Field and management tracking			X	X	X
Object libraries	X	X		X	
Online collaboration and project management	X	X	X	X	X
Phase planning (4D modelling)			X	X	X

Flow-on benefits

Improved productivity can help reduce the cost of design, construction and operation phases of an asset.

Metrics

• Labour intensity	• Speed of production
• Time per unit	• Cost per unit

Examples

Productivity improvements have been reported in the form of up to 20 per cent savings in man-hours, corresponding to up to 62 per cent cost savings, and from 2.6 per cent to 47.4 per cent reduction of work hours (Leite et al., 2011). Other reports have estimated that for the specific case of cast-in-place reinforced-concrete structures, productivity can be increased by 15–41 per cent during the drawing phase (Lu et al., 2013).

In a building design and construction project in Sydney, the architectural engineer stated that the first benefit realised from implementing BIM was higher productivity. This was achieved by using the 3D model to generate multiple views (e.g., floor plans, elevations and sections). The use of BIM also allowed having an embedded quality assurance process, which was seen as contributing to the higher productivity; 'all the views are coordinated by the system rather than manually generated in accordance with, but independent from, each other' (Fussel et al., 2009).

In an Australian rail project in New South Wales, the designers were able to save an estimated 1,500 man-hours by creating drawings directly from the model as opposed to designing them in CAD (Bentley, 2012a). In another transport infrastructure project in the US, the use of online collaboration and project management tools helped increase productivity by approximately 12 per cent. This was achieved by reducing the time required for travelling by designers, reducing the time required for searching and validating data, and reducing the amount of rework and duplication. This increment in productivity contributed to a total savings of US$850,000 for the designing firm (Bentley, 2012c).

Notes

- Sacks et al. (2005) suggested that a feasible target for large organisations implementing BIM is an increase of 2.3 per cent in productivity as a percentage of total cost.
- Please refer to Krygiel and Nies (2008) to read about the correlation between daylight use in buildings and user productivity.

Improved safety

Benefit description

This benefit refers to reduction of conditions that put staff at risk of injury or danger and it mainly applies to construction and operation phases of asset management. Improved data and information management, scenario and alternatives analysis, better use of supply-chain knowledge and rule-checking can, for example, reduce the risk of hazardous materials and improve safety management. Improved coordination, communication, visualisation and supply-chain knowledge use can also help identify, analyse and mitigate possible major and minor site hazards.

Visualisation tools can improve safety education and awareness by providing information about hazards and safety procedures in a more intuitive way to workers with fewer language barriers. Digital layout modelling can also help managers develop models that describe the movement of workers and equipment to improve safety management.

The use of 4D safety models can display safety risk levels visually for each task and team. This type of tool provides visual safety risk forecasts with detailed information. Accounting for short-term changes in schedules can contribute to improve weekly planning and increase awareness of specific risks among workers.

Construction

Digital modelling tools can include construction and safety constraints directly into the design, as well as considerations of assembly and disassembly of temporary structures. More pre-fabrication and improved programming of construction activities can also help improve on-site worker safety. Additionally, pre-fabrication and digital fabrication improve manufacturing safety and shift the site work towards assembly.

A constructions security sub-model can be developed to improve safety monitoring. This approach allows analysing and monitoring the safety of construction processes in a timely and dynamic way, while also providing early warnings of security risks. Field and management tracking can help prevent job-site hazards and at-risk behaviours, and ensure job-site safety.

Operations

In addition to the above description, BIM can be used to test the impact of different emergency scenarios on different design options that include evacuation and emergency response criteria. This can add another layer of information to the decision-making process that could help understand how the asset will hold during an emergency event, and how it will affect the level of disaster and the disaster risk map.

Beneficiaries

Beneficiaries include contractors, subcontractors, fabricators/manufacturers, asset managers and end-users.

Enablers

Table BD.23 Improved safety enablers

	Construction	*Operations*	*Decommissioning*
Automated rule-checking	X	X	X
Digital fabrication	X		
Construction system design (virtual mock-up)	X	X	X
Field and management tracking	X	X	X
3D control and planning	X		X

Flow-on benefits

Improved safety can help improve productivity during construction and operations by reducing the number of accidents and associated work stops as well as increasing worker morale.

Metrics

Safety measures such as:

- Number of injuries/casualties, accidents and near-misses
- Lost workday case incident rate
- Accident rate
- Recordable/reportable incident rate (RIR)
- Lost-time accident ratio

Examples

In an Australian highway upgrade project, the use of BIM contributed to having zero lost time due to injuries over a period of 1.175 million man-hours, which led to the project receiving a Health & Safety Environment (HSE) award (Wang, 2015).

In a case study of a hospital in Australia, the use of BIM to test alternative construction methodologies allowed the use of pre-fabricated sections of the mechanical riser, which improved the safety of the installation. The model reviews also prompted changes to certain features of the design that reduced work at heights as well as made easier future service reticulation in the ceiling space, all of which improved staff safety during construction works and future operations (Sanchez et al., 2015).

BIM was also used in Belgium to model, coordinate and plan a river-widening project as part of a flood prevention programme. Aside from using the model for coordination purposes, the team was able to carry out clash detection with a site-specific hazard: unexploded World War II bombs. This significantly increased the safety of the work crew (Autodesk, 2014).

The Department of Justice/Administrative Office of the US Courts has also expressed their interest in using BIM technological advances to improve the final asset's safety by ensuring measures are taken in complex facilities where competing security interests need to be addressed (Suermann, 2009).

Sacks et al. (2009) proposed a model that informs construction planners of safety risk levels for each team at all levels of planning resolution. Other 4D safety risk visualisation software application used colour codes to highlight dangerous locations according to activity types.

Notes

- Temporary structures such as scaffolding can have a significant impact on the quality, safety and profitability of construction projects. Compliance with safety standards have also been found to be an area of improvement. In Australia, for example, up to 40 per cent of all scaffolding projects do not comply with national safety and design standards (SBEnrc, 2014). Research has therefore been carried out into integrating scheduling and safety planning related to scaffolding. This area of research aims to use BIM to better understand workforce vulnerability to serious accidents and the level of danger and consequences that the hazards may result in. This type of system can support decision-making for proactive prevention of accidents and the improvement of job-site and workforce safety without sacrificing productivity (Hou et al., 2014).

- The construction industry is currently responsible for 36 per cent of all US workplace fatalities, 32 per cent of all worker fatal injuries in the UK and 25 per cent of fatal occupational accidents in Finland and Australia (Zhang et al., 2015; Safe Work Australia, 2015; Leigh, 2014).
- This benefit is starting to be recognised by universities such as the Australian University of Newcastle which is including the influence and application of BIM on fire safety and compliance in their course work (University of Newcastle, 2015).

Less rework

Benefit description

This benefit refers to the reduction of work generated due to errors, omissions or inefficient processes that requires a task to be done again. The use of interoperable and more accurate information reduces the rework associated with re-entering data; data generated and changes made at any stage can be readably transferred to subsequent phases and systems. For example, asset geometry does not have to be re-created for energy efficiency analysis.

Clash detection, integrated systems with interoperable data and 3D control and planning facilitate the reduction of errors and deficiency correction notices as well as duplication of efforts; all of which have the effect of reducing rework. BIM encourages a greater level of detail at earlier phases. This can help identify errors and omissions at earlier phases where there is less rework associated with correcting them.

Improved interoperability and information and data management helps reduce the cost of manual re-entry of data between different systems. Time spent duplicating data for different software packages and checking document versions can also be reduced as well as that required for information processing and data transactions.

Better coordination of disciplines also contributes to having less rework and doubling of tasks. Control points can be derived directly from the model reducing rework related to this task as well.

Design and construction

During the design and construction phases, the use of clash detection and rule-checking reduces the need for detailed checking during the creation of drawings by reducing the possibility of misaligned connections, incorrect architectural features and geometry conflicts. Interference checking and less field conflicts due to improved coordination can also significantly reduce rework.

Operations

The use of BIM enables easier and more accurate data-capturing during design, construction and operation phases. This in turn reduces the need for resurveying the asset for as-built documentation and operations and maintenance tasks.

Asset management databases can be populated faster from information-rich models than from traditional 2D drawings; data regarding spaces, equipment, systems, and zoning does not have to be re-entered into downstream systems.

Supply

As with asset managers, manufacturers can extract and introduce specifications of parts for off-site manufacturing directly from the model, reducing errors, rework and the need for cross-database checking.

Beneficiaries

Beneficiaries include clients/owners, designers, contractors, subcontractors, fabricators/manufacturers and asset managers.

Enablers

Table BD.24 Less rework enablers

	Planning	*Design*	*Construction*	*Operations*	*Decommissioning*	*Whole-of-life*
3D control and planning (digital layout)		X	X		X	
Automated clash detection	X	X	X	X	X	X
Automated rule-checking	X	X	X	X	X	X
BIM-based asset management systems				X	X	
Data-rich, geometrically accurate model components	X	X	X	X	X	X
Field and management tracking			X	X	X	

Table BD.24 (cont.)

	Planning	*Design*	*Construction*	*Operations*	*Decommissioning*	*Whole-of-life*
Integrated model management systems	X	X	X	X	X	X
Interoperable formats	X	X	X	X	X	X

Flow-on benefits

Less rework can have the following flow-on benefits: improved productivity, lower whole-of-life cost, lower design cost through reduced modelling time requirement and lower operational cost through more effective asset operation and maintenance.

Metrics

- Volume of rework
- Time per unit – rework
- Overall time – rework
- Cost of change
- Overall cost – rework

Examples

A facility assessment study carried out by the US Coast Guard reported that, in cases where BIM was used, a reduction of up to 98 per cent of the time and effort required for the production and updating of asset management databases was achieved (BSI and buildingSMART, 2010). On a different study, a project reported having had reduced rework by a factor of ten during design and construction (Thomas et al., 2004).

In study reported by Staub-French and Khanzode, the only rework observed occurred between trades that did not model their scope of work in 3D. In this project, 'the superintendent for the general contractor noted the "seamless" installation process for the 3D work' (Staub-French and Khanzode, 2007).

A prototype system developed in recent years called KanBIM was tested by staff working in a virtual construction site experimental set-up using virtual reality linked to a simulation engine. This was applied to a sixteen-apartments building to guide their performance of virtual work. 'The use of the system improved the process flow and eliminated waste. It gave the subjects situational awareness, leading them to perform the right work, reducing rework and time wasted' (Gurevich and Sacks, 2014).

Notes

- Rework can be defined as the unnecessary 'effort of redoing a process or activity that was incorrectly implemented the first time' (Love et al., 2009) and has been reported to be closely related to design changes, errors and omissions. In a project studied by Love et al. (2009), 79 per cent of the total rework costs experienced were generated by these factors and omissions errors alone accounted for 38 per cent.
- It should also be noted that, at the moment, there are still some issues to be resolved in the transfer from design to construction models. For subcontractors, design models are often started again, with the design model representing a *well-sorted brief* at best. This is sometimes due to issues of trust and liability as well as different software requirements for different phases.

Lower cost

Benefit description

This benefit refers to the reduction of cost across the different phases of an asset's life-cycle. Because the tools that generate cost savings and the flow-on benefits from each phase can be significantly different, this profile has been divided into five sub-profiles: design, construction, operations, decommissioning and whole-of-life.

Design

Drawings can be directly created from models with adequate levels of detail/development. This means that a larger number of drawings can be produced at a lower cost and faster, including shop fabrication drawings. The use of automated clash detection, rule-checking and integrated databases helps reduce rework and time needed for cross-checking and document coordination. This reduces the labour needed for these tasks and its associated cost.

Effort in terms of man-hours and number of staff employed during design can be reduced by, for example, allowing little or no division between design development and construction documentation tasks.

By using object libraries, the cost of designing and changing certain elements can be reduced, allowing more options to be reviewed with less time and effort. Design reviews also integrate a larger number of stakeholders, which can reduce cycle times and the associated cost.

Construction

The use of 4D modelling and visualisation tools to optimise construction tasks and their sequence can reduce construction time and increase accuracy, avoiding cost due to longer delivery and error correction. By using geometrically accurate information-rich digital objects, more elements can be pre-fabricated off-site and assembled by computer-controlled machines, therefore reducing the cost of on-site labour. Increasing the use of precast concrete during construction can, for example, reduce the engineering cost.

The use of integrated models has also been shown to have a positive effect in reducing the cycle time in bulk materials processes. Online collaboration and project management tools can also help reduce transaction costs by reducing the number of construction managers and coordinators needed. By using field collaboration tools and 4D, unforeseen schedule changes can be more effectively coordinated at the least cost. Additionally, all stakeholders have access to critical information when needed, reducing the number of bottlenecks during construction and increasing the ability to predict needed changes, thus reducing idle time and its associated cost.

The accuracy of mechanical and electrical components installation processes can be improved, reducing the time and resources needed, as well as number of errors.

Operations

Given the longevity of most assets, changes made in the design to increase the operational performance efficiency can have a great impact on the overall operational cost. BIM design reviews allow improving the asset's resource use efficiency such as electricity and water, reducing the cost of utilities. Better understanding of the heating, ventilating and air conditioning (HVAC) system parts, for example, can help reduce the cost of operations and commissioning.

Integration of asset management systems with BIM and environmental performance analysis allow a greater degree of control over the cost of operations. This more proactive savings-focused approach and dynamic performance monitoring can contribute to quickly identifying poor performing, faulty or about to fail equipment. Better maintenance information management allows more proactive preventive maintenance planning and procedures, increasing the lifespan of equipment and avoiding efficiency losses, as well as costly emergency repairs and impact on users. This reduces the cost of repair and replacement. This also enables better informed decision-making with respect to maintenance and repair expenditures. Better understanding of current conditions also helps increase the accuracy of bill management, and reduce the cost due to audits and surveys. BIM can also provide a framework to display real-time performance data to users, which may help improve behaviour and reduce running costs.

The use of integrated BIM asset management and maintenance systems allows staff to be more efficient through easier and faster access to key information required for the functioning of the asset. These systems also provide more control over asset management activity progress. The use of BIM can also help improve initial efficiency when changing asset managing companies through better change, data, information and knowledge management. Integrated life-cycle data and faster access to accurate information about management activities provide tools for more effective facility management with easier information exchange.

The use of BIM asset management systems provides the opportunity of dynamic monitoring and condition sensing. Building systems performance modelling and analysis provide a tool to measure and monitor the performance of the asset; opportunities to modify the system operations can be identified based on this information in order to improve overall performance.

Decommissioning

Having access to geospatial information about the location of objects and materials can help decrease the cost of demolition through better planning and management. Data-rich objects allow more accurate identification of those elements that can be recycled, are valuable or require special handling, thus improving decommissioning planning and costing.

Having fast and easier access to information about hazardous materials and other required information can also help increase the speed of decommissioning, decreasing the cost of crews and equipment. In conclusion, the deconstruction processes can be scheduled, sequenced, costed, tracked and optimised more cost-effectively. This includes rubble management.

Whole-of-life

Whole-of-life cost reductions are delivered by, for example, requiring teams to deliver a greater level of detail at earlier phases of the life-cycle, where changes can be made at a lower cost. The information-rich model allows assets to be modelled and changes agreed to early in the asset's life-cycle to produce better whole-of-life performance. This can help minimise change orders and costly variations due to unforeseen conditions after the fieldwork has started.

BIM can help improve cost performance and add value to assets over the whole life-cycle, by providing cost savings through different phases including design, construction, and operations and maintenance. Teams example, can leverage on better visualisation and collaboration platforms, and more efficient coordination and information retrieval to reduce cost. Asset performance can be better understood and optimised, reducing operational cost; information can be integrated across long-term and short-term asset management activities, producing further overall cost savings by identifying optimisation points across short- and long-term strategies.

The use of BIM can offer more control over whole-of-life cost and environmental data so they are more predictable and better understood; as the model becomes more comprehensive, costs can be better understood and reduced.

The high cost of inadequate interoperability can be reduced, allowing the transfer of accurate, complete and unambiguous information across stakeholders and life-cycle phases. This means that there are fewer costs associated with re-documenting as-built conditions, field surveys for renovation projects and destructive testing to confirm existing conditions. The use of BIM allows faster and more reliable validation of data, which contributes to higher efficiency and lower overall cost. Additionally, there is less cost associated with travel and document printing and shipping. When used by an experienced team, the staffing cost can also be reduced by improving the efficient delivery of individual tasks. The use of BIM can also help reduce overhead cost rates per unit of product due to increased utilisation capacity.

Finally, when used at the appropriate phases, BIM can enhance lean project delivery processes if implemented jointly.

Beneficiaries

Design

Beneficiaries include designers and clients.

Construction

Beneficiaries include contractors and clients.

Operations

Beneficiaries include asset-owners, managers and end-users.

Decommissioning

The main beneficiary is the asset-owner.

Whole-of-life

Main beneficiaries are asset-owners and managers. However, because whole-of-life reduced cost is driven by lower cost of design, construction and operation phases, designer, contractors and clients also benefit.

Enablers

Table BD.25 Lower cost enablers

	Planning	*Design*	*Construction*	*Operations*	*Decommissioning*	*Whole-of-life*
Automated clash detection	X	X	X	X	X	X
Automated rule-checking	X	X	X	X	X	X
Drawing generation	X	X				
Engineer analysis	X	X	X			
Integrated model management systems	X	X	X	X	X	X
Object libraries	X	X	X	X		
Barcoding/RFID			X	X	X	
Digital fabrication/off-site pre-fabrication			X			
Online collaboration and project management	X	X	X	X	X	X
Phase planning (4D modelling)			X	X		
BIM-based asset management systems				X	X	
Asset performance modelling, analysis and display				X		
Asset (preventative) maintenance scheduling				X		
Design reviews	X	X	X	X	X	X
Data-rich accurate BIM model elements	X	X	X	X	X	X
GIS-BIM		X	X	X	X	
Design authoring (3D visualisation)	X	X	X	X	X	X
Information delivery manuals (IDMs)	X	X	X	X	X	X
Lean construction			X	X	X	X
Cost planning (5D modelling)			X	X	X	X
Phase planning (4D modelling)		X	X	X	X	X

Flow-on benefits

- Lower design costs can provide a competitive advantage by improving profitability, and granting earlier access to the construction market.
- Lower construction costs provide competitive advantage gain through higher profitability.
- Lower operational costs contribute towards lower whole-of-life cost and higher customer satisfaction.

Metrics

Planning, design, and construction

• Cost per unit • Cost per defect-warranty • Cost avoidance/savings • Overall cost	• Profitability • Cost of change • Labour intensity

Operations and whole-of-life

• Cost avoidance/savings • Overall cost • Labour intensity	• Profitability • Cost of change • Asset/equipment useful life

Decommissioning

• Cost per unit • Cost avoidance/savings • Overall cost	• Cost recovery (decommissioning) • Cost of change • Labour intensity

Examples

Potential benefits from BIM have been estimated to be at least 2.3–4.2 per cent of total project cost for precast concrete companies, with reported savings of up to 10 per cent (Lu et al., 2013; Sacks et al., 2005). For example, one respondent to a survey by Thomas et al., attributed US$5 million savings in a US$230 million project to the use of BIM (Thomas et al., 2004). Leite et al. (2011) also reported cost savings of between US$200,000 and US$10 million in different projects.

Travel and document printing and shipping cost reductions across the life-cycle have also been reported. Although small as a percentage of total cost, these still contribute to a whole-of-life lower cost and higher return on investment (Becerik-Gerber and Rice, 2009).

In the UK, the net benefit of using BIM (if extended to all major projects) has been estimated to account for between £1–2.5 billion annually

(US$1.5–3.8 billion), just in the construction phase (BIM Industry Working Group, 2011).

Examples of specific cases where BIM has been shown to produce cost savings include a project reported by Azhar et al. (2011). They estimated cost benefits to be US$200,000 for a US$46 million project and almost US$2 million savings for a US$12 million project; both delivered under a construction manager at-risk contract. In another case reported by Thomas et al. (2004) the design time was reduced by 40 per cent and in a case reported by Leite et al. (2011), design man-hours were reduced by 20 per cent (corresponding to 62 per cent cost savings and up to 10 per cent of the contract value).

Examples of cost savings for specific stakeholders include a study where it was estimated that MEP installation contractors achieved savings of up to 20–30 per cent in labour cost (Leite et al., 2011). In a case reported by Kaner et al. (2008), the use of BIM helped produce practically error-free fabrication and erection in four projects and the effort required for checking drawings was significantly reduced.

In a highway upgrade project in Australia, the use of BIM tools contributed to savings of AU$24 million (US$18 million) – 14 per cent of the total project cost. This meant that the project cost AU$7 million less than the target budget and finished three months ahead of schedule. Some of these savings were achieved through reducing the construction cost associated with pipe relocation in the order of AU$2 million savings per kilometre (Wang, 2015).

A US government agency which manages and operates facilities across the country including 578 buildings with a current replacement value (CRV) of US$2.5 billion implemented a computerised maintenance management system. Thanks to this the agency was able to automate preventive maintenance programmes and optimise its capital asset replacement decision. They were able to extend the equipment asset useful life (EUL) by an average of 9.8 years over the average industry EUL value of 18.6 years (an increase of 53 per cent). This represented an estimated ownership savings of US$28.4 million per year or 1.12 per cent of the CRV per year (Teichholz, 2013). An accurate equipment inventory has also been shown to reduce operations and management contracting costs by 3–6 per cent by identifying and tracking facility equipment and facility square footage (GSA, 2011). Although these examples are not specifically about BIM, they showcase what is possible with more integrated digital asset management approaches.

The use of BIM in pilot projects has improved as-built documentation and provided major design savings by optimising the mechanical system design, uncovering design errors and omissions (e.g., envelope and coordination omissions in Houston Federal Office Building), and improving the means for communication. The US GSA reported that the cost savings they achieved from using BIM on a single pilot project already paid the start-up cost of the whole pilot program (GSA, 2007).

Notes

- It has been suggested that 60–90 per cent of project variations are the result of poor design documentation (CRC for Construction Innovation, 2007a).
- The asset's operations and maintenance cost is often the greatest contributor to whole-of-life cost (NATSPEC, 2014b).
- In the case of the precast sector, current project labour input due to drawings and checking typically accounts for 83 per cent of the total labour input. Precast CAD projects for a firm studied in 2008, ranged from 1,000 to 8,000 total labour hours. The breakdown of hours typically was: general and project management 7 per cent; erection layouts 25 per cent; engineering 10 per cent; shop drawings 38 per cent; and checking 20 per cent (Kaner et al., 2008).
- It should be noted that although lower cost during design and construction phases may be achieved, the fees charged to the client may not have been reduced.

More accurate quantity take-off

Benefit description

This benefit refers to fewer quantity take-off errors and a more reproducible estimating process. Data associated with the model elements can be directly used for quantity take-off. For example, dimensions and materials can be directly extracted from 3D models for the fabricators or suppliers. Space can also be quantified more easily based on the model. The use of accurate 3D models allows estimators to analyse the design in different ways and provide better visual representation of elements that must be estimated. Some software platforms also allow users to compare cost between design revisions, providing quantity differences as well as visual representations of the changes.

Quantity take-off can be made based on the most up-to-date information, providing more control and better forecasting. Quantities can also be automatically generated when introducing changes to design, reducing the risk of human error. Digital fabrication, for example, allows minimising the tolerances estimated through machine fabrication.

These processes can also be repeated more often within a specific time-frame than could be achieved with manual methods. For example, new cost estimates can be issued every two weeks instead of only at traditional milestones.

Beneficiaries

Beneficiaries include contractors, subcontractors, fabricators and asset managers.

Enablers

Table BD.26 More accurate quantity take-off enablers

	Design	*Construction*	*Operations*	*Decommissioning*
Cost estimation (quantity take-off)	X	X	X	X
Cost planning (5D modelling)	X	X	X	X
Data-rich, geometrically accurate model components	X	X	X	X
Digital fabrication		X		
Online collaboration and project management	X	X	X	X

Flow-on benefits

More accurate quantity take-off can have the following flow-on benefits:

- Better cost accounting through faster cost estimation of specific objects and easier estimation of cost associated with changes as well as more cost certainty. Quantity take-off software used for FF&E can also extract quantities directly from the model and compare against the brief as well as track changes, improving cost accounting and quality assurance.
- Better programming/scheduling through more efficient planning based on accurate quantity estimations.
- Lower construction costs due to more accurate and timely estimates for decision-making as well as more streamlined procurement orders. More accurate quantity take-off can also contribute to more proactive value engineering, where individual stakeholders can estimate costs within the authoring environment in order to work within cost constraints.
- Lower operational costs due to more accurate timely estimates for decision-making, more streamlined procurement orders and more accurate financial reporting, bidding and future cost estimations.
- Higher customer satisfaction through enhanced services to clients.

Metrics

- Accuracy and number of errors (quantity take-off)
- Cost predictability
- Cost avoidance/savings

Examples

One project reported by Thomas et al. (2004) achieved 30 per cent savings in electrical material alone through more accurate quantity take-off, which led to eliminating restocking chargers and leftovers. In a different study, Kang et al. (2013) reported that the use of BIM could help achieve more accurate quantity take-offs, leading to up to 25 per cent reduction of overall estimation effort.

In a case study of a medical centre in California, subcontractors were able to have a highly accurate bill of materials and use more pre-fabrication. This led to the reduction of cost due to increase productivity (Staub-French and Khanzode, 2007). At an organisational level, Turner and Townsend reported that accurate bulk quantification has allowed them to enhance their service to clients (NBS, 2014).

Note

- Some reports have estimated that 50 per cent to 80 per cent of the effort required for cost estimation is currently due to quantity take-off (Zhiliang et al., 2011).

More effective emergency management

Benefit description

This benefit refers to the more effective planning and management of responses to emergency situations during construction and operations. Information-rich BIM objects and cloud computing can, for example, facilitate emergency management and retrofit planning by providing emergency responders with real-time accurate information about asset conditions and component location. Access to up-to-date and accurate information about the asset and current conditions minimises the risk to emergency responders and improves the effectiveness of the response.

Improved data and information management allow access to more accurate information about piping, electrical systems and other relevant characteristics of the built asset. It also allows faster access to information that may be required in case of emergency and for emergency response planning such as security analyses and fire and life safety calculations. Better data-capturing additionally provides more accurate and visual representation of pre- and post-disaster conditions, which is useful for emergency management planning.

Better scenario analysis during different life-cycle phases allows the optimisation of the asset design based on emergency situation scenarios. These can also include disaster scenario simulations that can be used to analyse the potential causes, develop disaster protection plans, computer-based expert systems and evacuation programmes. Better use of supply change knowledge additionally facilitates the inclusion of emergency planning considerations during the design phase through more active involvement and feedback of asset managers. Finally, end-users and staff members can understand more easily disaster contingency plans based on 3D visualisation.

Beneficiaries

Beneficiaries include contractors, subcontractors, asset managers and end-users (tenants and emergency responders).

Enablers

- Data-rich, geometrically accurate model components
- Design reviews
- Disaster planning and response/disaster analysis software
- Common data protocol and environments

Flow-on benefits

More effective emergency management helps reduce the risk of material and human loss.

Metrics

- Emergency latency.
- Casualties and injuries during emergencies.
- Evacuation time predictability.
- Emergency plan and response effectiveness.
- Time required to access information during emergency drills and events.

Examples

A case study in Seattle modelled surface and sub-surface conditions to simulate the impact of an earthquake on the design of an elevated highway and the surrounding grade level improvements (McGraw Hill Construction, 2012).

A research group based in Germany developed an indoor emergency navigation system for complex buildings such as airports using BIM, wireless LAN (WLAN), ultra-wide-band (UWB) and RFID. This application allowed rescuers to find the shortest path and access relevant spatial information to improve the effectiveness of their response. The group highlighted the time and cost savings for emergency responders using Frankfurt Airport as an example. This airport experiences approximately 5,000 alarm calls a year, of which only 5 per cent are not false. In each case, the fire brigade is required to check the source of the alarm involving at least six fire-fighters who, 95 per cent of the time, waste 30–45 minutes just to locate the detector and return to the station. By implementing the new way-finding system, which could be accessed through mobile devices, it would be expected that the wasted time is reduced during false alarms and the effectiveness of the response during real emergency events would improve (Rueppel and Stuebbe, 2008).

Note

- It has been pointed out by the International Association of Emergency Managers 'that having interoperable communications equipment and systems would be of benefit in the response period following a hazard event' (Jensen and Duncan, 2011).

Optimisation of construction sequence

Benefit description

This benefit can be achieved during the construction phase and refers to identifying changes to the construction processes and task sequence in order to increase the speed and accuracy of construction. The use of 3D constructability analysis and digital layouts, for example, allows the construction process and different options to be visualised in order to select the optimal construction sequence. Improved visualisation through 4D prototyping/modelling can help practitioners to better understand the construction sequence in order to identify potential conflicts and optimise it. 4D location-based scheduling methods can also help optimise the construction sequence, reducing waste during the construction process (i.e., less time spent waiting by field crews, rework and disruptions).

Additional analysis can be carried out in order to better understand the consequences of sequencing and production strategies. Construction sequence optimisation planning can, for example, include human, equipment and material resources in the BIM model by linking model components to activities' characteristics and methods. Factors such as traffic, access and egress, site materials delivery and machinery can also be included in construction simulations.

Dynamic phasing plans of occupancy can offer multiple options and solutions to space conflicts. These can also be identified and resolved before construction. During construction, schedule changes due to unforeseen conditions can be addressed and mitigated more effectively and at the least cost. Web-based systems and field and management tracking additionally make this process faster and effortless.

Increased stakeholder involvement during phase planning can also help optimise the construction sequence by leveraging on their knowledge and experience. By being involved in this process and having access to better construction sequence visualisation, field teams can understand the assembling sequences better and help optimise its delivery.

Beneficiaries

Beneficiaries include contractors, subcontractors and fabricators/manufacturers.

Enablers

• 3D control and planning (digital layout) • Constructability analysis • Construction system design (virtual mock-up)	• Field and management tracking • Phase planning (4D modelling)

Flow-on benefits

Optimisation of construction sequence can have the following flow-on benefits:

- Improved output quality due to more readily constructible, operable and maintainable project.
- Lower construction cost achieved by improving site productivity and reducing wastage at construction site.
- Better programming.

Metrics

• Time per unit • Speed of production • Cost per unit	• Conflict (field) • Equipment stand-by time • Staff idle time

Examples

A case study reported in 2008 used a methodology developed by a Germany-based group to implement 4D simulations to optimise the construction sequence of a 28-storey building located in Munich, Germany. The construction spanned over three years and included a total floor area of 78,400m^2 with a maximum height of approximately 100m. The project team used 4D modelling to visualise 600 construction tasks out of the 800 included in the project schedule. One of the main benefits of the method used by this group included reducing the effort required to create and maintain the schedule. The authors suggested that future developments could further improve the process by including traffic conditions, site equipment, resource planning and payment schedule data (Tulke and Hanff, 2007).

In the design and construction of a train maintenance centre in Australia, the project team used simulation and other visualisation functions of Navisworks to improve understanding of and optimise the construction sequence. This process in turn helped identify high risk processes and address these issues before construction (Utiome et al., 2015).

Reduced execution times and lead times

Benefit description

This benefit refers to reduction of time required to complete a project delivery phase and the latency between the initiation and execution of a process. Lead times between the start of engineering design, contract and production can be reduced through improved information exchange and digital fabrication. Reduced engineering and delivery lead times for material procurement can also be achieved through better schedule management and clearer scope information available to engineers, fabricators, and contractors. The time required for shop drawings production, review and turnaround can also be reduced.

Supply chain management and certain design processes can be streamlined, expediting work packages and phase handover. Improved communications allow faster responses to project status, queries and budget, reducing latency and lead times.

Improved programming/scheduling and fewer errors allow construction works to be carried out faster and more accurately so there is also less time needed to correct defects. More elements of the asset being pre-fabricated off-site and improved coordination also allow paralleling of activities and reduction of execution time.

BIM models are also well suited to show the extent of different work packages in isolation or with regards to other works; both through the model and the derived 2D documentation. This and the higher degree of adaptability to programme and other changes can help improve construction planning and reduce execution times.

Beneficiaries

Beneficiaries include clients/owners, designers, contractors, subcontractors, fabricators/manufacturers, surveyors, asset managers, suppliers and end-users (tenants).

Enablers

Table BD.27 Reduced execution times and lead times enablers

	Design	*Construction*
Digital fabrication		X
Field and management tracking		X
Integrated model management systems	X	X
Online collaboration and project management	X	X
Phase planning (4D modelling)		X
Object libraries	X	X
Streamlined logistics		X

Flow-on benefits

Reducing lead times can help increase customer satisfaction by improving customer service quality and increasing responsiveness to client requirements.

Metrics

- Latency.
- Overall time.

Examples

Studies have reported up to 7 per cent reduction in project delivery time and a six-month reduction of the overall project schedule (Azhar, 2011; Lu et al., 2013; Leite et al., 2011). Other studies have reported significant reduction of the time required to deliver the design phase; one case reported a time reduction of 40 per cent (Thomas et al., 2004). The GSA has also reported a 19 per cent reduction of construction time on a renovation project that used Phase Planning (4D modelling) (Suermann, 2009).

A case study of a medical research lab in the US used BIM to continually update the model in order to match it to the existing conditions and be quickly adjusted based on user feedback. This process was estimated to have saved the project approximately more than 100 man-hours. The modelling effort itself took 78.5 man-hours, including all revisions. This meant approximately 20 per cent savings in man-hours 'for the existing division and department space calculations – which was equivalent to approximately 62 per cent cost savings'. These savings were attributed to different hourly rates and the learning curve required by the consultant 'to understand the current operational practices and organisational divisions before working through the existing space utilisation calculations' (Manning and Messner, 2008).

Note

- 'Components that are engineered to order, such as structural steel, precast concrete and curtain walling elements, often have long lead times in a process that is labour intensive and prone to error' (BSI and buildingSMART, 2010).

Reduced risk

Benefit description

This benefit refers to the reduction of the likelihood of adverse outcomes arising across the life-cycle of the asset. Reliable cost estimates and fewer change orders can, for example, help reduce financial risks. Additionally, the risk of financial claims due to variations and delay can also be minimised. In terms of reducing the risk during emergencies, evacuation plans can be studied to a greater detail, different scenarios simulated and action plans optimised during the design phase.

The risk of discrepancies between plans, drawings, sections and elevations is also minimised through higher levels of process automation, improved data and information management and better communication. The client and end-users can additionally be included during the design reviews to inspect the design options using 3D visualisation tools and provide feedback on the characteristics of the final asset. This can help reduce the risk of variations during construction and mitigate the risk of inadequate initial specifications or misaligned expectations/objectives.

There is also less uncertainty, more control over and better communication about the construction process, the constructed asset and its performance. Having access to more complete models in early phases of asset planning, design, construction and management reduces the risk for clients to overpay on contingency for unexpected change orders, and enables them to accurately define and review the scope of works and bid packages.

Planning

The use of BIM enables better front-end planning so owners can address risks at earlier phases and commit to mitigation activities. In early design/concept design, simple quantities can be extracted from schematic models to analyse the return on investment.

Construction

Project risk can be minimised by actively monitoring and managing construction progress through field management tools. Construction processes can also be simulated during the design or construction phase before being carried out in order to identify potential risks.

Automated clash detection and off-site manufacturing also reduce uncertainty and risk by reducing the likelihood of field conflicts and shifting site work towards assembly.

Operations

Performance risk can be better managed and mitigated. The use of BIM allows, for example, early forensic analysis to better assess the potential of asset failure. Additionally, having a long-lasting complete database of the materials used for the construction of the asset also reduces health hazard risks during operations and decommissioning. Most toxic and hazardous construction materials are only deemed so after the asset has been constructed; having these databases can assist asset managers to quickly locate the sections containing the material and device a mitigation plan.

Beneficiaries

Beneficiaries include client/owners, contractors, asset managers and end-users.

Enablers

Table BD.28 Reduced risk enablers

	Planning	*Design*	*Construction*	*Operations*	*Decommissioning*
Automated clash detection		X	X	X	X
Automated rule-checking	X	X	X	X	X
BIM-based asset management system				X	X
Construction system design (virtual mock-up)	X	X	X	X	X
Cost estimation and planning	X	X	X	X	X
Data-rich, geometrically accurate model components		X	X	X	X
Field and management tracking			X	X	X
Front-end planning	X	X	X		
Record model				X	X
Walk-through and animations		X	X	X	X

Flow-on benefits

Reducing risk can help reduce the cost of construction by reducing the contingency cost.

Metrics

- Risk.
- Contingency budget.

Examples

In a survey carried out by Gillian and Kunz, 18 per cent of respondents said they already allow less contingency in BIM-enabled projects and 28 per cent said they would do this in the near future. In the same survey, the majority of respondents saw less risk even with the same traditional contract structures. However, 30 per cent reported reduced risk only when collaborative contract models such as alliances are used (Gilligan and Kunz, 2007).

Bibliography

Ahmed, A., Ploennigs, J., Menzel, K. and Cahill, B., 2010. Multi-dimensional building performance data management for continuous commissioning. *Advanced Engineering Informatics*, 24(4), pp. 466–475.

Ajayi, S. O., Oyedele, L. O., Ceranic, B., Gallanagh, M. and Kadiri, K. O., 2015. Life cycle environmental performance of material specification: A BIM-enhanced comparative assessment. *International Journal of Sustainable Building Technology and Urban Development*, 6(1), pp. 14–24.

Allen Consulting Group, 2010. *Productivity in the Buildings Network: Assessing the Impacts of Building Information Models*, s.l.: BEIIC.

Allison, H., 2010. *How Swinerton Builders Deliver Virtual Design and Construction*. Available at: www.vicosoftware.com/vico-blogs/guest-blogger/tabid/88454/bid/12659/How-Swinerton-Builders-Deliver-Virtual-Design-and-Construction.aspx [Accessed 5 August 2014].

Arayici, Y., Coates, P., Koskela, L., Kagioglou, M., Usher, C. and O'Reilly, K., 2011. Technology adoption in the BIM implementation for lean architectural practice. *Automation in Construction*, 20(2), p. 189–195.

Arayici, Y., Onyenobi, T. and Egbu, C., 2012. Building information modelling (BIM) for facilities management (FM): the mediacity case study approach. *International Journal of 3D Information Modelling*, 1(1), pp. 55–73.

Australian Department of Infrastructure and Regional Development, 2014. *Trends Infrastructure and Transport to 2030*, Canberra: Australian Government.

Australian Government Productivity Commission, 2014a. *Public Infrastructure. Productivity Commission Inquiry Report: Volume 1*, Canberra: Commonwealth of Australia.

Australian Government Productivity Commission, 2014b. *Public Infrastructure. Productivity Commission Inquiry Report: Volume 2*, Canberra: Australian Government.

Autodesk, 2013. *Bringing Transportation Plans to Life: Multiconsult AS Uses InfraWorks for Early Planning and Visualization of Transportation and Infrastructure Projects*, s.l.: Autodesk.

Autodesk, 2014. *Making Room for Rivers: INFRANEA Uses BIM to Model, Coordinate, and Plan a River-Widening Project of the River Waal Nijmegen*, s.l.: Autodesk.

Azhar, S., 2011. Building information modeling (BIM): trends, benefits, risks, and challenges for the AEC industry. *Leadership and Management in Engineering*, 11(3), pp. 241–252.

Azhar, S. and Brown, J., 2009. BIM for sustainability analyses. *International Journal of Construction Education and Research*, 5(4), p. 276–292.

Azhar, S., Carlton, W. A., Olsen, D. and Ahmad, I., 2011. Building information modeling for sustainable design and LEED® rating analysis. *Automation in Construction*, 20(2), pp. 217–224.

Azhar, S., Khalfan, M. and Maqsood, T., 2012. Building information modelling (BIM): now and beyond. *Australasian Journal of Construction Economics and Building*, 12(4), pp. 15–28.

Barker, P., 2015. *Future Campus: Developing a Smart Low Carbon University.* Available at: www.saiwill.com/lcu/pdf/9.pdf [Accessed 3 June 2015].

Barlish, K., 2011. *How to Measure the Benefits of BIM: A Case Study Approach*, Phoenix, AZ: Arizona State University.

Becerik, B. and Pollalis, S. N., 2006. *Computer Aided Collaboration in Managing Construction*, Cambridge: Harvard University Graduate School of Design, Design and Technollogy Report Series 2006-2.

Becerik-Gerber, B. and Rice, S., 2009. *The Value of Building Information Modeling: Can We Measure the ROI of BIM?* Available at: www.aecbytes.com/viewpoint/2009/issue_47_pr.html [Accessed 20 May 2014].

Becerik-Gerber, B. and Rice, S., 2010. The perceived value of building information modeling in the US building industry. *Journal of Information Technology in Construction*, 15, pp. 185–201.

Bentley, 2012a. *Information Mobility Helps Deliver Award-Winning Project.* Available at: http://ftp2.bentley.com/dist/collateral/docs/case_studies/CS_Glenfield_Junction_LTR-EN_1113.pdf [Accessed 26 May 2015].

Bentley, 2012b. *Sweco's 'Live BIM' Railway Project Significantly Improves Efficiency on Hallandsås Project.* Available at: http://ftp2.bentley.com/dist/collateral/docs/case_studies/4653_CS_Hallandsas_LTR-EN_1213-s.pdf [Accessed 26 May 2015].

Bentley, 2012c. *AECOM Uses Projectwise to Complete a complex State Highway 50 Per Cent Faster and Under Budget.* Available at: http://ftp2.bentley.com/dist/collateral/docs/case_studies/CS-AECOM_SH130_s_08-12.pdf [Accessed 26 May 2015].

Bentley, 2013. *Bentley's LEAP Bridge Enterprise Saves Time in Analysis and Design of Prestressed Concrete Box Girder Bridges.* Available at: http://ftp2.bentley.com/dist/collateral/docs/newsletter/9414_GD_LEAP_Bridge_India_LTR-EN_0113-p.pdf [Accessed 27 May 2015].

Bentley, 2014. *Point Clouds and Scalable Terrain Models Support Network Rail's Great Western Rail Electrification Programme.* Available at: http://ftp2.bentley.com/dist/collateral/docs/point_clouds/DescartesV8iSS4_NetworkRail_0613_LTR_s.pdf [Accessed 26 May 2015].

BIM Industry Working Group, 2011. *A Report for the Government Construction Client Group Building Information Modelling (BIM) Working Party*, strategy paper, London: Department of Business, Innovation and Skills.

Botta, S., Comoglio, C. and Petrosillo, I., 2013. Implementing the environmental and an integrated management system: social policies of a municipality through theoretical framework and case study. *Journal of Environmental Planning and Management*, 56(7), pp. 1073–1095.

Brilakis, I., Lourakis, M., Sacks, R., Savarese, S., Christodoulou, S., Teizer, J. and Makhmalbaf, A., 2010. Toward automated generation of parametric BIMs based on hybrid video and laser scanning data. *Advanced Engineering Informatics*, 24(a), pp. 456–465.

Brown, D. M., 2013. Innovation is key to swinterton longevity. *Engineering News-Record*, 12 August, pp. 1–4.

Bryde, D., Broquetas, M. and Volm, J. M., 2013. The project benefits of building information modelling (BIM). *International Journal of Project Management*, 31, pp. 971–980.

BSI, 2014. *PAS 1192–3:2014: Specification for Information Management for the Operational Phase of Assets Using Building Information Modelling*, London: British Standards Institution.

BSI and buildingSMART, 2010. *Constructing the Business Case: Building Information Modelling*, London: British Standards Institution.

Building: Product Research, 2014. *CPD 30 2014: Ensuring Data Accuracy on BIM Projects*. Available at: www.building.co.uk/cpd-30-2014-ensuring-data-accuracy-on-bim-projects/5071499.article [Accessed 16 June 2015].

Cerovsek, T., 2011. A review and outlook for a 'Building Information Model' (BIM): a multi-standpoint framework for technological development. *Advanced Engineering Informatics*, 25(2), pp. 224–244.

Chan, A. P. and Chan, A. P., 2004. Key performance indicators for measuring construction success. *Benchmarking: An International Journal*, 11(2), pp. 203–221.

Charlesraj, V. P. C., 2014. *Knowledge-Based Building Information Modeling (K-BIM)*. Sydney: ISARC 2014.

Construction Industry Council, 2013. *BIM Factsheet: Strategic Implementation of Building Information Modelling (BIM) in Hong Kong's Construction Industry*, Hong Kong: Construction Industry Council.

Cox, R. F., Issa, R. R. and Ahrens, D., 2003. Management's perception of key performance indicators for construction. *Journal of Construction Engineering and Management*, 129(2), pp. 42–151.

CRC for Construction Innovation, 2007a. *Adopting BIM for Facilities Management: Solutions for Managing the Sydney Opera House*, Brisbane: Cooperative Research Centre for Construction Innovation.

CRC for Construction Innovation, 2007b. *FM as a Business Enabler*, Brisbane: Cooperative Research Centre for Construction Innovation.

CRC for Construction Innovation, 2009. *National Guidelines for Digital Modelling*, Brisbane, Australia: Cooperative Research Centre for Construction Innovation.

Dossick, C. S. and Neff, G., 2011. Messy talk and clean technology: communication, problem-solving and collaboration using building information modelling. *Engineering Project Organization Journal*, 1(2), pp. 83–93.

Eadie, R., Browne, M., Odeyinka, H., McKeown, C. and McNiff, S., 2013. BIM implementation throughout the UK construction project lifecycle: an analysis. *Automation In Construction*, 36, pp. 145–151.

Eastman, C., Lee, J.-M., Jeong, Y.-S. and Lee, J.-K., 2009. Automatic rule-based checking of building designs. *Automation in Construction*, 18(8), pp. 1011–1033.

Ecodomus, 2015. *Laser Scanning and BIM*. Available at: www.ecodomus.com/index.php/laser-scanning-for-fm [Accessed 3 June 2015].

Fallon, K. K. and Palmer, M. E., 2007. *General Buildings Information Handover Guide: Principles, Methodology and Case Studies*, Gaithersburg: US Department of Commerce, National Institute of Standards and Technology.

Furneaux, C. and Kivits, R., 2008. BIM –implications for government, in K. Brown (ed), *Case Study No. 5 [2004-032-A + Case study no. 5]*, Brisbane: CRC for Construction Innovation.

Fussel, T., Beazley, S., Aranda-Mena, G., Chevez, A., Crawford, J., Succar, B., Hainsworth, J., Hardy, S., McAtee, S., McCann, G., Rizzalli, R., Akhurst, P., Linning, C., Marchant, D., Law, J., Lord, P., Morse, D., Crapper, P. and Spathonis, J., 2009. *National Guidelines for Digital Modelling: Case Studies*, Brisbane: CRC for Construction Innovation.

Gao, J. and Fischer, M., 2008. *Framework & Case Studies Comparing Implementations & Impacts of 3D/4D Modeling Across Projects*, Stanford: Center for Integrated Facility Engineering (CIFE), Stanford University.

Gerbov, A., 2014. *Process Improvement and BIM in Infrastructure Design Projects – Findings from 4 Case Studies in Finland*, Aalto: Aalto University.

Giel, B. K. and Issa, R. R., 2013. Return on investment analysis of using building information modeling in construction. *Journal of Computing in Civil Engineering*, 27(5), pp. 511–521.

Gilligan, B. and Kunz, J., 2007. *VDC Use in 2007: Significant Value, Dramatic Growth, and Apparent Business Opportunity*, Stanford: Center for Integrated Facility Engineering, Stanford University.

Greenwood, D., Lockley, S., Malsane, S. and Matthews, J., 2010. *Automated Compliance Checking Using Building Information Models*, paper presented at the Construction, Building and Real Estate Research Conference of the Royal Institution of Chartered Surveyors, Paris, France, 2–3 September.

GSA, 2007. *GSA BIM Guide Overview*, Washington, DC: US General Services Administration.

GSA, 2011. *GSA BIM Guide for Facility Management*, Washington, DC: US General Services Administration.

GSA, 2012. *GSA BIM Guide Series 05*, Washington, DC: US General Services Administration.

Gurevich, U. and Sacks, R., 2014. Examination of the effects of a KanBIM production control system on subcontractors' task selections in interior works. *Automation in Construction*, 37, pp. 81–87.

Hakkarainen, M., Woodward, C. and Rainio, K., 2009. *Software Architecture for Mobile Mixed Reality and 4D BIM Interaction*, paper presented at 26th CIB W78 Conference, Istanbul, Turkey.

Hardin, B. and McCool, D., 2015. *BIM and Construction Management: Proven Tools, Methods, and Workflows*, 2nd edn, Indianapolis: Wiley.

Hartmann, T., van Meerveld, H., Vossebeld, N. and Adriaanse, A., 2012. Aligning building information model tools and construction management methods. *Automation in Construction*, 22, pp. 605–613.

Haymaker, J. and Fischer, M., 2001. *Challenges and Benefits of 4D Modeling on the Walt Disney Concert Hall Project*, Stanford: Center for Integrated Facility Engineering (CIFE), Stanford University.

Henrich, D., 2010. *Coordination Resolution – 40% Faster than Traditional Clash Detection*. Available at: www.vicosoftware.com/0/blogs/virtual-construction-lessons-and-insights/tabid/84142/bid/12818/Coordination-Resolution-40-Faster-than-Traditional-Clash-Detection.aspx.

Holzer, D., 2007. *Are You Talking To Me? Why BIM Alone is Not the Answer*, paper presented at 4th International Conference of the Association of Architecture Schools of Australasia, AASA, Sydney, Australia, 27–29 September.

Hou, L., Wu, C., Wang, X. and Wang, J., 2014. *A Framework Design for Optimizing Scaffolding Erection By Applying Mathematical Models and Virtual Simulation*, paper presented at Computing in Civil and Building Engineering, ASCE, Orlando, Florida, 23–25 June.

Irizarry, J., Karam, E. P. and Jalaei, F., 2013. Integrating BIM and GIS to improve the visual monitoring of construction supply. *Automantion in Construction*, 31, pp. 241–254.

Jensen, J. and Duncan, R. C., 2011. *Preparedness: A Principled Approach to Return on Investment*, Falls Church, VA: International Association of Emergency Managers.

Jiao, Y., Zhang, S., Li, Y., Wang, Y., and Yang, B., 2013. Towards cloud augmented reality for construction application by BIM and SNS integration. *Automation in Construction*, 33, pp. 37–47.

Jongeling, R. and Olofsson, T., 2007. A method for planning of work-flow by combined use of location-based scheduling and 4D CAD. *Automation in Construction*, 16(2), pp. 189–198.

Jongeling, R., Kim, J., Fischer, M., Mourgues, C. and Olofsson, T., 2008. Quantitative analysis of workflow, temporary structure usage, and productivity using 4D models. *Automation in Construction*, 17(6), pp. 780–791.

Kam, C., Senaratna, D., Xiao, Y. and McKinney, B., 2013. *The VDC Scorecard: Evaluation of AEC Projects and Industry Trends*, Stanford, US: Center for Integrated Facility Engineering (CIFE), Stanford University.

Kam, C., Senaratna, D., McKinney, B., Xiao, Y. and Song, M., 2014. *The VDC Scorecard: Formulation and Validation*, Stanford: Center for Integrated Facility Engineering (CIFE), Stanford University.

Kaner, I., Sacks, R., Kassian, W. and Quitt, T., 2008. Case studies of BIM adoption for precast concrete design by mid-sized structural engineering firms. *ITcon*, 13, pp. 303–323.

Kang, J. H., Anderson, S. D. and Clayton, M. J., 2007. Empirical study on the merit of web-based 4D visualization in collaborative construction planning and scheduling. *Journal of Construction Engineering and Management*, 133(6), pp. 447–461.

Kang, Y., O'Brien, W. J. and Mulva, S. P., 2013. Value of IT: indirect impact of IT on construction project performance via best practices. *Automation in Construction*, 35, pp. 383–396.

Kasprzak, C. and Dubler, C., 2012. Aligning BIM with FM: streamlining the process for future projects. *Australasian Journal of Construction Economics and Building*, 12(4), pp. 68–77.

Kassem, M., Kelly, G., Dawood, N., Serginson, M. and Lockley, S., 2015. BIM in facilities management applications: a case study of a large university complex. *Built Environment Project and Asset Management*, 5(3), pp. 261–277.

Khanzode, A., 2010. *An Integrated, Virtual Design and Construction and Lean (IVL) Method for Coordination of MEP*, Stanford: Centre for Integrated Facility Engineering, Stanford University.

Khosrowshahi, F. and Arayici, Y., 2012. Roadmap for implementation of BIM in the UK construction industry. *Engineering, Construction and Architectural Management*, 19(6), pp. 610–635.

Kim, C., Park, T., Lim, H. and Kim, H., 2013. On-site construction management using mobile computing technology. *Automation in Construction*, 35, pp. 415–423.

Kim, H., Stumpf, A. and Kim, W., 2011. Analysis of an energy efficient building design through data mining approach. *Automation in Construction*, 20, pp. 37–43.

Kivits, R. A. and Furneaux, C., 2013. BIM: enabling sustainability and asset management through knowledge management. *The Scientific World Journal*, pp. 1–14.

Kreider, R., Messner, J. and Dubler, C., 2010. *Determining the Frequency and Impact of Applying BIM for Different Purposes on Projects*, paper presented at Proceedings of the 6th International Conference on Innovation in Architecture, Engineering & Construction (AEC), University Park, PA, 9–11 June.

Krigsvoll, G., 2007. Life Cycle Costing as part of decision making-use of building information models, in L. Bragança, M. Pinheiro, S. Jalali, R. Mateus, R. Amoeda, and M. Correia Guedes (eds.), *Portugal SB07 Sustainable Construction, Materials and Practices: Challenge of the Industry for the New Millennium*, Amsterdam: IOS Press, pp. 433–440.

Krygiel, E. and Nies, B., 2008. *Green BIM: Succesful Sustainable Design with Building Information Modeling*, Hoboken: Wiley.

Kunz, J. and Fischer, M., 2012. *Virtual Design and Construction: Themes, Case Studies and Implementation Suggestions. CIFE Working Paper #097*, Stanford: Sanford University.

Kwon, O.-S., Park, C.-S. and Lim, C.-R., 2014. A defectmanagement system for reinforced concrete work utilizing BIM, image-matching and augmented reality. *Automation in Construction*, 46, pp. 74–81.

Lagüela, S., Díaz-Vilariño, L., Martínez, J. and Armesto, J., 2013. Automatic thermographic and RGB texture of as-built BIM for energy rehabilitation purposes. *Automation in Construction*, 31, pp. 230–240.

Lee, G., Lee, J., Jones, S. A., Uhm, M., Won, J., Ham, S. and Park, Y., 2012a. *The Business Value of BIM in South Korea: How Building Information Modeling Is Driving Positive Change in the South Korean Construction Industry*, Bedford: McGraw Hill Construction.

Lee, J.-K., Lee, J., Jeong, Y-S. Sheward, H., Sanguinetti, P., Abdelmohsen, S. and Eastman, C. M., 2012b. Development of space database for automated building design review systems. *Automation in Construction*, 24, pp. 203–212.

Leigh, D., 2014. *Statistics on Fatal Injuries in the Workplace in Great Britain 2014*, Pontefract: Health and Safety Executive, National Statistics.

Leite, F., Akcamete, A., Akinci, B., Atasoy, G. and Kiziltas, S., 2011. Analysis of modeling effort and impact of different levels of detail in building information models. *Automation in Construction*, 20(5), pp. 601–609.

Lin, Y.-C., 2014. Construction 3D BIM-based knowledge management system: a case study. *Journal of Civil Engineering and Management*, 20(2), pp. 186–200.

Liu, F., Jallow, A. K., Anumba, C. J. and Wu, D., 2013. *Building Knowledge Modelling: Integrating Knowledge in BIM*. Beijing: CIB, pp. 1–10.

Love, P. E., Edwards, D. J., Irani, Z. and Walker, D. H., 2009. Project pathogens: the anatomy of omission errors in construction and resource engineering project. *IEEE Transactions on Engineering Management*, 56(3), pp. 425–435.

Love, P. E., Edwards, D. J., Han, S. and Goh, Y. M., 2011. Design error reduction: toward the effective utilization of building information modeling. *Research in Engineering Design*, 22(3), pp. 173–187.

Love, P. E., Matthews, J., Simpson, I., Hill, A. and Olatunji, O. A., 2014. A benefits realization management building information modeling framework for asset owners. *Automation in Construction*, 37, pp. 1–10.

Lu, W., Peng, Y., Shen, Q. and Li, H., 2013. Generic model for measuring benefits of BIM as a learning tool in construction tasks. *Journal of Construction Engineering and Management*, 139(2), pp. 195–203.

Manning, R. and Messner, J. L., 2008. Case studies in BIM implementation for programming of healthcare facilities. *ITCon*, 13, pp. 446–457.

Marzouk, M., Hisham, M., Ismail, S., Youssef, M. and Seif, O., 2010. *On the Use of Building Information Modeling in Infrastructure Bridges*, paper presented at 27th International Conference–Applications of IT in the AEC Industry (CIB W78), Cairo, Egypt, 16–19 November.

McGraw Hill Construction, 2012. *The Business Value of BIM for Infrastructure: Addressing America's Infrastructure Challenges with Collaboration and Technology SmartMarket Report*, Bedford, MA: McGraw Hill Construction.

McGraw Hill Construction, 2014a. *The Business Value of BIM for Construction in Major Global Markets: How Contractors Around the World are Driving Innovation with Building Information Modeling*, Bedford, MA: McGraw Hill Construction.

McGraw Hill Construction, 2014b. *The Business Value of BIM in Australia and New Zealand: How Building Information Modeling is Transforming the Design and Construction Industry SmartMarket Report*, Bedford, MA: McGraw Hill Construction.

Meža, S., Turk, Ž. and Dolenc, M., 2014. Component based engineering of a mobile BIM-based augmented reality system. *Automation in Construction*, 42, pp. 1–12.

Migilinskas, D., Popov, V., Juocevicius, V. and Ustinovichius, L., 2013. The benefits, obstacles and problems of practical BIM implementation. *Procedia Engineering*, 57, pp. 767–774.

Model the Planet Corp., 2015. *Live BIM™ Capabilities*. Available at: www.modeltheplanet.com/what-is-live-bim/live-bim-capabilities [Accessed 23 August 2015].

NATSPEC, 2014a. *Introduction to BIM*. Available at: http://bim.natspec.org/index.php/resources/introduction-to-bim [Accessed 25 August 2014].

NATSPEC, 2014b. *Getting Started With BIM*, Sydney: NATSPEC.

Nawari, N. O., 2012. *Automated Code Checking in BIM Environments*, paper presented at the 14th International Conference of Computing in Civil and Building Engineering, Moscow, Russia, 27–29 June.

NBS, 2014. *NBS National BIM Report*, Newcastle upon Tyne: RIBA.

Northumbria University, 2014. *Modern Technology Explores Medieval Room*. Available at: www.northumbria.ac.uk/about-us/news-events/news/2014/07/modern-technology-explores-medieval-room [Accessed 3 June 2015].

O'Connor, J. T. and Yang, L.-R., 2004. Project performance versus use of technologies at project and phase levels. *Journal of Construction Engineering and Management*, 130, pp. 322–329.

OpenBIM, 2012. *BIM for Facilities Management – University Campus*. Available at: www.openbim.org/case-studies/university-campus-facilities-management-bim-model [Accessed 26 May 2015].

Park, C.-S., Lee, D.-Y., Kwon, O.-S. and Wang, X., 2013. A framework for proactive construction defect management using BIM, augmented reality and ontology-based data collection template. *Automation in Construction*, 33, pp. 61–71.

Penn State, 2011. *BIM Execution Plan*. Available at: http://bim.psu.edu/Uses/Resources/default.aspx [Accessed 5 September 2014].

Peppard, J., Ward, J. and Daniel, E., 2007. Managing the realization of business benefits from IT investments. *MIS Quarterly Executive*, 6(1), pp. 1–11.

Peterson, F., Fischer, M. and Tutti, T., 2009. *Integrated Scope-Cost-Schedule Model System for Civil Works*, Stanford: Center for Integrated Facility Engineering, Stanford University.

Philip, D., 2015. *Level 2 'Package'*, Sydney: Australian Construction Industry Forum/Australasian Procurement and Construction Council.

Raheem, A. A., Issa, R. R. and Olbina, S., 2011. *Environmental Performance Analysis of a Single Family House Using BIM*, paper presented at 2011 Computing in Civil Engineering Conference Miami, Florida, 19–22 June.

Read, T., 2014. *Case Study: CMB BIM Pilot Project*. Available at: http://blog.solibri.com.au/case-study-cmb-bim-pilot-project [Accessed 3 June 2015].

Redmond, A., Hore, A., Alshawi, M. and West, R., 2012. Exploring how information exchanges can be enhanced through cloud BIM. *Automation in Construction*, 24, pp. 175–183.

Roper, K. and McLin, M., 2005. *Key Performance Indicators Drive Best Practices for General Contractors*, Raleigh, NC: FMI, Management Consulting, Investment Banking for the Construction Industry, Microsoft Corporation.

Rowe, J., 2013. *Analyzing Embodied Environmental Impacts with Building Information Modeling Tools*. Available at: http://autodesk.typepad.com/bpa/2013/03/analyzing-embodied-environmental-impacts-with-building-information-modeling-tools.html [Accessed 29 April 2015].

Rueppel, U. and Stuebbe, M. K., 2008. BIM-based indoor-emergency-navigation-system for complex buildings. *Tsinghua Science & Technology*, 13, pp. 362–637.

Sacks, R., Eastman, C. M., Lee, G. and Orndorff, D., 2005. A target benchmark of the impact of three-dimensional parametric modeling in precast construction. *PCI Journal*, 50(4), pp. 126–138.

Sacks, R., Treckmann, M. and Rozenfeld, O., 2009. Visualization of work flow to support lean construction. *Journal of Construction Engineering and Management*, 135, pp. 1307–1315.

Sacks, R., Koskela, L., Dave, B. A. and Owen, R., 2010. Interaction of lean and building information modeling in construction. *Journal of Construction Engineering and Management*, pp. 968–980.

Safe Work Australia, 2015. *Worker Fatalities*. Available at: www.safeworkaustralia.gov.au/sites/swa/statistics/work-related-fatalities/pages/worker-fatalities [Accessed 22 June 2015].

Salazar, G., Mokbel, H., Aboulezz, M. and Kearney, W., 2006. *The Use of the Building Information Model in Construction Logistics and Progress Tracking in the Worcester Trail Courthouse*, paper presented at Joint International Conference on Computing and Decision Making in Civil and Building, Engineering Montréal, Canada, 14–16 June.

Sanchez, A. X., Kraatz, J. A. and Hampson, K. D., 2014a. *Document Review. Research Report 2*, Perth: Sustainable Built Environment National Research Centre.

Sanchez, A. X., Kraatz, J. A., Hampson, K. D. and Loganathan, S., 2014b. *BIM for Sustainable Whole-of-Life Transport Infrastructure Asset Management*, paper presented at IPWEA Sustainability in Public Works Conference, Tweed Heads, Australia, 27–29 July.

Sanchez, A. X., Hampson, K. D. and Mohamed, S., 2015. *Perth Children's Hospital Case Study Report*, Perth: Sustainable Built Environment National Research Centre.

Sawyer, T., 2014. Dynamic models for safer sites. *Engineering News-Record*, 2 June, pp. 34–38.

SBEnrc, 2014. *Sustainable Built Environment National Research Centre Project 3.27 Using Building Information Modelling (BIM) for Smarter and Safer Scaffolding Construction*. Available at: www.sbenrc.com.au/research-programs/3-27-using-building-information-modelling-bim-for-smarter-and-safer-scaffolding-construction [Accessed 10 November 2014].

Shou, W., Wang, J., Wang, X. and Chong, H. Y., 2015. A comparative review of building information modelling implementation in building and infrastructure industries. *Archives of Computational Methods in Engineering*, 22(2), pp. 291–308.

Smith, P., 2014. *BIM Implementation – Global Initiatives & Creative Approaches*. Prague: Creative Construction Conference.

Staub-French, S. and Khanzode, A., 2007. 3D and 4D modeling for design and construction. *ITCON*, 12, pp. 381–407.

Succar,. B., Sher, W. and Williams, A., 2012. Measuring BIM performance: five metrics. *Architectural Engineering and Design Management*, 8(2), pp. 120–142.

Su, Y. C., Lee, C. Y. and Lin, Y. C., 2011. *Enhancing Maintenance Management Using Building Information Modelling in Facilities Management*, Seoul: International Association for Automation and Robotics in Construction, pp. 752–757.

Suermann, P. C., 2009. *Evaluating the Impact of Building Information Modelling (BIM) on Construction*, doctoral thesis, Gainesville: University of Florida.

Tan, X., Hammad, A. and Fazio, P., 2010. Automated code compliance checking for building envelope design. *Journal of Computing in Civil Engineering*, 24(2), pp. 203–211.

Tang, A., Owen, C., Biocca, F. and Mou, W., 2003. *Comparative Effectiveness of Augmented Reality in Object Assembly*, paper presented at SIGCHI Conference on Human Factors in Computing Systems, Lauderdale, FL, 5–10 April.

Tang, P., Huber, D., Akinci, B., Lipman, R. and Lytle, A., 2010. Automatic reconstruction of as-built building information models from laser-scanned point clouds: a review of related techniques. *Automation in Construction*, 19, pp. 829–843.

Tang, P., Anil, E. B., Akinci, B. and Huber, D., 2011. *Efficient and Effective Quality Assessment of As-Is Building Information Models and 3D Laser-scanned Data*, paper presented at 2011 ASCE International Workshop on Computing in Civil Engineering, Miami, FL, 19–22 June.

Teichholz, P., 2013. *BIM for Facility Managers*. Hoboken: Wiley.

Thomas, S. R., Lee, S.-H., Spencer, J. D. and Tucker, R. L., 2004. Impacts of design information technology on project outcomes. *Journal of Construction Engineering and Management*, 130(4), pp. 586–597.

Tsai, M.-H., Mom, M. and Hsieh, S.-H., 2014. Developing critical success factors for the assessment of bim technology adoption: Part I. methodology and survey. *Journal of the Chinese Institute of Engineers*, 37(7), pp. 1–14.

Tulke, J. and Hanff, J., 2007. *4D Construction Sequence Planning – New Process and Data Model*, paper presented at the CIB-W78 24th International Conference on Information Technology in Construction, Maribor, Slovenia, 27–29 June.

Tuttas, S., Braun, A., Borrmann, A. and Stilla, U., 2014. *Comparision of Photogrammetric Point Clouds with BIM Building Elements for Construction Progress Monitoring*, paper presented at 2014 ISPRS-International Archives of the Photogrammetry, Remote Sensing and Spatial Information Sciences, Zurich, Switzerland, 5–7 September.

University of Newcastle, 2015. *Building Fire Safety and Compliance*. Available at: www.newcastle.edu.au/course/ARBE3306 [Accessed 10 July 2015].

Utiome, E., Mohamed, S., Sanchez, A. and Hampson, K., 2015. *Case Study Report – New Generation Rollingstock*, Perth: Sustainable Built Environment National Research Centre.

Volk, R., Stengel, J. and Schultmann, F., 2014. Building information modeling (BIM) for existing buildings – literature review and future needs. *Automation in Construction*, 38, pp. 109–127.

Wang, W.-C., Weng, S.-W., Wang, S.-H. and Chen, C.-Y., 2014a. Integrating building information models with construction process simulations for project scheduling support. *Automation in Construction*, 37, pp. 68–80.

Wang, X., 2015. *National BIM Guidelines and Case Studies for Infrastructure*, Perth, Australia: Sustainable Built Environment National Research Centre.

Wang, X., Love, P. E., Kim, M. J., Park, C., Sing, C. and Hou, L., 2013a. A conceptual framework for integrating building information modeling. *Automation in Construction*, 34, pp. 37–44.

Wang, X., Truijens, M., Hou, L., Wang, Y., and Zhou, Y., 2014b. Integrating augmented reality with building information modeling: onsite construction process controlling for liquefied natural gas industry. *Automation in Construction*, 40, pp. 96–105.

Wang, Y., Wang, X., Wang, J., Yung, P. and Jun, G., 2013b. Engagement of facilities management in design stage through BIM: framework and a case study. *Advances in Civil Engineering*, pp. 1–8.

Williams, J., Amor, R., Apleby, S., Boyden, G., Davis, S., Greenstreet, N., Hawkins, J., Jowett, G., Read, H., Hunter, F. and Reding, A., 2014. *New Zealand BIM Handbook: A Guide to Enabling BIM on Building Projects*, s.l.: Building and Construction Productivity Partnership.

Xiong, X., Adan, A., Akinci, B. and Huber, D., 2013. Automatic creation of semantically rich 3D building models from laser scanner data. *Automation in Construction*, 31, p. 325–337.

Xu, X., Ma, L. and Ding, L., 2014. A framework for BIM-enabled life-cycle information management of construction project. *International Journal of Advanced Robotic Systems*, 11, pp. 126 I.

Yan, H. and Damian, P., 2008. *Benefits and Barriers of Building Information Modelling*, paper presented at 12th International Conference on Computing in Civil and Building Engineering, ITcon, Beijing, China, 16–18 October.

Yang, Q. Z. and Xu, X., 2004. Design knowledge modeling and software implementation for building code compliance checking. *Building and Environment*, 39(6), pp. 689–698.

Yates, K., Sapountzis, S., Lou, E. and Kagioglou, M., 2009. *BeReal: Tools and Methods for Implementing Benefits Realisation and Management*, in paper presented at 5th Nordic Conference on Construction Economics and Organisation, Reykjavík, Iceland, 10–12 June.

Yoders, J., 2009. BIM + IPD three success stories. *Building Design & Construction*, 50(4), pp. 26–40.

Yun, S.-h., Jun, K.-H., Son, C.-B. and Kim, S.-C., 2014. Preliminary study for performance analysis of BIM-based building construction simulation system. *KSCE Journal of Civil Engineering*, 18(2), pp. 531–540.

Zhang, S. Sulankivi, K., Kiviniemi, M., Romo, I., Eastman, C. M. and Teizer, J., 2015. BIM-based fall hazard identification and prevention in construction safety planning. *Safety Science*, 72, pp. 31–45.

Zhiliang, M., Zhenhua, W., Xiude, Z., Shixun, Q. and Pengyi, W., 2011. *Intelligent Generation of Bill of Quantity from IFC Data Subject to Chinese Standards*, paper presented at the 28th International Association for Automation and Robotics in Construction (ISARC), Seoul, South Korea, 29 June–2 July.

Enablers dictionary

Adriana X. Sanchez and Will Joske

This dictionary contains the definitions of those tools, actions and processes that facilitate the realisation of and maximise benefits of implementing BIM. These can often be implemented in non-BIM projects but, when used as part of the BIM implementation strategy, help maximise the value of BIM through the associated benefit. Each enabler profile:

- defines the enabler;
- provides information about how it can be used in relation to BIM;
- provides examples of how it has been used when case studies are available;
- when appropriate, examples of commercial tools that fall in this category are also indicated.

The profiles have been organised in two categories:

- **Intrinsic/core:** These are enablers that were considered by the authors to form the basis of BIM and maximising benefits from its use across different life-cycle phases. These may include some processes that are not standard practice in many countries yet. However, it was considered that these processes should be an intrinsic part of the BIM implementation strategy in order to maximise its value for all stakeholders.
- **In use:** These are enablers that are either commonly used nowadays and/ or that, although having had limited use in common practice, either are already growing in use or have the potential to do so in the near future and provide significant benefits.

This dictionary should be updated by the framework implementers as technology advances and new tools and uses are developed. This is a key part of the ongoing learning stage of the value realisation framework and a way of staying up to date with the fast pace of technology progress. Internal dictionaries should also include notes from user experience.

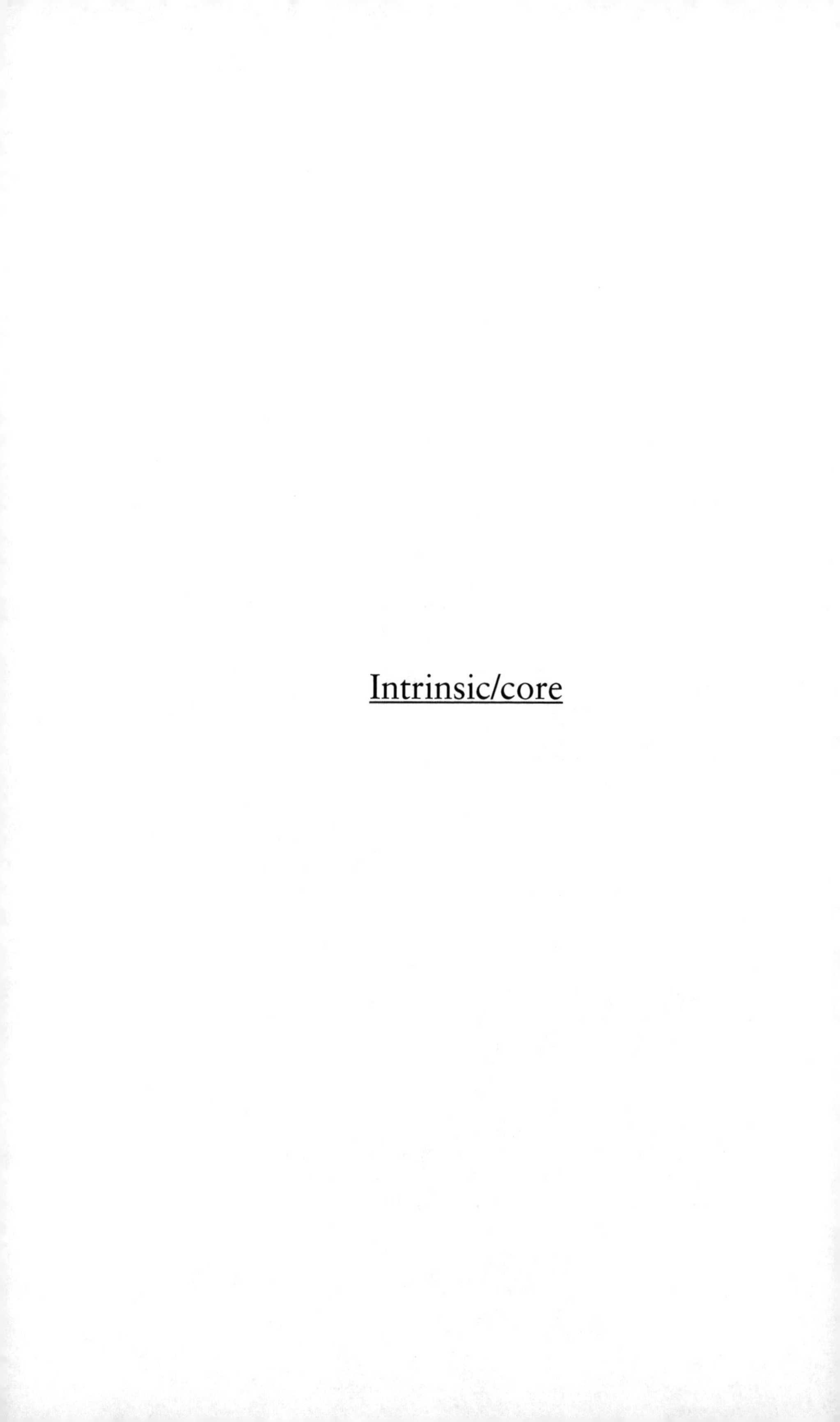

Intrinsic/core

Design authoring and data-rich accurate models

Even at its most basic level, a BIM model is characterised by containing 3D objects with associated data beyond their dimensional and positional information (CRC for Construction Innovation, 2009). These objects are sometimes referred to as *intelligent* because they have the ability to contain a significant amount of information about the object itself as well as about other objects associated with it (Conover et al., 2009). Having a BIM data-rich model with static and real-time information expands its application from a 3D design image into a tool that can be used to carry out a number of design, construction and operations tasks more efficiently and effectively. This requires reliable and accurate data acquisition sources that can be used for updating the model frequently based on the project needs (Caldas and O'Brien, 2009). For example, one of the main limitations of using BIM for automated rule-checking in fall hazard modelling is that if the information in a BIM model 'is incomplete, incorrect, or inaccurate, the correctness of the safety analysis will be largely affected' (Zhang et al., 2015). There has been a significant level of research into the use of laser scanning and radio-frequency identification (RFID) to automatically create data-rich models (Xiong et al., 2013; Hajian and Becerik-Gerber, 2009). However, as it will later be discussed in the laser scanning profile, these methods are not commonly used yet.

Design authoring is the process by which 3D modelling software is used to develop a BIM model based on asset design criteria. Design authoring is central to the development of a BIM model as well as defining quantities, means, methods, costs and schedules associated with individual objects (Penn State, 2011; Bloomberg et al., 2012). These tools are focused on the use of parametric modelling to create a BIM model and their goal is to increase the effectiveness of design (Kreider et al., 2010; Marzouk et al., 2010). Commercial design authoring tools include Autodesk's Revit, Graphisoft's ArchiCAD, Gehry Technologies' Digital Project (Ku and Mills, 2010) and Bentley's BIM and BrIM software (Bentley, 2015a; 2015d). Team competencies required to carry out these processes are: 'ability to manipulate, navigate and review a 3D model, knowledge of construction means and methods, and design and construction experience' (Penn State, 2011).

A 2009 United States survey about benefits from BIM linked to its uses answered by 175 professionals from across the supply chain showed that, at that time, design authoring was the third most popular BIM process (Kreider et al., 2010).

Parametric modelling

> If a diffuser is associated with a certain low-pressure duct, and that diffuser is moved, the associated duct will automatically relocate to the appropriate new position relative to the diffuser. Thus, not only can

> design changes be made earlier, they can also be made much faster and easier. This provides the designer greater certainty that all views have been updated with current information.
>
> (Conover et al., 2009)

The word *parametric* means that there are rule-based relationships created between intelligent objects that enable related attributes to be updated when one attribute changes (Conover et al., 2009). 'Parametric objects automatically re-build themselves according to the rules embedded in them' (Apollonio et al., 2012). Parametric attributes allow teams to quickly compile information with a high degree of confidence in its accuracy. In the past, these comparisons would have been largely based on assumptions and only spot-checked due to the amount of time involved relative to updates (Manning and Messner, 2008).

Parametric modelling allows users to coordinate changes and maintain consistency across documentation at all times (Azhar et al., 2009). In an early case study of a hospital designed and built using BIM, having parametric attributes allowed the modeller to spend only 30 minutes verifying changes before reissuing updated concept drawings for review. These changes, which would have normally taken the project team days to verify and cross discussed across all documentation, were updated instantaneously. The effect of this process was estimated to have reduced the time required for this task from weeks and months to days in a similar case study of a medical research lab (Manning and Messner, 2008).

One study carried out in 2008 to test the impact of 3D parametric modelling on the productivity of structural engineering practices concluded that this approach could increase productivity of drawings production by 15 to 41 per cent based on conservative estimates. When other activities were included into the considerations the upper limit of this range increased to 48 per cent (Sacks and Barak, 2008).

Record modelling (as-built BIM)

Record modelling is the process of creating an accurate representation of physical conditions, environment and elements of built assets. This model contains information related to architectural, structural and mechanical, electrical and plumbing (MEP) elements and is normally handed over to the asset manager as the as-built model. The as-built BIM model may also contain other required element types such as equipment and space planning systems. It combines elements from all previous models that may be of use to the asset manager such as design, construction and manufacturing elements. The record model therefore substitutes the as-built drawings (Penn State, 2011). In this case, data quality is also crucial to any applied functionality of record models. This process can be greatly improved if the initial data capture and creation was planned to be as accurate as possible and the data itself is updated and maintained as changes arise (Volk et al., 2014).

Record models provide data that is essential for effective management of operations and maintenance of a built asset, its equipment and utility systems (Jung et al., 2014). 'Having an accurate as-built model of the existing structure allows owners to visualize and analyze proposed retrofit solutions and ensures that the retrofit meets the owner's requirements and provides the best value' (Woo et al., 2010). These models can also be created for existing buildings either through 3D imaging and laser scanning techniques, by converting from 2D drawings or through traditional surveying methods (Hichri et al., 2013b). The Sydney Opera House, for example, has used a mix of these three options (Sanchez et al., 2015a). In terms of speed, 3D laser scanning is the most efficient technique (Hichri et al., 2013b), at the time of writing, however, there were no commercial BIM design authoring tools that could do this automatically. Nevertheless, as observed in the 3D laser scanning profile, there have been significant strides towards automating this process in experimental settings. Photogrammetry has also been used to verify existing record models and ensure they accurately represent as-built conditions (Klein et al., 2012).

Drawing generation

Drawing generation refers to the process of using a BIM model to create drawings and drawing sets, including schematic, design development, construction and shop drawings (Penn State, 2011). 'Because all drawings are derived from a single model, many of the errors caused by a lack of coordination between documents created separately by 2D CAD systems are eliminated' (NATSPEC, 2014b). Until recently, BIM design authoring software offered different degrees of automation for generation of drawings and documents. Some also include full bi-directional editing capabilities; this means that model objects can be edited from links embedded in drawings (Sacks et al., 2010a). AGA CAD's Smart Assemblies tool, for example, allows manufacturers and on-site personnel to directly generate shop drawings based on a Revit model (AGA CAD, 2014).

In 2008, there was a series of experiments carried out to study the impact of BIM versus traditional 2D drafting methods in structural engineering. This study showed a productivity gain in drawing production of 21–61 per cent for buildings with fairly repetitive floors, with a mean greater than 41 per cent. It also showed gains of 55 per cent for a large public building with more varied geometry from floor to floor (Sacks and Barak, 2008). Previous research had also pointed to high productivity gains in this area for structural precast concrete projects (82–84 per cent) (Sacks et al., 2005).

A case study on a precast air-raid shelter design and construction located in southern Israel showed the impact of formal training when transitioning to BIM for drawing production. In this case, staff formal training increased the productivity by more than 600 per cent. In this project, the drawings and pieces were completely error-free, which had never been experienced by the erection team (Kaner et al., 2008).

3D visualisation and walk-through

This enabler refers to the process of using BIM models and associated data to either directly visualise the interior of an asset through predefined views and model navigation or to provide a virtual environment to *walk-through* the space. This can be used to explore sections of the design in a more intuitive 3D environment during sign-off processes or during tender presentation and other review processes (Eynon, 2013). An integrated 3D model also makes visible issues and attributes that had been traditionally difficult to see in 2D documentation such as plinths for plant, detailed structure to support other systems and outlets for serviced equipment (Fussell et al., 2009).

In a 2012 survey that focused on infrastructure, 33 per cent of designers considered that 3D visualisation helped foster greater client engagement (Bernstein et al., 2012b). In a more recent global survey by McGraw Hill, 60 per cent of those surveyed identified better multiparty communication and understanding based on 3D visualisation as one of the most important factors improving return on investment from implementing BIM (McGraw Hill Construction, 2014a). The Manchester City Council, for example, highlighted that the use of 3D visualisation to provide clarity about and improve health and safety standards was one of the top benefits of using BIM during the refurbishment of the Manchester Town Hall Complex project completed in 2013 (Manchester City Council, 2013).

In a case study published in 2008 of a hospital designed and built using BIM, the use of the 3D model allowed the team to produce multiple 3D walk-through videos and perspective views for different design configurations within 16 hours. At this point, previous design reviews using 2D drawings had taken hours and produced few outcomes in terms of decisions. After having access to the visualisations created from the model, the team was able to come to a conclusion within less than 30 minutes (Manning and Messner, 2008). The benefits of 3D visualisation are highlighted throughout the *Benefits dictionary* and therefore will not be discussed in depth here.

Notes

- Design authoring and record modelling are two of four BIM uses mandated by the US Army Corps of Engineers (US Army Corps of Engineers, 2012).
- In order to properly exhaust all opportunities for utilising record models, dialogue must take place before the project begins. The critical factor to the success of creating a record model is properly documenting the owner's intended use of the information stored within the model.

Teams are also encouraged to outline the delivery strategy, implementation plan and level of document detail early on (Penn State, 2011).

- If the record models continue to be updated and maintained during operations, it can serve as a living document containing an accurate representation of the completed space (Penn State, 2011).
- Volk et al. (2014) provide a comprehensive literature review on record modelling of existing buildings covering more than 250 references.
- Depending on the circumstances of an individual project, an as-built model may be used to create an Asset Information Model (AIM). An AIM will have irrelevant objects and information removed so as to be better suited and more efficient for an asset or facilities management solution.
- Drawing production consumes approximately 60 per cent of the hours spent in structural engineering of buildings; figures, however, vary across different building types. In a series of case study projects published in 2008 this value was 52 per cent for commercial buildings, 52 per cent for public/educational buildings and 71 per cent for residential (Sacks and Barak, 2008).
- A study based on a mid-sized structural engineering firm in Calgary, Canada, found that 83 per cent of the labour input in typical projects was due to drawing and checking. A typical project was defined as precast CAD project, ranging in magnitude from 1,000 to 8,000 total labour hours. The typical breakdown of hours was: 7 per cent for general and project management tasks; 25 per cent for erection layouts; 10 per cent for engineering; 38 per cent for shop drawing; and 20 per cent for checking (Kaner et al., 2008).
- Implementation of walk-throughs and animations should include a platform and a systematic process for feedback from relevant stakeholders in order to inform decision-making (Peavey et al., 2012).

Early and effective stakeholder engagement

> The early involvement of stakeholders could identify any potential problems early-on and continue providing input throughout the construction phase. Therefore, there is a need to engage the stakeholder in the project as early as possible to determine the need and capture stakeholder inputs in such projects.
>
> (Baharuddin et al., 2013)

The main causes of inadequate or ineffective engagement of key stakeholders at early phases often leading to project failure are confusion, complexity and lack of communication. This can result in disbenefits such as reworks, disputes and variations, and cost overruns (Baharuddin et al., 2013; Hardin and McCool, 2015). This enabler refers to the process of using BIM models and tools to ensure early and effective stakeholder engagement. This process may include design reviews with all relevant stakeholders and the use of groupware, BIM server and cloud computing, mobile devices and augmented reality and other communication media (Volk et al., 2014).

This enabler is often encouraged by international BIM implementation guidelines used to develop client briefs and information requirements. The AEC (UK) BIM Protocol V2.0 for example, states that the Project Leader should initiate a *kick-off* meeting where the BIM goals and the project-wide execution plan are defined. This meeting should involve key stakeholders and consider the BIM requirements for the full life-cycle of the project (Coombes et al., 2012). This implies that disciplines that would normally be involved in later phases are engaged earlier in the life-cycle of the asset to ensure downstream requirements are met. Engaging stakeholders in the decision-making process can also enhance the sense of ownership of a project or asset (Mathur et al., 2008).

The US Army Corps of Engineers' BIM roadmap also includes the goal of implementing eco-charrettes in all applicable projects starting by 2015. These are stakeholder group meetings focused on ideas that may increase the efficiency of resource use of the final asset (US Army Corps of Engineers, 2012; Good, 2015). These meetings involved all relevant stakeholder groups from owners, designers, and contractors to maintenance staff, potential tenants and especially concerned neighbours (Department of Ecology, State of Washington, 2015). Using BIM for this purpose can facilitate effective stakeholder engagement that considers end-user needs throughout the asset's life-cycle (Bryde et al., 2013).

To ensure constructability, requirements of disciplines that would normally be involved at later stages are realised in earlier phases of design. In BIM for transport infrastructure design of new highway sections that, for example, include a bridge, disciplines such as bridge designers may need to be involved in the early design phase for cases where multiple options are available or any problem or risk exists regarding their technical area (Gerbov, 2014). This early engagement allows the design to be optimised at a stage when changes are the least expensive to analyse and apply (Conover et al., 2009).

In addition to examples of improved stakeholder engagement described in the AR, walk-through, animation and design review sections, a UK-based group of researchers studied the use of BIM-based automatic virtual environments (video-based method) and their effect on client engagement. This experiment allowed clients and end-users with no technical expertise to quickly identify issues such as light switches and socket outlets located in the wrong place (Tutt and Harty, 2013).

Note

- In addition to ensuring that all relevant supply chain stakeholders are involved at a time where their input can be most effective, team leaders should develop plans that describe the tools to be used, file formats and information drop schedule (Hardin and McCool, 2015).

Integrated model and programme management systems

A fully integrated model and programme management system is the ultimate goal for BIM systems where multidisciplinary models are integrated through model servers or other network-based technology. This can be done within a local server or online through a web server (Fussell et al., 2009). It requires an integrated model development and management approach and an effective stakeholder engagement process that considers user needs throughout each phase (Bryde et al., 2013). These systems provide a framework to describe, track and manage output, organisational and process changes over time, allowing these changes to be visualised and managed (Kunz and Fischer, 2012).

Integrated management systems:

- minimise documentation and record duplication;
- help create integrated and univocal strategies and policies as well as non-conflicting objectives;
- help produce cost savings by optimising time, resources and responsibilities;
- simplify internal and external audits, interaction and cooperation with and across stakeholders;
- improved management reliability; and
- help ensure more sustainable management (Botta et al., 2013).

The integration of all objects and aspects that can be managed within a single source enables teams to find coordinated solutions to complex interrelated problems. When a fully integrated system is used, changes in quantities, for example, are passed to the process and cost models. Likewise, changes to the cost model results in a revised process model and changes to the process model results in a revised cost model, avoiding rework and errors (Peterson et al., 2009). Integrated model and programme management systems with mobile access in the field 'allow access to all documentation without making trips back to the office' (GSA, 2011).

Commercial software providers with tools that are examples of this enabler include Newforma, Aconex and Union Square. Newforma Building Information Management is an extra module of their Project Center software that aims to improve BIM workflows management (Newforma Inc., 2015). Aconex Project-Wide Solutions includes Connected BIM, which allows merging large models from different authors and all stakeholders to have access to a single integrated model (Aconex, 2015). Union Squares has similar functions and claims to be 'the leading information management software for the UK Construction industry' (Union Square, 2015a). This provider has also recently joined the Open BIM Alliance, which has the goal of 'providing certainty for Open Standard BIM exchange' (Union Square, 2015b).

There are also several software tools commercially available for integrated programme management of complex projects such as dRofus and CodeBook. dRofus is a software for integrated programme management that supports planning and administration of areas, rooms and departments/functions; room datasheets; registration and monitoring of the requirements for each room; FF&E planning, cost control and procurement of FF&E; checking and visualisation of designed model through IFC; and bi-directional plug-ins for Revit and ArchiCAD (dRofus, 2015). The developer of this software platform, Nosyco now dRofus SA, states that this tool provides a data-centric approach to BIM; data input from every stakeholder can be planned, created and managed through this platform while capturing relevant information for all objects across domains, disciplines and models, including data outside of the BIM platform (dRofus, 2015). CodeBook has a series of data management products that support BIM across asset life-cycle, from inception to handover. They state that the tool allows different stakeholders to create 'a single source of BIM truth about the origins, design, construction and use of any building type and size' (CodeBook, 2015).

There have also been tools developed in experimental settings to address shortfalls of commercial software platforms such as:

- the need for manual data collection and entry for certain functionalities;
- the lack of some of the important functions of an integrated project control system, such as forecasting performance, analysing variances, and recommending corrective actions;
- flexibility and interoperability issues;
- data-sharing issues; and
- limited parameters that can be optimised within a limited domain (Chen et al., 2013).

Beyond current levels of integration, a fully integrated model management system during the operations phase can be further integrated with asset performance and structural health monitoring sensors as well as simulation systems. This was being done, for example, by Enemetric, a small Scottish manufacturer of volumetric building systems, as part of their strategy for modular housing developments (Seeam et al., 2013).

Common data environments and protocols

> The responsibility for data integrity, including backups, periodic and milestone archiving, and logs of interaction, is critical to maintaining the value and integrity of the data within the model.
>
> (AIA, 2013a)

A common data environment is a 'web-based server that enables multiple users to collaborate in managing digital information in accordance with an agreed protocol' (CIOB, 2013b) and is recommended by the Chartered Institute of Building (CIOB), AEC (UK) and the American Institute of Architects (AIA) (CIOB, 2013a; Coombes et al., 2012; AIA, 2013a). The complexity of the systems can vary wildly from simple databases that allow users to view, use and modify digital data to document management products with document tracking, version control and other features (Sanchez et al., 2014b). If such a system is to be used, AIA also recommends that parties document 'what the system is intended to achieve and how the system will impact the role of the Project Participants' (AIA, 2013a).

Common data environments and protocols facilitate collaboration and data management. 'Developing precise methods for exchanging project data in initial stages eases the transition of information throughout the project' (Fillingham et al., 2014). The classification proposed by AEC (UK) (Work in Progress (WIP); Shared; Published; and Archive) offers a structured and easy way of organising information. User manuals and information such as what the system is intended to achieve and how it will impact the role of project participants can also improve acceptance by the project team (Sanchez et al., 2014b).

In a case study about the implementation of BIM in the construction of the Birmingham City University Parkside Development for whole-of-life asset management, common data environments allowed design staff members to always access the same set of shared modelling data. This provided 'clarity of understanding when it came to discussing and resolving issues' and enabled simpler and better collaboration (Fillingham et al., 2014).

Some common data environments use cloud-based systems; this is sometimes called cloud-BIM.

> Cloud computing refers to the services over the Internet, and the hardware and software in the data servers that provide these services. It is quickly becoming an innovative model for delivering IT infrastructure, applications, and data management… 'CloudBIM' shifts the emphasis from static stand-alone applications, to dynamic shared environments, dynamically allocated tasks, and access via a network… Because the Internet operates in real time, computer devices with cloud-based tools can make information more accessible to users.
>
> (Chi et al., 2013)

Cloud-BIM is considered to be almost BIM 2.0, BIM's second generation expected to create a new wave of change (Wong et al., 2014). Using more traditional BIM methods, files are exported from one application and imported into another; this creates multiple copies of the data (Chi et al., 2013). Cloud-BIM allows information to be readily exchanged between parties using BIM software on a cloud platform (Redmond et al., 2012).

It also allows the use of mobile devices that communicate with traditional computers (Wong et al., 2014). Innovations in this field have, for example, a particular impact on a team's ability to complete cost estimates with platforms such as Onuma System allowing 'multiple users to collaborate in real time in the cloud with the same datasets' (Hardin and McCool, 2015).

There are also BIM-cloud computing frameworks, which include:

- services within specific software applications for basic cloud storage (e.g., Autodesk Revit Server);
- within a vendor suite (e.g., Autodesk BIM360 solutions and A360 Collaboration);
- high-level general purpose cloud frameworks; programme and data run in the cloud for general applications; this requires more advanced IT skills (e.g., J2EE, Stratus and Microsoft Azure); and
- low-level system software and hardware virtual machine (e.g., Amazon Elastic Compute Cloud (EC2), Citrix and VMware systems) (Zhang and Issa, 2012).

In the construction of a major hospital in Norway worth €670 million, designers were able to choose either to work co-located or distributed by establishing a cloud-based framework (Merschbrock and Munkvold, 2014). In the construction of another hospital in Australia, the use of BIM-cloud framework applications also allowed staff members to work off-site. In this case, Panzura was used as the cloud storage system for working on the model remotely and Buzzsaw as the cloud-based file transfer system to send and receive all model documentation. These tools were seen to help increase efficiencies by allowing modellers to be located off-site, including other countries with different time-zones. These systems also allowed tracking which staff members have logged in, when and for how long as well as clearly identifying changes made by that person. This allowed the team to produce an auditable report and improve change management; although it also created concerns regarding data security (Sanchez et al., 2015b).

BIM360 was also used to create field reports and was planned to be used for commissioning. This tool was seen to have provided better communication and improved data management benefits, as well as reduced conflict and improved efficiency. BIM360 Glue was planned to be used for immersion technology and access to the model through handheld devices. Together with the use of RFID in every room, this tool was expected to allow the team to use their handheld devices to recall information related to the objects in a room. Eventually, the team expected to implement a system that would allow staff to scan the RFID tag on an object and then populate any reports or introduce changes directly into the model (Sanchez et al., 2015b).

Centralised document management systems

The use of centralised document management systems as the only source of truth is becoming more common among projects (AIA, 2013a). AIA's *Project Building Information Modeling Protocol Form*, for example, establishes clauses that assign responsibility to specific roles for managing and maintaining a centralised electronic document management system (AIA, 2013b). The *Project Digital Data Protocol Form* then establishes the requirements to establish such system (AIA, 2013c). The use of such centralised data management systems can avoid duplication of data, which often results from planning authorities insisting on using traditional drawings (Sherman, 2013).

Notes

- Examples of commercial software currently available for integrated programme management are NavisWorks from Autodesk (2015m) and ProjectWise from Bentley (2015b).
- A common strategy for Cloud-BIM is to use systems such as Citrix that connect team members running less sophisticated hardware to a server at a central location where the design authoring software is running from. These arrangements, however, can experience time lags from one networked point to another, particularly when working from geographic locations that are far apart. The latency can make this strategy difficult to implement.
- As of 2014, most cloud-BIM applications were 'centred on building design and construction phases' (Wong et al., 2014).
- While more data can be moved and shared among the project team via the web, this type of cloud-based collaboration will require more powerful web-based operating systems, file-sharing platforms and hardware controllers to support the consolidation and archiving of and increased access to project data (Wong et al., 2014).
- Although cloud-based system providers such as Autodesk assure their clients that these systems incorporate 'advanced data security technology to deliver secure cloud technology to users' (Autodesk, 2015d), data integrity and security is often a concern for users. The McGraw Hill Global Markets survey published in 2014, for example, highlighted that 62 per cent of contractors expressed a moderate or high level of concern, with South Korea having the highest number at almost 80 per cent and Germany the lowest at 31 per cent. This number was also highest among firms with high levels of BIM engagement (McGraw Hill Construction, 2014a). Data security can be addressed in the File Transfer Protocol and managed by a specific role such as a data security manager (CIOB, 2013a). Another alternative is to develop a data security protocol and set adequate user rights to prevent data loss or damage

during file exchange, maintenance and archiving (NATSPEC, 2011). AEC (UK) also strongly recommends that all BIM project data resides on network servers subject to regular back-ups and that staff access to BIM project data held on the network servers is done only through controlled access permissions (Coombes et al., 2012).

- Ding and Xu (2014) explore requirements for cloud storage for BIM in more detail and provide a framework for a cloud-based BIM life-cycle management system.

Online collaboration and project management (OCPM)

This enabler refers to the process of using web-based collaboration and project management software platforms to communicate project information among team members in a more accurate, effective and timely manner. This process has been shown to produce improved results in construction projects using computer-aided project management. A study published in 2006 analysing nine case studies covering a range of organisations found that implementing OCPM tools reduced request for information (RFI) turnaround to an average of 5.44–6.04 workdays instead of the industry average of 14 days, saving up to hundreds of thousands of dollars to each organisation (Becerik and Pollalis, 2006).

This process is particularly useful when working with BIM and aiming to improve collaboration and coordination across multiple disciplines. There are a number of BIM web-based collaboration systems available with functionalities such as content management and viewing and reporting, which are complementary to model creation or system administration tools (Volk et al., 2014). The BIM360 suite from Autodesk, for example, is a collaborative construction management software used for remote access to project data throughout the building construction life-cycle that supports project online collaboration (Autodesk, 2015h). In the Perth Children's Hospital in Australia, the use of this tool provided benefits such as better communication, improved data management, less conflict and improved efficiency (Sanchez et al., 2015b).

Note

- Implementation strategies of OCPM, as with cloud-based model management, should address data security early on when using any web-based collaboration and project management software (Singh et al., 2011).

Interoperability and data formats

Interoperability provides the ability to exchange information across different systems or components and to use the information that has been exchanged

(Williams et al., 2014). This depends on the format used to store the data and the software requirements to read it. At a project level within a single phase, the challenge is to ensure interoperability between the software used across different disciplines. When this enabler is considered for the complete life-cycle of an asset, the challenge becomes ensuring interoperability across phases and rapidly developing BIM models, functionalities and technology (Volk et al., 2014). BIM design authoring software providers, for example, typically release new versions every year or two and sometimes stop providing support to older versions, but facility and asset management legacy systems can be used for the complete lifespan of the asset (Kassem et al., 2015). Chapter 5 explored this issue to some extent and how this has been dealt with at the Sydney Opera House.

Data exchange is then made possible either by using directly interoperable formats or through the use of proprietary or non-proprietary exchange formats (Volk et al., 2014). This is achieved by mapping sections of the data structures of the different applications to a universal data model and vice versa. Non-proprietary universal data models are commonly called *open data exchange formats* and the idea is that they can be used by any application independently of who developed it (Grilo and Jardim-Goncalves, 2010).

The use of interoperable file formats in BIM when using different platforms allows:

- faster informed decision-making;
- faster iteration of performance and task sequencing simulations;
- streamlined information flow and reduced time-to-complete in certain supply chains (e.g., steel);
- substantially reduced field problems and material waste;
- makes feasible off-site fabrication of larger percentages of the asset components and assemblies, increasing their quality and longevity; and
- reduced on-site construction activities and materials staging, creating a less crowded and safer site (Fallon and Palmer, 2007).

Ensuring the use of interoperable formats across life-cycle phases and functionalities is, however, still a challenge in many projects due to incomplete, differently or ambiguously used IFC and other format attributes, denotations or contents.

> Recent developments focus on the implementation of specifiers' Properties Information Exchange (SPie) and semantic web technologies in open formats and ontologies like HTML, XHTML, bcXML, gbXML, e-COGNOS, COBie, IFCXML, IFC, ifcOWL or CIS/2 to enable expert software applications.
>
> (Volk et al., 2014)

The application of some of these are, however, often limited to experimental and academic settings. Nevertheless, organisations such as the US Army Corps of Engineers have shown continued advocacy and support for the use of interoperable formats and exchange standards 'in order to maintain the broadest options for software selection based on cost-effectiveness and end-user satisfaction' (US Army Corps of Engineers, 2012). This is a particularly important consideration for asset managers who accumulated data over long periods of time and use systems that tend to lack compatibility with data exchange standards. Most facility and asset management systems tend to support simpler formats such as text, spreadsheets and relational database files with highly heterogeneous data (Kang and Hong, 2015).

Data exchange standards

Data exchange standards are standardised electronic file format specifications used to exchange data between different software applications. These have been developed and evaluated by standard-developing organisations (Conover et al., 2009). These organisations include international standard institutes such as the International Organization for Standardization (ISO) as well as regional or national organisations such as the European Committee for Standardization (CEN) and the German Standards Institute (DIN). There are also a number of standards that have been developed by independent groups such as Electronic Business Extensible Markup Language (ebXML) (Grilo and Jardim-Goncalves, 2010).

Industry Foundation Classes (IFC), originally developed in 1995 by a group of American and European firms through the International Alliance for Interoperability (IAI), have now been standardised through ISO/PAS 16739 and are maintained by buildingSMART International. IFC is the dominant non-proprietary (vendor-neutral and open) data exchange format for BIM files and uses object-oriented file formats (Volk et al., 2014; Conover et al., 2009). This standard was developed to represent asset information across its life-cycle and facilitate data transfer between BIM design authoring tools, IFC viewers and expert software applications such as performance simulation software and rule-checking applications.

> Data exchanges between source and receiving software systems are performed through mainly proprietary translators with own data structures. Certifications like 'openBIM' of buildingSMART and NBIMS award software solutions with high IFC interoperability.
>
> (Volk et al., 2014)

Construction Operations Building Information Exchange (COBie) is a subset of IFC developed by the US Army Corps of Engineers and has now been mandated by the UK government. It is an 'information exchange specification for the life-cycle capture and delivery of information needed by facility managers' (East, 2012). The standard BS 1192–4:2014 outlines COBie data structure code of practice for the exchange of information across life-cycle phases of buildings and infrastructure assets (BIM Task Group, 2014).

At an information requirement level rather than specific formats, the ISO Standard 10303 (also known as Standard for the Exchange of Product Model Data; STEP) contains part 225, *Application Protocol (AP): Building Elements Using Explicit Shape Representation.* This part provides specifications for requirements related to the information exchange of object shape, property, and spatial configuration between systems 'with explicit shape representations, specifically the physical parts of which a building is composed, such as structural elements, enclosing and separating elements, service elements, fixtures and equipment, and spaces' (Grilo and Jardim-Goncalves, 2010).

Notes

- BSi's PAS BIM standards such as the PAS1192 series specify BIM data and strategic requirements to achieve level 2 BIM across design, construction and operations (BSI, 2013; 2014).
- Although open formats are certainly available, common design practices tend to support using proprietary formats as much as possible in order to retain fidelity of models and associated data. Open formats are then mostly left for cases where specific stakeholders are using incompatible software that is required to make exchanges to the model data. Open formats are also valuable for creating archives and handover deliverables in a way that minimises the risk of future interoperability.

Object libraries

An object is an item that has a state, behaviour and unique identity (NBS National BIM Library, 2015). A BIM object is a model element that contains:

- information that defines the product or element;
- geometric characteristics of the physical product or element;
- visualisation data giving the object a recognisable appearance; and
- functional data that allows the object to be positioned or to behave in a certain manner (Waterhouse et al., 2014).

An object library is a repository that contains these objects and can be internally developed by an organisation for private use or centrally developed by a public, private or umbrella organisation for open, public use (Duddy et al., 2013).

> The construction industry needs access to BIM objects that can be used freely, safe in the knowledge that they contain the right levels of

> information with the appropriate geometry, all wrapped up in a consistent, yet structured and easy to use format.
>
> (NBS National BIM Library, 2015)

Developing libraries commonly requires separate objects to be produced for all of the different software platforms. This increases the cost associated with populating object libraries and the difficulties associated with maintaining consistency between representations across different software platforms and versions. To address these issues and maximise the efficient use of resources industry-wide, several countries are developing open national libraries that contain vendor neutral objects (Duddy et al., 2013). In the UK, the National Building Specification (NBS) has developed a set of standards for BIM objects and a national object library to help increase the level of adoption industry-wide. The object standard is based on the British Standard series BS 8541, COBie and IFC schema, and it contains both generic and manufacturer-specific objects. This library won an award in 2013 as BIM Initiative of the year (NBS National BIM Library, 2015).

Standardisation of the information requirements across objects allows different elements to be collated and compared. Having common approaches to the representation of the physical characteristics of products and elements makes the BIM objects simpler to use, more affordable and reliable (Waterhouse et al., 2014). In Australia, SBEnrc started to develop a national open access collaborative object library, which was based on generic object descriptions and non-proprietary formats (SBEnrc, 2012). This initiative has then been continued by buildingSMART Australasia and NATSPEC who are now aiming to align practices with NBS (NATSPEC, 2014a; NATSPEC BIM, 2015). Also in Australia, BIM-MEP^{AUS} is developing a set of standards and content for MEP models. Plant, equipment and fitting content, for example, include Level of Development (LOD) 300 IFC model objects and LOD 400/500 Manufacturer's Certified objects (BIM-MEP AUS, 2014). Other countries developing national libraries for specific sectors of the industry include Sweden, where the Swedish Transport Administration has made available an object template and tutorial library for road and rail assets (Trafikverket, 2015).

Libraries developed by software providers include 3D Warehouse, which contains millions of objects modelled with SketchUp (Trimble, 2015a) and Autodesk's Seek BIM Model Catalogue (Autodesk, 2015c). Libraries based on proprietary objects from specific product manufacturers include SWEETS BIM Catalogue (Dodge Data and Analytics, 2015) and SMARTBIM Library (SmartBIM, 2015).

Object libraries have also been developed for specific purposes such as safety planning, in the case of TurvaBIM developed by the Technical Research Centre of Finland (VTT) (Sulankivi et al., 2009), and building energy simulations, in the case of ModelicaBIM library (Kim et al., 2015).

Well-structured data

This enabler refers to the process by which information within model objects is classified in a standardised way in order to be exchanged and integrated across systems in an efficient and effective manner. A well-defined data structure that covers information types and responsibility for compiling the data and is itself defined across the different asset types is critical in order to maximise benefits from implementing BIM. Without an agreed comprehensive data classification, it is impossible to ensure direct interoperability between information manipulation and management systems and tools.

> A construction classification system must include buildings, infrastructure and integrated project and office management. It must be able to map project information from the initial concept through brief, detailed design, construction, handover and facility operation and maintenance. To fully adopt the BIM process it is important to organise such information systematically and consistently.
>
> (BIM Task Group, 2013)

Well-structured data in a single database makes information more accessible and makes possible filtering and mining data. A structured data scheme can be developed early on by the project team to be used throughout the life-cycle of an asset in order to have consistency across the database that will later be used for asset management (Sanchez et al., 2015b). This enabler allows immediate machine interpretation and helps

> eliminate the cost of manipulating and interpreting the data in the receiving system each time information is handed over to another application for analysis. Structured form is the key to highly optimized design, supply chain streamlining and the ability to use information captured during design and construction in downstream operations and maintenance applications without additional cost.
>
> (Fallon and Palmer, 2007)

Object classification systems

Object classification systems provide a standardised classification structure to organise information (OmniClass™, 2006). Applied to defect management, for example, standard classifications and data compositions need to be structured in such a way that defect management plans can be conveniently prepared while also ensuring that defect causations and impacts are reflected precisely. At the same time the classification used needs to comply with industry standards of information classification structure in addition to project- and task-specific data requirements. The classification must also allow data to be conveniently and efficiently collected, searched and reused. Well-known industrial standards include MasterFormat, UniClass and OmniClass Construction Classification System (Park et al., 2013)

OmniClass was developed by the Construction Standards Institute (CSI) and is a construction classification system. UniClass and MasterFormat were developed by the Construction Specifications Institute in conjunction with Construction Specifications Canada (CSC). UniClass is a 'classification system for building elements (including designed elements) that forms the basis of Table 21 of the OmniClass System' (Williams et al., 2014). MasterFormat, like UniClass, organises information according to the type of work performed and the type of object. The use of these classifications allows, for example, for cost data to be easily derived from the model (Weygant, 2011).

Most BIM software platforms are able to use data organised according to more than one classification. Trimble, for example, can uses UniClass or MasterFormat. According to this software provider, UniClass is more commonly used for preliminary project description, performance specifications and cost estimation while the MasterFormat is primarily used for detailed cost information (Vico Software, 2015c).

Note

- Data exchange standards such as COBie require that each component is assigned a classification code from classification systems such as those mentioned above.

In use

3D laser scanning – point cloud manipulation

Laser scanning is a form of 3D imaging that has been popular in geospatial and survey industries for years now (Gleason, 2013; GSA, 2009). This is the practice of using specialised instruments to rapidly measure or capture existing conditions in the built or natural environment. This is done by measuring 'the range and bearing to and/or the 3D coordinates of points on an object or within a region of interest' (GSA, 2009). Other examples of 3D imaging include triangulation-based systems and other approaches based on interferometry. Information capture by these systems in addition to coordinates can include colour, intensity and texture (Brilakis et al., 2010; GSA, 2009).

Laser scanning allows capturing the detailed geometry of constructed assets for as-built BIM models (Tang et al., 2011). This technology has been used in conjunction with geographic information systems (GIS), underground radar and other tools in infrastructure asset planning, construction and management. This allows firms to develop more complete models of the existing civil conditions that can be used for a variety of analyses, simulations and visualisations to optimise project delivery (Bernstein et al., 2012b). Laser scanning has also been used to test the accuracy of BIM models of built assets by overlaying the model with point cloud data (Fussell et al., 2009).

Very few constructed facilities have complete as-built records; this type of tool can be useful to address this issue and develop BIM models of existing assets. Laser scanners are increasingly being accepted by the construction industry for this purpose and can help shorten the time required to develop an as-built BIM model, which would normally take months of modelling for an average-sized building (Brilakis et al., 2010; Tang et al., 2010).

Creating a BIM model from laser scanning data has three main steps:

1. **Data collection:** Dense point cloud data is collected with laser scanners positioned in key areas of the asset (point clouds can have up to billions of data points). These measurements can have an error of millimetres if highly accurate representations are required and be taken in short periods of time. In recent times, mobile scanners have also helped to greatly reduce the scanning time when higher accuracy is not required (Bosché, 2010; Tang et al., 2010; Xiong et al., 2013; Gleason, 2013).
2. **Data pre-processing:** Sets of point measurements (point clouds) are filtered to remove errors and combined (or stitched) into a single surface representation with a specific coordinates system. This process is known as registration (Tang et al., 2010; Xiong et al., 2013).
3. **Modelling the BIM model:** Point cloud data *is transformed into a semantically rich BIM*. This process is commonly known as scan-to-BIM and has traditionally been mostly a manual, labour-intensive process. Recently, however, there have been several reports of automation of this process leading to trials being able to do all three steps within one to two hours for a space that would take up to a day to be modelled manually.[1]

> One of the approaches is having libraries of established taxonomies of common objects to automatically compare against. Commercial registration software packages such as Trimble RealWorks use this approach but have some limitations regarding accuracy and acceptance of modelled objects' metadata (Tang et al., 2010; Xiong et al., 2013; Bosché, 2010; Brilakis et al., 2010; Gleason, 2013).

Although scan-to-BIM can be an expensive option, this initial cost can be later offset by 'speed, quality and quantity of the visual information produced' (Page, 2012). One US-based high-definition surveying firm, for example, states that the cost savings achieved by using laser scanning to create BIM models have been three to ten times the cost of this service (Darling Geomatics, 2015). The economic decision of using laser scanning or other 3D imaging systems versus traditional methods will be based on project requirements, schedule and cost. For example, projects involving simple geometries and readily accessible work-sites may not need 3D imaging systems but projects with safety or time restrictions may be more suitable (GSA, 2009).

The US Office of the Chief Architect (part of the Office of Design and Construction) is encouraging the use of '3D laser scanning technologies on a project-by-project need basis' (GSA, 2015). This office has also partnered with other US agencies and organisations such as the National Institute of Standards and Technology (NIST), National Institute of Building Science (NIBS), the American Society for Testing and Materials (ASTM) and Fiatech. Together, they are using 'funding awarded by the Public Buildings Services Office of the Chief Information Officer's Venture Capital program to develop best practices and standards for 3D laser scanning'. This effort includes pilot projects in Brooklyn, New York; Atlanta, Georgia; and Miami, Florida (GSA, 2015).

A bridge infrastructure project in the US used 3D laser scanners as part of the accelerated bridge construction process that led to minimising traffic disruptions that would otherwise have had an added economic impact of up to US$4 million (Jenkins, 2007). A few years earlier, the US Federal Highway Administration had directed a central artery/tunnel project in Boston to investigate the use of 3D laser scanning. In this project the initial investment in scanning equipment was about US$300,000 but the savings provided by its use were around US$2 million (Jenkins, 2004).

'Estimation of damages caused by a disaster is a major task in the post disaster mitigation process' (Dash et al., 2004). 3D laser scanning has also been used to assess damages after natural disasters. In 2004, this technology was used to shorten the repair schedule of a platform rig on the Gulf of Mexico by 26 days (SPAR Point Group Staff, 2005). This type of technology can help provide near real-time damage models through, for example, the use of airborne laser scanning (ALS) to determine the type and extent of damage and compare to pre-disaster models (Dash et al., 2004). Laser scanning has also been proposed for 3D project progress status tracking and dimensional quality control (Bosché, 2010).

Notes

- Teams planning on implementing scan-to-BIM in their project should plan the effort carefully based on clear objectives and future uses of the data. The choice between commercial registration software and external modelling applications will depend on the scope of the desired outcome (accuracy level and meta-data acceptance) (Gleason, 2013).
- Interferometers are optical instruments that measure 'distances based on the interference phenomena between a reference wave and the reflected pattern' (GSA, 2009).

3D control and planning (digital layout)

3D control and planning, also called digital or field layout, is defined as the process of using a BIM model to create detailed control points for assembly layout or automate control of equipment movement and location (Penn State, 2011). This enabler allows heavy field equipment to adjust automatically to, for example, make cuts and fills in accordance with design plans (LandTech Consultants, 2014). This process has been used to produce lift drawings that are utilised by foremen on-site (Anumba et al., 2010) and aims to decrease site operator errors while improving communication during construction (Marzouk et al., 2010). This in turn can help reduce survey layout cost and construction time (LandTech Consultants, 2014).

A 2009 US survey about benefits from BIM linked to its uses was answered by 175 professionals from across the supply chain and showed that, at the time, only a small percentage of the industry used BIM for this purpose. However, this study's largest respondents were from the architecture and design sector, which would have less need to use this kind of tool (Kreider et al., 2010). It has also been noted that contractors have been slow in adopting 3D control and planning due to this technology normally being used only for the final finish or grade; usually final 5–10 per cent of the project schedule (Jackson, 2014). In Australia and New Zealand, however, model-driven field layout and pre-fabrication have recently been shown to be 'highly rated construction phase BIM activities' (Scott et al., 2014).

> Imagine being able to download the instructions to assemble your bike and then upload them to your own personal robot, which would assemble it for you. That might sound like science fiction, but something similar to that is common practice in the land development industry. Models... are being uploaded to GPS-guided earthmoving machines. These giant 'robots' synchronize GPS-based locations of themselves and their digging implements with the dimensions of Civil 3D model until the real dirt and rock are a match to the model. Without a model, there is no GPS-guided machine control.
>
> (Chappell, 2014)

In order to use BIM for this purpose, the model requires a rather high level of accuracy regarding size and location from the onset, which should be taken into consideration (Marzouk et al., 2010). The higher accuracy required will guide the blades, buckets, wheels and hoes to the precise location planned (Hannon, 2007). This enabler also requires machinery with global positioning system (GPS) capabilities, digital layout equipment and model transition software. The project will need to have staff that can create and manipulate 3D models as well as interpret whether 'the model data is appropriate for layout and equipment control' (Penn State, 2011). It has also been noted that in order to use 3D control most efficiently, 'agencies should reengineer or adapt the design phase involvement of construction contractors and consultants in the creation of digital 3D project plans'. Otherwise, establishing model constructability reviews as a best practice may be a good alternative. Additionally, not having a 3D model developed during the design phase has been identified as a barrier to cost-effective adoption by contractors (Hannon, 2007).

In 2006, Pennsylvania Department of Transport was already developing specifications to use GPS systems in bulldozers and motor graders for automated precision grading. This organisation was using digital information and GPS technology to carry out *stakeless* grading in road construction and stated that one of the benefits they expected to receive from it was faster completion dates (Grading and Excavation Contractor, 2007). The head of the Minnesota department of transport for example stated in 2005 that their goal was to have machine control in 100 per cent of their projects by 2007 (Hannon, 2007).

> GPS machine control combines two fantastic technologies, Satellite Positioning and Digital 3D models, so that your machine operators can see themselves moving on the site plan in real time, with cut/fill information constantly displayed anywhere.
>
> (Topcon Positioning Systems, 2008)

A river shore restoration project in the US in 2014 used 3D control to place approximately 30 boulders across two sections of the river in order to create a tranquil area for trout and salmon conservation purposes. The assistant project manager stated that using traditional methods would have made it impossible to drive stakes in the river in the same way as it was easily achieved using 3D control. This project also used 3D controlled bulldozers to finish the grading with a tolerance of ±0.64cm, which allowed them to complete that part of the project eight days ahead of schedule. The president of the general contractor firm estimated that using this technology avoided 30–35 per cent cost growth and time overruns. Additionally, the project team was able to include topographical details into the model that showed the location of specific plants that required different extraction methods (Winke, 2014).

Industry leaders in this area include organisations such as Trimble, Topcon, Leica and Caterpillar (Hannon, 2007). In 2014, Topcon developed a 3D positioning device specially designed to be used with BIM and digital layouts and cost 50 per cent less than regular automatic machine control systems (Jackson, 2014). In 2015, Autodesk and Leica announced they were partnering to provide a BIM-to-field solution that would allow users to access sensor data directly from the Design Authoring software platform. This means that users can access sensor data and control the robotic stations through a BIM360 app for smart handheld devices (Autodesk, 2015j).

Although the uptake across the industry has been relatively sparse, 3D control and planning has significant potential for future development and greater level of automation. Systems such as that patented by Google in early 2015 showcase this point. The patent proposes a cloud-based method to connect and direct robotic devices that may use GPS and RFID systems as well as optical, infrared and biosensors to complete tasks (GPS World staff, 2015).

Note

- Utah Council of Land Surveyors published a presentation in 2013 about GPS machine control that stated that this technology can help deliver tasks 30–80 per cent faster and with 70 per cent less rework. It also stated that it has helped reduce lost time injury frequency (LTIF) by up to 40 per cent and change orders by approximately 70 per cent (Herring, 2013). The authors contacted the person who created this presentation but were not able to confirm how these figures had been calculated.

Animations and simulations

This enabler refers to the process of using BIM models and associated data to create animations and simulations that provide a virtual environment to review a space. This can be used to explore sections of the design in a more intuitive 3D environment during sign-off processes or during tender presentation and other review processes (Eynon, 2013). Experiential simulation models (ESMs), for example, allow future asset users, researchers, clients, contractors and designers to test room fitting and configuration scenarios. In hospitals, this may include care scenarios and critical events such as emergency codes. This process tends to be carried out only for key spaces and complex construction tasks (Peavey et al., 2012).

Some virtual prototype simulation tools allow directly importing BIM models into the simulation software. Virtual prototype simulations can include avatars to allow users to explore the model virtually and better understand spatial relations (Peavey et al., 2012). In the Perth Children's Hospital in Australia, for example, the design team developed accurate models of the rooms and included avatars that could *walk* through the room. The avatars were adjusted to fit the height and build of some end-users such as doctors and nurses so they could provide feedback on the location of fixtures and equipment. This allowed demonstration of line-of-sight and movement patterns within the model for an area that was disputed based on a misunderstanding of the 2D drawings, which then led to the almost immediate sign-off of individual spaces (Sanchez et al., 2015b).

Specific BIM virtual environments have also been developed for particular objectives such as improving fire emergency evacuation preparedness and training. A group from Cardiff University in the UK, for example, developed a BIM-based virtual environment that used a virtual reality and serious games engine for emergency awareness training and emergency management two-way information updating. This system was a prototype and was only able to work with Revit and specific data formats but it is a good example of what can be achieved. The system, denominated BIM-VE, was able to create dynamic emergency scenarios where the aim was to test the user response under different circumstances. The fire scenarios (fire propagation speed and other characteristics) were also to change when variations to the building materials were introduced in the model (Wang et al., 2014a).

Another prototype system recently reported created a Smart-BIM where objects within the model could directly react to interactions with users based on pre-defined tasks. The idea of this prototype was that eventually the model could also interact with smart technology (Heidari et al., 2014).

Augmented reality (AR)

> Augmented reality (AR) is an environment where virtual elements are embedded in a live picture of real surroundings... AR is an environment where additional data generated by the computer is fed into the user's view of a real scene... In effect, AR brings the virtual computer world to us and relates it in a way, mostly geometrically and visually, to the real world. AR requires a technique to capture the world around us and the mode to experience the computer world (typically by overlaying computer graphics in the camera window). Because the entry requirements are minimal, many of today's smart phones are suitable for AR devices.
>
> (Meža et al., 2014)

AR in the simplest terms is an environment where real-world objects and computer generated relevant information co-exist by one being superimposed on the other through the use of machine interfaces (Chi et al., 2013; Wang, 2009). This means that users can access additional information about current conditions in a mixed reality, ideally, through seamless interaction cycles between the real world and digital information (Jiao et al., 2013; Hakkarainen et al., 2009). This technology has been proposed to be used in the construction industry for equipment operation training (Wang and Dunston, 2007), displaying position and layout of underground infrastructure during maintenance, planning and surveying (Schall et al., 2009), remotely operate equipment (Chi et al., 2012), and assembly task sequencing and coordination (Tang et al., 2003).

AR can aid architecture, engineering, construction and owner-operated (AECO) users 'with their routine tasks because their live view of a space can be supplemented by the information they need, all in one interface' (Williams et al., 2015). When jointly implemented, this technology is thought to enhance benefits achieved from implementing BIM in construction and maintenance projects such as improving visualisation and comprehension (Hakkarainen et al., 2009; Meža et al., 2014). For example, one early prototype was able to help reduce performance time of simple conflict detection by 71 per cent (Wang and Dunston, 2006).

Most AR processes follow two steps (Chi et al., 2013):

1 Tracking and registration: defines where to display the digital content.
2 Determine media content: defines what digital content to display.

Tracking is used to follow the user's orientation and position (Meža et al., 2014). It is usually:

- Sensor-based: based on acoustical, optical, mechanical, inertial, or magnetic sensors; such as GPS, Wi-Fi and Bluetooth.
- Vision-based: for example, feature-based (natural or artificial) and model-based.
- A hybrid: for example, a system that uses natural edges and 3D models coupled with inertial sensors to track location (Williams et al., 2015).

This information has then to be calibrated with the position within the virtual environment and processed to merge digital and real environment data in order to display it (Meža et al., 2014). One issue in BIM-AR is that there needs to be a conversion platform from BIM objects to AR elements (Jiao et al., 2013). Existing commercial AR systems use proprietary models based on custom software packages. This means they are not standardised and may not be compatible with, for example, Industry Foundation Classes (IFC), and may require a great deal of coding to develop a fully functional system (Meža et al., 2014).

AR has been seen as an important area of opportunity for the construction industry since the 1990s (Wang, 2009). In more recent years, 'advances in computer interface design and hardware capability have fostered a number of AR research prototypes or test platforms to be developed for application in construction' (Wang et al., 2013a). There are, however, few mature AR industrial applications in the construction industry, with most existing work being prototypes with limited commercialisation (Jiao et al., 2013). Nevertheless, promises made by companies such as Google with Glass 2

(Allsopp, 2015) and Microsoft HoloLens (Microsoft, 2015) as well as the proliferation of AR mobile applications such as Google Goggles (Google, 2014), Oculus Rift (Oculus VR, 2015), Blippar (2015), Aurasma (2015) and Wikitude (Wikitude GmbH, 2015), to mention just a few, may be an indication of a general move towards AR becoming more commonly used.

At the current rate of technology development, AR technologies may be mature enough to be implemented across a number of industry fields over the next decade. Specifically, advances in location, user interface, cloud computing environments and mobile display devices will help make BIM-AR more accurate, intuitive, information-rich and cost effective (Chi et al., 2013). AR image display, for example, is commonly done through either handheld devices, such as phones and tablets, or wearable devices, such as Google Glass and the iHelmet (Meža et al., 2014; Yeh et al., 2012). The popularisation of smartphones and other mobile devices means that developing integrated mobile augmented reality (MAR) environments could 'provide users in the AECO industry with the opportunity to efficiently access their augmented layers of information through natural interaction with their mobile devices' (Williams et al., 2015).

As mentioned earlier, BIM-AR has been tested in a number of fields within construction. For example, the AR4BC project (Augmented Reality for Building and Construction) carried out by the Technical Research Centre of Finland (VTT) and industrial partners from the Finnish building and construction sector combined AR with BIM 4D modelling functionalities. Expected benefits included improved visualisation, mobile access and feedback to BIM data from the work-site, improved communications and a more versatile use of BIM. This project used an ultra-mobile PC to allow users to compare BIM-based schedules with job-site conditions in real time, as well as attaching visual content to the BIM model from this location (Hakkarainen et al., 2009).

One research group developed a video-based AR platform for multiple-user collaboration that integrated as-built BIM models with business social network services to establish an integrated AR cloud environment. This prototype was tested in Shanghai by a general contractor during a construction project of a 16th-floor for pipes and water supply drainage ducts. The system was used to carry out and document on-site tasks and to access more extensive views from the off-site office that allowed the project manager to confirm acceptance standards had been met (Jiao et al., 2013).

An earlier research project investigated the feasibility of using AR for rapid earthquake-induced building damage evaluation by being able to compare current conditions with the digital model. This research showed that this use was feasible but, at the time, the GPS sensors and wearable display were bulky and had low precision, which hindered its use in practice (Kamat and El-Tawil, 2007). Another research project proposed a BIM-AR system for defect detection, which was argued to be the primary cause of schedule and cost overruns in construction. This system aimed to allow carrying out quality inspections off-site as well as automatically detecting dimension errors and omissions at the job-site (Kwon et al., 2014). A similar lab-base study developed two automatic field inspection BIM-AR systems based on markers and image-matching techniques (Park et al., 2013).

A different BIM-AR system reported in 2015 as BIM2MAR acted as a low-cost '"transparent window" that interactively provides facility managers with the required information for performing their tasks in a single interface'. This system was piloted in a healthcare facility in the US for facility management tasks. The trial concluded that the system was easy to use and intuitive, but there were registration errors and workflow issues due in part to the lack of as-built BIM models (Williams et al., 2015).

Recent work carried out at Curtin University in Australia has also suggested that 'the integration of BIM with AR can provide a platform for a site management team and subcontractors to effectively interact and utilize data contained within a BIM model' (Wang et al., 2013a). This group suggests that BIM-AR can help improve assembly tasks efficiency, comparison of as-planned and as-built as well as on-site design reviews. AR could, for example, enable subcontractors to virtually walk into the models during design reviews enabling better visualisation of constructability and sequencing issues (Wang et al., 2013a, 2014b).

In asset management, a group of German researchers developed a mobile BIM-AR application based on *markerless tracking technology* for building life-cycle management that allows manager to add semantic annotations to the model. Maintenance employees exposed to this platform stated that

> such a system would remarkably simplify their work because they would not need to carry on heavy, printed manuals anymore and that they would not need to tediously search the required information in the paperwork any longer.
>
> (Olbrich et al., 2013)

Note

- A 2003 study where AR was used to instruct field workers in how to carry out certain assembly tasks found that overlaying 3D instructions on the actual work pieces reduced the error rate for an assembly task by 82 per cent, particularly diminishing cumulative errors; errors due to previous assembly mistakes (Tang et al., 2003).

Automated clash detection (spatial and 3D coordination)

Clash detection, also known as spatial coordination and collision control, commonly refers to use of BIM models for the automated detection of clashes that occur between objects in the 3D model (Gijezen et al., 2009; Bernstein et al., 2012b; Marzouk et al., 2010). In 3D coordination, clash detection software is used to detect field conflicts between disciplines during the coordination process. This BIM use can help reduce or eliminate field conflicts and RFIs, increase productivity and speed of construction and improve the construction project coordination (Penn State, 2011).

> A common use is in the identification of clashes in elements under the control of different disciplines, with clashes between ductwork and structure being one of the more common instances.
>
> (Fussell et al., 2009)

Clash detection software combines multiple systems into a single model to automatically search for conflicts. Clash detection and resolution is an iterative process where clashes are first identified by the software and then need to be resolved by the project team by, for example, assigning priorities (Khanzode, 2010). Clash filters are normally established in order to carry out smart automated clash detection. This means differentiating between acceptable and inappropriate clashes based on a selected group of systems to analyse (Gijezen et al., 2009; Fussell et al., 2009).

> Some elements can show as a clash, yet be not only acceptable, but necessary. For example, a power point needs to be located within a wall, but will show as a clash unless the settings in the clash detection software recognise it to be an acceptable clash and ignore it. The same power point sitting in a window needs to be identified and highlighted. Another example of this is a pipe running through a footing.
>
> (Fussell et al., 2009)

Clashes are commonly visualised by colour coding the model view. An Australian hospital construction project established as project objective to find and resolve all major spatial coordination issues before the start of construction. This project reported having reduced the time used for clash detection from an estimated 40 hours with traditional methods to two hours with BIM. In order to expedite the resolution of these clashes, the team developed a hierarchical classification to visualise which systems had priority (Sanchez et al., 2015b). A similar method was used in another Australian hospital through a clash-detection priority matrix and resolution process. This was reported to have helped the team answer RFIs earlier and more efficiently, while also reducing the total number. It should be noted, however, that in this case, the team reported that the time spent coordinating and modelling using the new workflow was probably twice the amount saved in answering RFIs (McGraw Hill Construction, 2014b).

Another case study of the construction of a US$100 million plant facility used clash detection to identify more than 500 significant clashes. These included ten items that were estimated could have costed up to US$10 million in rework had they not been found prior to construction (Bernstein et al.,

2012a). In Australian iron ore port facility construction, clash detection was reported to have been fundamental for the coordination of steel placement and finding the safest and most efficient way to remove temporary steel. Also in Australia, a 28-storey high-rise used clash detection to allow a tighter placement of equipment, resulting in smaller plant rooms and less excavation. This project reported more than 10,000 clashes resolved through clash detection and coordination (McGraw Hill Construction, 2014b).

A report by McGraw Hill on the value of BIM in infrastructure pointed out that automated clash detection, which is seen as a staple in building construction, can produce even larger savings in infrastructure construction. This was explained to be due to the fact that 'conflicts tend to be more expensive in treatment plants and heavy construction than in commercial work'. In the same report, surveyed infrastructure practitioners identified clash detection as one of the top value-adding applications of BIM for contractors and consultants (Bernstein et al., 2012b). Another report by McGraw Hill of a survey carried out in South Korea also identified spatial coordination as one of the top two value-adding uses of BIM (Lee et al., 2012a).

In the design and construction of a mixed-use hotel in the US, clash detection was used to identify 55 spatial conflicts during the design development phase. This resulted in estimated savings of US$124,500 based on appraisals of the cost that would have been incurred due to rework had the collision not been detected earlier. A total of 590 clashes were identified and resolved before the start of construction, resulting in an estimated net adjusted cost saving of US$200,392. This calculation is based on the assumption that 75 per cent of clashes would have been identified using traditional methods (Azhar, 2011).

Notes

- Mechanical, electrical and plumbing (MEP) systems can account for up to 40–60 per cent of the total value of complex projects (Khanzode, 2010).
- One of the lessons learned in an Australian hospital designed and built using BIM was the importance of maintaining the same naming conventions across all consultants in order to ensure effective clash detection. To make sure this was happening, the team implemented simple audits selecting all search sets by changing their colour set, quickly identifying those that were not being recognised due to incorrect naming (Sanchez et al., 2015b). In a different medical facility built in the US, the team also highlighted the fact that it is paramount that all teams use the same reference point in order to correctly integrate all the models (Staub-French and Khanzode, 2007).
- 3D coordination during design and construction for interference management is one of four BIM uses mandated by the US Army Corps of Engineers (2012).

- Commercial software packages such as NavisWorks from Autodesk (2015j) and ProjectWise from Bentley provide clash detection functionalities (Bentley, 2015b).

Automated rule-checking

Automated rule-checking refers to a software application 'that does not modify a building design, but rather assesses a design on the basis of the configuration of objects, their relations or attributes' (Eastman et al., 2009). Rule-checking has been acknowledged as an area of opportunity well before BIM became common, with early research into this topic dating back to the 1960s (Solihin and Eastman, 2015). However, until recent years, the only means to deal with complex topics such as building codes and safety was human cognition and review processes (Eastman et al., 2009). In addition to codes and standards, designers normally use previous experiences and perform analogies (Nawari, 2012). This manual process to ensure compliance is 'complex and prone to human error with significant cost implications' (Greenwood et al., 2010). The use of BIM has opened the door for automated checking of designs and code plans, which are often costly bottlenecks in the project delivery process, potentially providing significant time and cost savings (Eastman et al., 2009).

Automated rule-checking relies upon an initial interpretation and logical structuring of rules for the software. It may also be necessary for the objects within the BIM model(s) to have data structured in a certain way. The software then applies the rules to the model and creates reports that, similar to the process of clash detection, can be prioritised and assigned to the relevant stakeholders. Procedures can then be set up to automatically correct failed rules (Eastman et al., 2009; Solihin and Eastman, 2015).

This enabler has been used for an extensive range of applications such as:

- well-formedness of a building: syntactic aspects required for model views;
- building regulatory code; compliance with well-defined building regulation and code;
- specific client requirements;
- constructability and other contractor requirements;
- safety and other rules with possible programmed corrective action: potential hazards during construction and operations;
- warranty approvals; post-construction model check against warranty issues that may generate cost – this is normally accompanied with site inspections and acts as guide for potential issues; and
- model completeness for handover (Solihin and Eastman, 2015).

Challenges for the successful application of automated rule-checking are mostly related to the complexities inherent to the rules and the conditions to which they apply (Solihin and Eastman, 2015). Another challenge is that standard model-checking software may not populate the objects with the specific information required to check against local or project specific rules (Greenwood et al., 2010). This means that the strategy to ensure all required information is included in the model needs to be addressed in early project phases to avoid downstream rework. Depending on the intended use of automated rule-checking, this can be done by explicitly requiring parties to provide specific information as part of the model, deriving new data from the model through computer processing, and/or application of analysis models to derived and input data (Eastman et al., 2009).

Automated rule-checking was, for example, used in a recent project to identify fall hazard and prevention in construction safety planning. The prototype algorithm was applied to an office and residential building project and a multi-storey precast concrete apartment building in Finland. In the first project the team compared manual versus automated fall modelling and found the following benefits: (1) significantly less time requirement (only seconds or minutes with automated rule-checking); (2) lower level of safety expertise required for the modeller; and (3) less effort required to obtain new safety reports after design and schedule changes (Zhang et al., 2015).

Code validation

Building codes aim 'to organize, classify, label, and define the rules, events, and patterns of the built environment to achieve safety, efficiency and economy' (Nawari, 2012). Pennsylvania State University defines code validation 'as a process in which code validation software is utilized to check the model parameters against project specific codes' (Penn State, 2011). This is a special type of automated rule-checking which has been extensively researched over the past 40 years (Dimyadi and Amor, 2013). Some groups have, for example, worked in developing country code-specific software and IFC extensions. A group at Northumbria University proposed a domain extension within the IFC model for England and Wales Building Regulations (Greenwood et al., 2010). This work has been later extended by NBS in partnership with Solibri and Butler & Young 'to demonstrate how currently available tools can be used to automate checking of Building Regulations Approved Documents compliance in a Building Information Model' (NBS, 2014).

In Australia, the Cooperative Research Centre (CRC) for Construction Innovation developed DesignCheck; 'an automated rule checking engine capable of checking building designs for compliance against the BCA [Building Code of Australia] ' (CRC for Construction Innovation, 2006). In the US, the SmartCodes project focused on transforming paper-based codes into machine-interpretable rules (Greenwood et al., 2010). An automated rule-checking system for design building requirements on occupant circulation was also developed in a project sponsored by the US GSA based on the *US Courts Design Guide* and reported in 2010. This tool allowed different rules to be applied to different design stages (Lee, 2010). The International Code Council (ICC) founded in 1994 used to publish building fire and safety codes through documents such as the International Building Code (Nguyen and Kim, 2011). The ICC started producing SMARTCodes in 2006 that provided official representations of a few standards that could be used by authoring tools to manage amendments to the code. This effort was, however, discontinued in 2010 due to lack of funding (Dimyadi and Amor, 2013).

There are also a number of commercial model-checking software that include building code validation, such as Solibri Model Checker, Jotne Express Data Manager suit (EDM, formerly EDModelChecker) and Fornax Plan Checking. Solibri reads/writes IFC and ifcXML, and provides built-in functions such as fire code-checking and a rule library. EDM uses *express language*, which is the same language used for IFC, to support open development by users. Fornax is a 'C++ object library that derives new data and generates extended views of IFC data' (Nawari, 2012; Jotne IT, 2015; ZSL, 2012; Solibri, 2015).

Note

- It is recommended to derive any complex properties required for rule-checking directly from the model whenever possible. This can be done directly through the rule-checking software or the design authoring software to avoid introducing human errors (Eastman et al., 2009).

Asset knowledge management

> Managing knowledge can be a challenge, especially in the [architecture, engineering and construction] AEC industry where short-term working contracts and temporary coalitions of individuals can inhibit knowledge sharing... To make use of tacit knowledge for competitive advantage, it needs to be articulated and utilized by companies and their partners... thus while knowledge management (KM) is fairly new to the AEC industry, it is emerging as a significant concept in management science.
>
> (Becerik and Pollalis, 2006)

Knowledge management systems are IT-based systems 'developed to support and enhance the organizational processes of knowledge sharing, transfer, retrieval, and creation' (Vorakulpipat and Rezgui, 2008). Although to date few building (or asset) knowledge management tools have been developed and even fewer commercialised, 'one of the areas growing in importance is the ability for a team to collect and aggregate best practices in a central location that is accessible' (Hardin and McCool, 2015). The ability to add semantic information to model objects in a single source and easily share it with other parties within and across organisations can greatly facilitate asset knowledge management (Kivits and Furneaux, 2013). A BIM model can be used on its own to retain knowledge in a digital format associated to specific objects and facilitate easy updating and transference of this knowledge in a 3D environment (Lin, 2014).

The National Taipei University of Technology in Taiwan developed a BIM-based knowledge management system for contractors. This system was then trialled by a local general contractor and monitored for six months during a construction project on a 12-floor commercial building. The total cost of implementing the system, including hardware and BIM software, was TW$120,000 (US$3,882). The system was then evaluated through a questionnaire completed by 13 users. The main benefits identified by the users were: (1) automatic correction and notices when knowledge was updated; (2) easy and effective integration between 3D CAD-based knowledge and knowledge management tools; and (3) being able to clearly identify available knowledge and experience when requested for a project (Lin, 2014).

This case study also pointed to general advantages of model-based asset knowledge management tools such as: (1) generates visual 3D illustrations with knowledge; (2) it allows viewing knowledge and information easily and effectively; and (3) enables engineers to trace and manage acquired CAD-based knowledge. The surveyed staff also found barriers to wider adoption: (1) if BIM is not used for all projects, then this knowledge management approach is difficult to maintain across projects; (2) high hardware requirements; (3) time requirements for staff to use BIM software to edit and update the system; (4) internet speed limitations; and (5) unwillingness of participants to share knowledge and experience (Lin, 2014).

K-BIM was a framework developed in 2014 for the same purpose but applied to asset and facilities management, integrating computer-aided facility management (CAFM), BIM and knowledge management systems (Charlesraj, 2014). Another asset knowledge management system developed in the UK also focused on the operation and maintenance phase and aimed to facilitate 'full retrieval of information and knowledge for maintenance work' (Motawa and Almarshad, 2013). This system had a case-based reasoning approach that classified knowledge according to maintenance issues associated with individual objects. However, as with K-BIM, it is unclear the extent to which these systems have been used in practice.

Note

- K-BIM should not be confused with the kBIM Template and Library developed by Karpinski Engineering (ASHRAE, 2015).

Asset performance assessment, modelling and displays

This enabler refers to the use of a BIM model to assess the future performance of a future or existing built asset. Maintainability analyses, for example, focus on desired performance and can be carried out at various stages of an asset's life-cycle. This can address issues such as accessibility to objects for maintenance, sustainability of materials (avoiding materials that will cause defects or lower performance), and preventive maintenance (Becerik-Gerber et al., 2012). All performance assessments and models are based on systematic procedures and verifiable data. Environmental design tools used in Australia, for example, can model performance indicators for various impacts, such as relative embodied energy and CO_2 emissions. The results are commonly provided in relative scores to a standard or reference asset of the same type. 'The greater the score relative to the benchmark, the better the design decision is for the environment' (Downton, 2011).

Performance analysis tools can be classified in two groups: physical calculation models (use precise calculation of detailed tasks) and statistics calculation models (simplified models for the estimation of total demand) (Schlueter and Thesseling, 2009).

There have also been BIM-based frameworks developed to model and monitor performance during operations. In 2014, for example, such a system was developed and applied to a network of subway stations to monitor indoor environmental quality through a wireless sensor network linked to the record model (Marzouk and Abdelaty, 2014). In 2013, an Irish research group also developed a BIM-based integrated toolkit 'to assist energy managers at different stages of their activity relating to systematic energy management in buildings' (Costa et al., 2013). This system, however, still presented some interoperability issues that needed to be addressed before it could be linked to the building automation systems (BAS).

Although not without challenges, the integration of BIM with real-time power monitoring systems can help reduce energy consumption by creating detailed energy-consumption databases and load profiles, suitable for effective demand-side management. This information can be made available to users through display devices with a more intuitive 3D visualisation interface helping tenants to control their own energy use behaviour. It can also be integrated with smart grid systems to optimise energy management (Seeam et al., 2013).

Engineering analysis

Engineering analysis is a term that encompasses other enablers profiled in this dictionary. It is the process where 'intelligent modeling software uses the BIM model to determine the most effective engineering method based on design specifications' (Penn State, 2011). Examples of engineering analyses are structural, energy, disaster prevention, emergency evacuation planning during design phase, construction planning and scheduling, and project control and safety (Wang et al., 2014b).

> Varying degrees of human effort are needed to adapt the exported data to the forms required by the analysis tools, and different degrees of rework are required to change the analysis models whenever the building model is changed. Nevertheless, the procedures are more productive, less error prone, and quicker than compilation of the analysis models from scratch.
>
> (Sacks et al., 2010)

Most BIM software platforms can export relevant pre-processed data into engineering analysis tools; some of these software solutions additionally include engineering analysis functionalities such as finite-element and energy analyses (Sacks et al., 2010). Energy modelling tools, for example, commonly involved the use of gbXML files, which can be exported from BIM design authoring software and have the ability to carry asset environmental sensing information (Bahar et al., 2013). This process, however, can sometimes be difficult due to the different modelling strategies the analysis software may require.

In structural engineering analysis, on the other hand, there are a number of software applications that have been specifically developed to interact with BIM design authoring tools. Bentley, for example, offers a suite of interoperable software solutions, including AutoPIPE (pipe stress analysis), SACS (finite-element analysis for off-shore structures), and the Integrated Structural Modeling suite (Bentley, 2015e, 2015f, 2015g). Autodesk has Robot, a BIM-based structural analysis that includes aspects such as wind and structural loads (Autodesk, 2015a).

Environmental performance and sustainability evaluation

> The use of BIM to support sustainability goals is increasingly valuable. Tools for analysis and simulation are helping design professionals to generate higher performing design solutions, and contractors can leverage models in a variety of ways to improve the quality and reduce the environmental impact of their work. In addition, an emerging area of BIM activity relates to owners using models to improve building performance by optimizing facilities management.
>
> (McGraw Hill Construction, 2014a)

This refers to the process of assessing and evaluating the environmental performance of an asset through its BIM model based on sustainability criteria (Azhar et al., 2009; Penn State, 2011). Environmental performance assessment and modelling are routinely carried out in most building, and often in infrastructure, construction projects during the design phase. This process aims to understand what the resource use is expected to be for the constructed asset during operations (Becerik-Gerber et al., 2012). This can be done, for example, through sustainability ratings such as the US Leadership in Energy and Environmental Design (LEED), the UK Building Research Establishment Environmental Assessment Methodology (BREEAM) or Australia's Green Star and Infrastructure Sustainability rating analysis.

This evaluation can be done at any life-cycle phase of a built asset, but it is often most effective when used during planning and design stages (Penn State, 2011). This enabler is also recognised as *Green BIM* and in 2010, the top sustainability criterion used by the industry included energy performance, lighting analysis, HVAC design and green certifications (Bernstein et al., 2010). Commercial BIM model authoring software providers such as Autodesk have also highlighted this as a prominent BIM use that provides substantial benefits (Malkin, 2010).

The New York City BIM guidelines state the value of BIM sustainability evaluations is:

- accelerated design review and rating certification process through the efficient use of a single database where all sustainability features are archived;
- improved communication between stakeholders in order to achieve sustainability rating credits and therefore reduce rework for redesign;
- aligned scheduling and material quantities tracking for more efficient material use and better cash flow analysis; and
- optimised asset performance by tracking energy use, indoor air quality and space planning are in accordance to rating standards (Bloomberg et al., 2012).

A UK-based group at Northumbria University highlighted that for this enabler to be used most effectively, asset and operations managers need to be more involved during the design phases and develop a framework to support this interaction. They also highlighted the importance of integrating COBie (or the exchange format in use) with the sustainability evaluation data requirements, which could also be tied to BAS, computerised maintenance management system (CMMS) and CAFM system requirements (Alwan and Gledson, 2015).

Three commonly used BIM-based environmental analyses software are based on physical calculation models: Ecotect and Green Building Studio from Autodesk and Integrated Environmental Solutions (IES) Virtual Environment (VE). Ecotect, allows designers to work in a 3D environment while covering a wide range of simulation and analysis functions (Wong and Fan, 2013). In March 2015, however, Autodesk took Ecotect out of the market and integrated its functionalities directly into their design authoring product family (Autodesk, 2015b).

Energy simulation tools

Energy simulation software is used to predict the energy performance of assets, inform design development and define equipment size. These software packages provide insight into which aspects of the 'design, materials and features contribute to or undermine energy efficiency and comfort, so that improvements can be made' (Downton, 2011). Building energy simulation tools can use 'a properly adjusted BIM model to conduct energy assessments for the current building design' (Penn State, 2011). The goal of using such tools is to inspect the asset's 'energy standard compatibility and seek opportunities to optimize proposed design to reduce structure's life-cycle costs' (Penn State, 2011).

> A quick energy analysis by using a simple BIM model during early design stage could help select best building orientations and configurations to improve building load and energy consumption profiles. Detailed BIM energy analysis is typically done in late design phase by using more powerful energy simulation tools, most of which are currently capable of behaving an hourly building load, system and plant energy simulation with economic analysis based on building location and local utility rates, and supporting BIM model files as inputs.
>
> (Penn State, 2011)

One of the main advantages of integrating BIM with energy simulation platforms is that the time and cost associated with geometry modelling for energy simulations can be significantly reduced by using BIM-based tools and thus reducing rework (Ahn et al., 2014).

Sustainability ratings

Sustainability rating systems provide guidance to improve the sustainability of the final asset throughout planning, design, construction and operations. LEED® is one of the methods most commonly used in the US to rate the environmental performance of buildings. It was originally developed by the US Green Building Council in 1998 'to provide building owners and operators a concise framework for identifying and implementing practical and measurable green building design, construction, operation and maintenance solutions' (Azhar et al., 2011). The Building Environmental Assessment Method (BEAM Plus) is one of the ratings most used in Hong Kong and is a similar green building rating system based on credits (Wong and Kuan, 2014). The LEED v4 assigns credits across six categories: location and transportation, sustainable sites, water efficiency, energy and atmosphere, materials and resources, and indoor environmental quality (USGBC, 2014). BEAM assigns credits across five categories: sites aspect, water use, energy use, materials aspect, indoor environmental quality and innovation and additions. A study published in 2014 found that there were at least 26 BEAM Plus credits that can be attained with the support of BIM-based documentation and 13 LEED credits plus one prerequisite that could be directly calculated and documented using design authoring tools such as Revit (Wong and Kuan, 2014). A prior study had also suggested that all documentation required for LEED can be 'directly or indirectly prepared using the results of BIM-based sustainability analyses software' and this could help streamline the certification process (Azhar et al., 2011).

Over the past 15 years, there have been a number of case studies about BIM implementation that have used sustainability ratings certifications to evaluate the environmental performance of the built asset. In the design and construction of the Living/Learning Center (2001), the project achieved a LEED silver certification with a university standard construction budget. In this project, the use of a BIM model

> allowed concepts for daylighting the 'underground' building to be studied quickly, and the final concept proved to be the most effective and least expensive. The savings are estimated to be approximately $900,000 over the next ten years at current electricity rates.
>
> (Gleeson, 2005)

BREEAM has a special in-use assessment scheme that uses initial pre-assessment questionnaires and later questionnaires, which are then verified by an independent assessor. This rating assesses the overall performance of the asset against BREEAM in-use criteria and classifies the asset's performance as: unclassified, acceptable, pass, good, very good, excellent and outstanding (Alwan and Gledson, 2015).

Building automation systems

> Building Automation Systems (BAS) are centralized, interlinked, networks of hardware and software, which monitor and control the environment in commercial, industrial, and institutional facilities. While managing various building systems, the automation system ensures the operational performance of the facility as well as the comfort and safety of building occupants.
>
> (KMC Controls, 2015)

Data monitored by these systems can be displayed for users in order to promote awareness and behaviour changes that improve asset environmental performance (Staub, 2014). BIM models can be used as a virtual framework to display this performance data in real-time (Teichholz, 2013).

Notes

- It has been noted that although interoperability between design authoring software and engineering analysis platforms is still an area of improvement for some tools, a major requirement to allow automated transaction between them is the information included. This 'must be evaluated through a series of interoperability tests to ensure consistency between all participants using AEC software' (Bahar et al., 2013).
- In the US in 2013, 72 per cent of electricity consumption, 39 per cent of energy use and 38 per cent of all carbon dioxide were produced/consumed by buildings (Wang et al., 2013b); up from 70 per cent of electricity and 30 per cent of greenhouse gas emissions in 2008 (Azhar et al., 2009).
- Before a BIM model is used for energy analysis, the responsible party (mechanical engineers or energy analysts) should review the model and make proper adjustments if the BIM model is not ready for simulation. The reviewing work includes checking model integrity and ensuring all parameters needed are not missed (Penn State, 2011).
- Levy (2012) provides a good overview of BIM-based environmental sustainability analysis for small and medium scale projects. Although a few years older, Krygiel and Nies (2008) also provide a broader insight into environmental performance assessment and modelling with BIM.
- It has been noted that as of April 2015 there was 'a lack of [an] all-inclusive green BIM tool that provides a "cradle to grave" management of a building's environmental sustainability, including the building materials, products and energy required over the building's full life cycle' (Wong and Zhou, 2015).
- The US Army Corps of Engineers has set the objective to 'conduct analysis to meet LEED and/or appropriate energy optimization' as part of their BIM roadmap to be used in all applicable projects by 2016. This aims to contribute to the achievement of the roadmap's second goal: 'Integrate: establish policies and procedures for measuring process improvement' (US Army Corps of Engineers, 2012).

Asset (preventive) maintenance scheduling

Asset maintenance scheduling is the process by which the functionality of the asset structure and equipment serving the asset are maintained over its operational life in the model. This process requires access to design authoring software, BAS or CMMS linked to the model and/or user-friendly interfaces to 'provide performance and/or other information to educate users' (Penn State, 2011).

Preventive maintenance is based on establishing maintenance requirements that provide a foundation 'for planning, scheduling, and executing scheduled maintenance, planned versus corrective for the purpose of improving equipment life and to avoid any unplanned maintenance activity/ minimize equipment breakdowns' (Hunt, 2013). This can be done within a BIM environment in new or existing buildings, although the latter may be more challenging. This process can be especially effective for environmentally sensitive indoor environments such as hospitals (BIMEX, 2015).

Data such as equipment tag information, procedures, replacement parts, special tools, lubrication requirements, service providers and warranty information can be initially added to the BIM model and then transferred directly into the CMMS or CAFM software platforms (Hunt, 2013). Ideally, the BAS can be linked to the record model to provide location-based maintenance scheduling. These systems control and monitor the use of MEP equipment during operations and may be integrated with system analysis platforms that include energy and lighting to monitor asset performance (Hergunsel, 2011).

Unfortunately, although Becerik-Gerber et al. (2012) provide some insight about the data requirements to integrate these systems, not many case studies of successful implementation of BIM for this purpose have been documented. In infrastructure, New South Wales Roads and Maritimes Services in Australia is currently developing BIM capabilities for asset management and maintenance for operational (reactive) maintenance as well as planned (preventive) maintenance. This system is being developed based on a trial on the Pyrmont Bridge, which is a swing bridge located across Cockle Bay in Darling Harbour, Sydney (Sahlman, 2015). Some commercial software packages, however, include functionalities that relate to this enabler such as Ecodomus FM (Ecodomus, 2015b) and Zuuse (2015).

Notes

- 48 per cent of preventive maintenance data can be available at design completion (Hunt, 2013).
- 25–40 per cent of the energy consumed by HVAC systems can be saved by mitigating component faults and improving control strategies (Liu et al., 2012).

BIM-based asset management

> When one considers the extensive documentation of information needed for effective maintenance and operation of most facilities, it is clear that finding efficient ways to collect, access and update this information is very important… an inordinate amount of time is spent locating and verifying specific facility and project information from previous activities.
>
> (Teichholz, 2013)

BIM-based asset (and facilities) management is 'a process in which an organized management system is bi-directionally linked to a record model to efficiently aid in the maintenance and operation' of an asset[2] (Penn State, 2011). To do this, an interface may be used to allow asset management systems to be populated by or directly utilise the data contained in the BIM model. This application can be used, for example, to determine the cost of potential changes to the asset, segregate cost of assets for tax purposes or maintain a current asset management database (Penn State, 2011). The interface also allows users to visualise information related to the asset in a more intuitive manner than referring to 2D CAD-based information (Su et al., 2011).

Although this is a rapidly advancing field through BIM for facilities management (FM) in buildings, to date there are few well-documented BIM for asset management (AM) or FM case studies. This may be due to challenges faced by this sector to implement BIM. These include skills requirements for staff to update and maintain the model, asset managers normally having limited input into early life-cycle stages regarding AM systems requirements, and interoperability issues between BIM platforms and AM systems (Kassem et al., 2015). To deal with some of these challenges, some owners have started to require information from new assets in formats that allow them to seamlessly integrate it into their asset management systems (Hardin and McCool, 2015). The US Army Corps of Engineers developed the Construction Operations Building Information Exchange (COBie) as a way of dealing with this issue (East, 2015). COBie is 'a system for capturing information during the design and construction of projects that can be used for Facilities Management purposes including operation and maintenance' (Williams et al., 2014). Making sure that the information contained in the model can be integrated with the asset management systems reduces the risk of falling out of warranty and the amount of man-hours required to input this information (Hardin and McCool, 2015).

Managers of existing assets also face challenges related to the costs associated with the creation of the model. Advances in technologies such as laser scanning are expected to help reduce these costs and given the longevity of certain assets, the investment may be well justified by the potential gains. Some areas that may help build this case are applications on management of maintenance work orders, emergency service requests, space planning and management, inventory and inspections, move and real estate portfolios (Azhar et al., 2012). The University of Southern Carolina's School of Cinematic Arts, for example, found that using BIM for the management of their new complex containing six buildings allowed its staff to have an immediate high-level understanding of MEP and fire safety systems, with their interrelations and connections. This group also developed a portal interface to access the model and asset management information (Teichholz, 2013).

Northumbria University's campus, based in Newcastle upon Tyne in the UK, has been used to investigate the possibilities of using BIM to manage 32 non-residential buildings. These facilities have a gross area of more than

120,000m^2. The first steps towards developing the record model were taken in 2010 when five developers were commissioned to produce 32 separate models with a focus on space management. These were completed in five weeks at a cost of approximately £0.33/m^2 (US$0.52/m^2) (OpenBIM, 2012). Before implementing BIM, the asset management staff had to update their drawings and information in two separate environments. The resulting efficiency gains associated with avoiding this rework were estimated to allow the university to reduce the need for a full time CAD technician and provide further cumulative savings (Kassem et al., 2015).

> Additional information relating to statutory compliance such as integrated asbestos register, emergency equipment, escape routes, accessibility and essential maintenance, can be easily traced, updated and reported in schedules.
>
> (Kassem et al., 2015)

Another case study in the UK, based on the MediaCityUK project, was used to explore the direct use of BIM to support asset and facility management tasks. This case showed that the visualisation capabilities provided by BIM through, for example, virtual walk-throughs assisted staff during relocation. Cost and time target setting was also facilitated by automated quantification and scheduling capabilities (Arayici et al., 2012).

Computerised maintenance management systems (CMMS), computer-aided asset management systems (CAAMS) and computer-aided facility management (CAFM)

CCMS, CAAMS and CAFM systems are tools used for managing assets during the operations and maintenance phase that have existed for decades (Motamedi et al., 2014; Jones and Collis, 1996). They automate and help organise most of the logistic functions performed by maintenance and management staff. These systems help eliminate paperwork and manual activity tracking. As with BIM, the benefits achieved and functionalities of these systems depend on the ability to collect and store accurate information that can be easily retrieved (Sullivan et al., 2010; Elmualim and Pelumi-Johnson, 2009).

Owners who have implemented these tools need to transfer all the equipment and asset management relevant information from the construction phase into digital files that can be used by the system. This is commonly done manually by asset or facility management staff (Teichholz, 2013). A major concern for owners and operation managers that are to use BIM-based asset management is the interoperability between the BIM platform and the CCMS/CAFM systems. This 'can be exacerbated by the huge difference amongst the lifecycle of updates of BIM technologies, FM technologies and buildings' (Kassem et al., 2015).

CCMS/CAFM systems can also be integrated directly with the BIM model through BIM application programming interfaces (API). There are also some facility management platforms that have been developed specifically for this purpose. Ecodomus FM, for example, is a software platform that is advertised to provide

> real-time integration of BIM with Building Automation Systems, like Honeywell, Johnson Controls, or others, with CMMS/CAFM/IWMS software like IBM Maximo, ARCHIBUS, AssetWorks AiM, Accruent FAMIS, TMA Systems' webTMA, Corrigo, or others, and GIS from ESRI (ArcGIS) and Google (Google Earth).
>
> (Ecodomus, 2015b)

This software utilises a 3D interface to allow 'interactive virtual preventive maintenance inspections, and quicker response to service requests' (Ecodomus, 2015b). If the BIM asset management requirements have been thought through and the correct file exchange format has been implemented in earlier phases of the life-cycle, information can be directly extracted from the model into the CMMS/CAFM platform and vice versa (Teichholz, 2013). Within the industrial management context, there are great potential gains. 'Integrated maintenance management when properly implemented can lessen emergencies by 75%, cut purchasing by 25%, increase warehouse accuracy by 95% and improve preventative maintenance by 200%' (O'Donoghue and Prendergast, 2004).

Research prototypes include one developed in 2014 by a Canadian research group. The proposed system directly integrates the CMMS and BIM systems to help technicians identify possible causes of 'too hot' incidents at the Genomics Research Center of Concordia University. The solution designed was able to 'create custom visualizations through an interactive user interface... for field experts to find root causes of failures' in the building (Motamedi et al., 2014).

Notes

- In buildings, the average life-cycle cost has been estimated to be five to seven times higher than the initial investment and three times the cost of construction (Kassem et al., 2015).
- 'Organisations wishing to implement BIM for FM [and asset management] in the immediate term should take a long-term view (e.g. minimum five years) and be willing to work with different standards and information formats' (Kassem et al., 2015).
- 'Around 80 per cent of an asset's cost is incurred during its operational phase' (buildingSMART, 2010).

- 'The number of owners requiring BIM on over 60 per cent of their projects surged from 18 per cent in 2009 to 44 per cent in 2014' (Hardin and McCool, 2015).
- Commercial software for BIM asset management include: Bentley's *BIM for Operational Management of Space and Assets* (Bentley, 2015c), Ecodomus FM (Ecodomus, 2015b), Onuma Systems (Onuma, 2011) and Archifm.net (Archifm, 2015).
- The US Army Corps of Engineers' BIM roadmap includes COBie requirements in all projects by the end of 2017 (US Army Corps of Engineers, 2012). The UK's PAS1192 also requires the use of COBie (BIM Task Group, 2014). It should also be noted that the UK has developed its own COBie standard called COBie-UK-2012 (BIM Task Group, 2012).
- As of March 2014, appropriate data for metrics such as resource use (e.g., energy), waste production, and predicted maintenance were not often adequately captured for the purpose of effective asset management and CMMS/CAFM systems. They are also not commonly considered within BIM tools and processes (Alwan and Gledson, 2015).

Constructability analysis

> Constructability is the effective and timely integration of construction knowledge into the conceptual planning, design, construction and field operations of a project to achieve the overall project objectives in the best possible time and accuracy, at the most cost-effective levels.
>
> (Kang et al., 2013)

Constructability analysis aims to review the construction logic from beginning to end during pre-construction phases in order to identify constraints and potential issues (Hardin and McCool, 2015). Many construction companies are now conducting constructability reviews at different design stages to improve project outcomes (Jiang and Leicht, 2015). In traditional constructability reviews, construction experts go through the design and use previous experience and knowledge to detect potential constructability issues. These are commonly classified according to the following categories: 'design inadequacies, design omissions, design ambiguities, design coordination, unforeseen conditions, resource constraints, and construction performance' (Jiang and Leicht, 2015).

In a BIM context, constructability analysis refers to the use of BIM models to 'perform detailed constructability analysis to plan sequence of operations at the jobsite' (Azhar et al., 2012). 'The use of BIM during constructability for spatial coordination was a catalyst for the rapid adoption of BIM in the construction industry' (Hardin and McCool, 2015). This is because the enhanced 3D visualisation provides a more intuitive understanding of the design and site layout (Lin, 2014).

Although this is already an improvement with respect to constructability analysis based on 2D drawings, it is still a time-consuming task that relies on experts who may or may not find the issues (Jongeling et al., 2008). More recently, applications such as clash detection and automated rule-checking have been developed to automate this process (Hardin and McCool, 2015; Jiang and Leicht, 2015).

> It is not uncommon to see clash detection reviews now include installation and maintenance, clearance objects or 'blobs', in front of equipment. These blobs allow team members to understand that even though there is nothing physically in the space, the space is actually needed for accessibility and maintenance purposes. For example, a mechanical subcontractor may place a 12" clearance blob in front of a piece of equipment to ensure that the filters in a VAV box can be replaced.
>
> (Hardin and McCool, 2015)

A recent case study reported the use of automated rule-checking for the constructability analysis of a new educational facility in Baltimore, US. This 12,500m^2, four-storey building was built using cast-in-place concrete superstructures and braced structural steel frames. The team used Solibri Model Checker as the software platform to apply a set of constructability rules to identify the most suitable horizontal formwork for the floor system. The team concluded that the method is consistent with practical project decisions and saves time and effort normally dedicated to manual checking (Jiang and Leicht, 2015).

BIM can also be used to test different construction methodologies within a virtual environment as part of this process. In the Perth Children's Hospital in Australia, the BIM model was used to redesign structural steel elements to ensure constructability of the risers, and to review and test geometrically different construction methodologies. This led to the decision of constructing the mechanical riser in pre-fabricated sections within steel cages that could be delivered to site and lowered into position. This change also produced time savings and improved safety (Sanchez et al., 2015b).

Construction and phase planning (4D modelling)

Construction planning is a process in which construction professionals break the planned structure into smaller packages, build a logical network between them and estimate the time required to complete each package, while identifying the critical path (Kang et al., 2007). Phase planning is the process by which a 4D model is used 'to effectively plan the phased occupancy in a renovation, retrofit, addition, or to show the construction sequence and space requirements' on a construction site (Penn State, 2011).

4D modelling refers to a process in which 3D model objects are linked to time or activity-based scheduling data (Jongeling et al., 2008). This can be used for construction scheduling analysis and management, as well as to create animations of construction processes (Williams et al., 2014). 'Visualization of construction plans allows the project team to be more creative in providing and testing solutions by means of viewing the simulated time-lapse representation of corresponding construction sequences' (Huang et al., 2007). This removes model abstraction and enables project teams to understand schedules quickly and identify potential issues based on this understanding (Yeh et al., 2012).

In an experiment reported in 2007, 84 project management students were asked to carry out construction planning tasks in 2D and 4D, co-located as well as web-based. The results showed that teams using 4D representations 'detected more logical errors, more accurately, and faster, with less need for intra team communication than teams using 2D computer graphics'. It also showed that 4D web-based construction visualisation can enhance team performance and may also improve team collaboration in construction planning and scheduling (Kang et al., 2007).

Multiple 4D models can be produced to simulate different production options and allow project stakeholders to compare construction alternatives (Jongeling et al., 2008). In a case study based on the construction of the Australian Queensland State Archives Extension Project in 2005, 4D modelling was seen as one of the key uses of BIM, delivering benefits such as improved communications and better construction planning. The model was able to represent 5,000 lines of tasks in the Gantt chart as well as their progress status using colours and other visual attributes. Construction planning is closely related to constructability analysis, where the former is established first so other planning and cost factors can be added in order to enable cost claims analysis based on site progress information (Fussell et al., 2009).

4D modelling can also be combined with location-based scheduling to further improve construction planning. A case study based on a construction project of a cultural centre in Sweden showed that this combination has the potential to 'significantly contribute to the reduction of waste in the construction process' and facilitates work-flow planning (Jongeling and Olofsson, 2007).

A larger study carried out in 2008 analysed 32 construction projects completed in 1997–2008 in North America, Europe and Asia. This group found that the use of 4D modelling early in the design phase enabled both immediate and late benefits. These ranged from improved communication to earlier detection and resolution of potential site logistics and accessibility constraints, as well as interference between trades. The benefits were then maximised when all the key stakeholders were involved in creating and using 4D models, and the models were created just-in-time and at the appropriate level of detail (Gao and Fischer, 2008).

Site utilisation planning and analysis

> Because the 3D model components can be directly linked to the schedule, site management functions such as visualized planning, short-term re-planning, and resource analysis can be analyzed over different spatial and temporal data.
>
> (Penn State, 2011)

Site utilisation planning is the process of using a BIM model to graphically represent on-site permanent and temporary structures throughout different construction phases. The model may also be linked to 'the construction activity schedule to convey space and sequencing requirements' (Penn State, 2011). Site analysis links the model to GIS data in order to determine whether potential sites meet certain criteria based on project requirements as well as technical and financial factors (Azhar et al., 2012).

Site utilisation planning and analysis enables early detection of logistic and site accessibility issues (Gao and Fischer, 2008). To do this, the project team can include additional information such as labour resources, materials with associated deliveries and equipment location. They can also produce site usage digital layouts more efficiently, which include assembly areas, temporary structures and material delivery to identify critical space and time conflicts as well as improve the HS&E assessment (Penn State, 2011).

Site utilisation planning was used to test and review the construction schedule of the Perth Children's Hospital in Australia. In one example provided by the case study, the construction team had concerns regarding site logistics associated to the reach and manoeuvring of booms within project boundaries. The booms were modelled and incorporated into the BIM model. This allowed the team to visualise and schedule the use and location changes of equipment required for the installation of the façade. This in turn reduced idle time of equipment and labour while increasing the accuracy of site logistics (Sanchez et al., 2015b).

This enabler can help reduce time–space conflicts that have been cited as a major source of productivity loss (Akinci et al., 2000). The Salford Medical Centre in the UK, for example, was completely designed in BIM, including quantities, location and schedule of material movement and disciplines' workflow. This led to the project being delivered six weeks ahead of schedule (with a 28 per cent increase in productivity), reduced rework (between 60 and 90 per cent), and achieving all the sustainability targets set by the client (Fischer, 2013).

Site coordination

Site coordination refers to the process by which 3D and 4D modelling applications are used for site coordination purposes in order to plan site logistics, develop traffic layouts and identify potential job-site hazards (Azhar et al., 2012). In the 2014 Global Markets BIM report by McGraw Hill, 82 per cent of contractors with very high levels of BIM engagement and 60 per cent of all contractors used the model for this purpose (McGraw Hill Construction, 2014a).

Expanding the use of BIM tools for site management and coordination requires changes to the level of collaboration between BIM model authors and site managers, as well as changes to the way site managers carry out their daily tasks. Site managers need to trust the design is correct, closely collaborate with designers and be able to interrogate the model in order to find the information they require as well as potential sources of error, hazard or conflict (Mäki and Kerosuo, 2015). For this to happen, all contributors to the model, including the head contractor and subcontractors, should have agreed during earlier phases that the model would later be used for this purpose. Site coordination through BIM models can also help address challenges originating from the reduction of core head-contractor staff and the use of large numbers of subcontractors often seen in construction projects nowadays. The use of 3D and 4D models can facilitate fine-grained site coordination required for multidiscipline project teams based on more intuitive visualisation of the job-site and project scope (Sacks et al., 2010b).

A group from the University of Reading proposed to use an application they called SiteBIM, which effectively provided mobile access to a read-only coordinated BIM model and document management system. This tool was tested during the construction of a hospital in the UK, where the contractor identified significant benefits such as 'waste reduction, a lower than usual cost growth in packages for service installation, and significant savings in spending on administrative and coordinating staff time' (Davies and Harty, 2013).

BIM360 Glue is a commercial cloud-based BIM application by Autodesk used for construction site coordination, which provides mobile access to a model and project data viewer. It also has site layout and workflow functionalities (Autodesk, 2015g).

Notes

- A recent global survey by McGraw Hill showed that 29 per cent of contractors consider 4D modelling one of the most important BIM uses. The report highlighted that 'these activities are likely to grow as the technical and interoperability challenges of integrating model data with contractors' legacy scheduling and costs systems become easier to manage' (McGraw Hill Construction, 2014a).
- The US Army Corps of Engineers' BIM roadmap states that 4D modelling is to be a requirement in all projects starting 2015 (US Army Corps of Engineers, 2012).
- The use of BIM models for site coordination has been noted to be hindered by insufficient construction information included in models and

lack of access to mobile (handheld) devices as well as job-site staff with basic BIM skills (Mäki and Kerosuo, 2015).

Construction system design (virtual mock-up)

> A picture paints a thousand words, and for an industry that is so concerned with the visual and experiential, it is surprising how little we use this aspect to clarify requirements... A mock-up of a particular detail can help resolve buildability or sequencing issues, which can be difficult to do with just drawings.
>
> (Eynon, 2013)

Pre-construction mock-ups are full-size representations of a proposed design commonly produced before the design is finalised in order to test appearance, performance and other details of the final structure. Physical mock-ups 'are time consuming, costly, and are limited to one design alternative with only minor variations' (Vico Software, 2015d). Virtual mock-ups are the result of a process in which a BIM design authoring software is used to design and analyse the construction of a complex asset or system in order to improve planning. This analysis may include aspects such as formwork, glazing, tie-backs and other elements, and can be carried out at different levels of detail based on project needs (Penn State, 2011). Virtual mock-up has also been known as virtual prototyping where mock-ups and simulations are used to 'address the broad issues of physical layout, operational concept, functional specifications, and dynamics analysis under various operating environments' (Huang et al., 2007). As such, this enabler is a special use of 3D visualisation that is sometimes combined with animations and simulations.

This practice has now become commonplace in complex projects using BIM in planning and design phases. For example, during the design and construction of the MIT Koch Institute 33,445m^2 new cancer research building completed in 2010 in Cambridge, US, full mock-ups of animal and procedure rooms were created. These included furniture, fixture and equipment (FF&E) features, switch locations and animal cages. The virtual mock-up provided clear understanding of the design intend for the owners and allowed confirmation of the location of the FF&E and electrical switches (Hergunsel, 2011).

In the Mission Hospital Patient Care Tower in California, which included tunnels, four storeys and a central plant, virtual mock-ups were used after having had to halt construction in order to address a particular glass curtain wall, steel exoskeleton and fireproofing intersection. The owner had first requested a physical mock-up of the building costing US$150,000. However, after finding that there were significant fire and infection risks, the project team developed virtual mock-ups of the building and the steel structure to detail steel and glass intersections, as well as discuss and resolve constructability issues in real time. This work required an investment of US$80,000, which included engaging an on-site project manager with modelling expertise. The outcomes led to man-hour savings in the order of ten project work-weeks and rework on, more than 1,000 panels. The constructability analysis based on the virtual mock-ups saved the project an estimated US$1.8 million (Vico Software, 2013).

In another case study reported by Mortenson Construction in 2012, an estimate of 40 hours were required to produce interior virtual mock-ups of a baby changing station of a day-care centre in Wisconsin, US. The project superintendent estimated that, had the mock-ups not been used to address different scenarios and design changes as well as FF&E issues, there would have been at least two and half weeks of downtime in addition to rework due to redesigning (Filkins, 2012).

Cost estimation (quantity take-off) and planning (5D)

Cost estimation within a BIM environment refers to the process of using the BIM model to assist in the efficient production of accurate quantity take-offs and cost estimates throughout the life-cycle of an asset. Project and asset management teams can use this process to better understand the cost implications of changes made during any phase of the life-cycle of an asset. This in turn 'can help curb excessive budget overruns due to project modifications' (Penn State, 2011).

In this process, estimators can extract quantities directly from the BIM model and use this information in downstream cost estimation applications. '3D visualizations generated from the information in a BIM can also provide important insights because it enables estimators to analyze the design in different ways' (Hartmann et al., 2012). Cost planning or 5D BIM normally refers to a BIM model that has been linked not only to schedule information (4D) but also to cost data; this data may also be included in the model objects themselves. This allows users to map cost against project programme when doing cash flow analysis (Williams et al., 2014; Stanley and Thurnell, 2014).

It is important to highlight that for cost estimation to be accurate, the team must ensure that the correct naming conventions are being used throughout the model as well as the correct space area calculation standards (Matta et al., 2015). In the design and construction of a fire station and headquarter building, the project team was able to use the BIM model to automatically generate cost estimates that were accurate to within 0.6 per cent of the guaranteed price (Gilligan and Kunz, 2007). Other reports have estimated up to 80 per cent reduction of the time required to generate cost estimates with an accuracy of 3 per cent (Australian Government Productivity Commission, 2014). These time savings allow reports to be generated more frequently, helping improve cost control and earlier detection of potential issues that may generate cost overruns as well as signal other unplanned events.

In a railway renovation project carried out in 2012–2013 under a design-bid-build contract in Finland, BIM-based cost estimation was applied to different geotechnical solutions and track designs in order to compare the cost impact and select the best option (Gerbov, 2014). In the design and construction of a polyethylene terephthalate (PET) plant in Lithuania, the economic benefits of using this process to produce accurate quantity take-offs was estimated to be at least ten times the investment required to implement BIM. This process also helped minimise supplier overpayment while also reducing disputes with this stakeholder group (Migilinskas et al., 2013).

Results of a survey carried out in New Zealand across a small sample of quantity surveyors suggests that 5D BIM may provide advantages for this stakeholder group such as higher efficiency, improved visualisation of construction details and earlier risk identification. This survey also pointed to some barriers such as interoperability between software platforms, lack of coding protocols, high set-up cost and lack of integrated models (Stanley and Thurnell, 2014). Research carried out in Australia showed similar barriers within the quantity surveying sector (Smith, 2014).

Notes

- In a global survey carried out by McGraw Hill, 49 per cent of trade contractors and 30 per cent of head contractors used the BIM model for determining quantities (McGraw Hill Construction, 2014a).
- ASTM E1557 (Uniformat) is one of the most common standards currently used for cost estimation during design and construction (Matta et al., 2015).
- Software platforms that support this process include Trimble's Vico Software (2015a; 2015b), CostX (Exactal, 2015a; 2015b) and Nemetschek (Nemetschek Group, 2014).
- The US Army Corps of Engineers' BIM roadmap states that 5D BIM is to be required in all projects starting 2015 (US Army Corps of Engineers, 2012).

Design reviews

> These graphically explicit applications also support strong teaming environments and help ensure that the project vision is shared among all stakeholders. The result includes less risk for the team, the client, and the authorities having jurisdiction (AHJs), as well as safer, more effective structures for the owners and occupants.
>
> (Sullivan, 2007)

Design review in a BIM environment refers to the process by which different stakeholders view the model and provide feedback to validate design and other aspects such as construction programme, space aesthetics and layout, and criteria such as sightlines, lighting, security, acoustics, textures and colours (Penn State, 2011). This process can be carried out throughout the design and construction process and is critical when moving between design stages (Bloomberg et al., 2012). This process can make use of design review software and/or special virtual environment and immersive facilities to review the complete design or portions of it (Penn State, 2011).

Design-review tools allow different BIM software platforms to communicate seamlessly. They are also commonly easier to use than design authoring tools, which allows non-technical stakeholders to engage with the model data and participate more actively in the review process (Sullivan, 2007). Conventional design review is performed by engaging experts from different stakeholder groups who provide feedback based on their own experience. However, different research groups have worked on automating this process for specific tasks or assets (Lee et al., 2012b). See automated rule-checking for more information on this.

In a survey carried out in 2009, this process was one of the two BIM uses with the most perceived benefit and most often implemented (Kreider et al., 2010). A case study published in 2012 about the Wisconsin Department of Transport's use of BIM highlighted that this enabler improved the efficiency of the design review process. A survey presented in the same report also highlighted that infrastructure owners assign high value to this enabler during the planning phase because it allows them to visually convey complex engineering solutions to non-technical stakeholder groups. Owners also stated that the review and approval cycles were improved (Bernstein et al., 2012b).

In the Perth Children's Hospital case study in Australia, design reviews were carried out often and included end-users, clients and contractors as well as the design group. This process prompted changes to the design to reduce work at heights, make future service reticulation in the ceiling space easier, selection and layout of FF&E features, and reduce the lead time required for feedback and approval (Sanchez et al., 2015b).

Note

- It has been noted that in order to maximise the benefits from this process for all relevant stakeholders, formal design review methodologies should be developed (Dunston et al., 2010).

Digital fabrication

Digital fabrication refers to the process of using digitised information 'to facilitate the fabrication of construction materials or assemblies' (Penn State, 2011). This can be done based on: computer numerical control (CNC) processes (creating objects by removing material such as laser cutting) or rapid prototyping (RP) processes (creating objects by building layer-by-layer such as 3D printing) (Seely, 2004).

Digital fabrication is also known as BIM-driven pre-fabrication, which in turn can increase safety, reduce delays due to weather and unforeseen conditions, and increase speed of construction (Hardin and McCool, 2015). Interestingly, in the recent past, although industry professionals generally recognise the benefits from digital fabrication, this has not been a common use of BIM (Kreider et al., 2010). This may have been because the resources required include machine-readable data for fabrication software (Penn State, 2011), which may not be directly interoperable with BIM formats. Such was the case of the Perth Children's Hospital in Australia, where the incompatibility between software packages created rework for the consultant who had to re-create the information in the suppliers/fabricators' required formats (Sanchez et al., 2015b). Nevertheless, according to the most recent McGraw Hill Global Markets report, this may be changing; 43 per cent of contractors cited BIM-driven pre-fabrication as one of the top ways their firm uses BIM and 40 per cent of trades chose this as a top driver (McGraw Hill Construction, 2014a).

Digital fabrication, however, may prove impractical for some specific applications such as roof building where as-designed dimensions often cannot be used for fabrication. Using digital fabrication in this case would require the project team manually creating an XML file based on manual measurements. Some research has been done in creating these models automatically by using videogrammetry (by capturing stereo video streams of the roof area). The accuracy of the outcomes has however not yet been sufficient to be acceptable (Fathi and Brilakis, 2013).

Some firms have created proprietary software and machinery to allow direct on-site digital fabrication from BIM models. A London-based housing firm, for example, created a CNC-based machine that can be used on-site from a portable computer to cut raw wood panels to be immediately assembled. The tolerance of this system is 0–2mm (*AEC Magazine*, 2013). A more common application of digital fabrication is perhaps structural steel. To do this, a BIM model that contains all the shop drawings required for steel detailing and fabrication is used directly for the structural fabrication process (Avsatthi, 2015).

> Reusing the design model in this fashion is inherently more efficient (time normally spent creating a fabrication model is eliminated) and produces higher quality results (discrepancies between the design and fabrication models are eliminated). In addition, the source of the information used in the steel detailing and fabrication software is digital design data based on a highly accurate, coordinated, consistent building information model – data worth sharing for related building activities... Using the design model directly for fabrication will create a natural feedback loop between fabricators and designers – and bring fabrication considerations forward into the building design process. Sharing the design model with fabricators for bidding will shorten

> the bid cycle and lead to more uniform bids based on consistent steel tonnages.
>
> (Rundell, 2008)

Most case studies of digital fabrication are based on CNC systems. This may, however, soon change with the rise of 3D printing in the construction industry. The Chinese company WinSun, for example, claims to have printed ten houses at a cost of US$5,000 per house with side benefits of 50–70 per cent of production time savings, 30–60 per cent reduction of construction waste and 50–80 per cent reduction of labour cost. More recently, they printed a five-storey apartment building and a 1,100m^2 villa costing US$161,000 (Starr, 2015; Costrel and Rega, 2015). The Dutch firm DUS is also building a 3D-printed canal house and opened the construction site to visitors who can also pay for an audio tour (3D Print Canal House, 2015).

Disaster planning and response/disaster analysis

BIM-based disaster planning and response is here defined as a process in which a BIM model is used for at least one of the four disaster cycle phases: prevention and mitigation, preparation, response, and recovery (Drogemuller, 2013). Using BIM during the first two stages aims to minimise the human and economic loss risk by using the model for prediction and preparedness planning processes and combine them with an effective post-disaster management program (Dash et al., 2004).

During an emergency, responders would

> have access to critical building information in the form of a model and information system. The BIM would provide critical building information to the responders that would improve the efficiency of the response and minimize the safety risks. The dynamic building information would be provided by a building automation system (BAS), while the static building information, such as floor plans and equipment schematics, would reside in a BIM model. These two systems would be integrated via a wireless connection and emergency responders would be linked to an overall system. The BIM coupled with the BAS would be able to clearly display where the emergency was located within the building, possible routes to the area, and any other harmful locations within the building.
>
> (Penn State, 2011)

This process provides emergency responders with information to support better decision-making during the event response (Autodesk, 2013). Until recently (2010), at least in the US, the use of BIM for disaster response was one of the least frequent uses (Kreider et al., 2010). However, McGraw Hill's market report on the value of BIM for infrastructure highlights the potential of using BIM for disaster simulations to improve the resilience of the final asset. This was done in a project in Seattle, US, to assess the impact of earthquakes on an elevated highway and the surrounding grade level improvements (Bernstein et al., 2012b). 'BIM is a powerful tool for analysis, and can play a central role in disaster preparedness when leveraged during the very early schematic design phase of a project' (Stone, 2013). For example, when designing tunnels, fire safety and ventilation can be considered in more detail by integrating the 3D model into simulations of different scenarios, which can then be used to develop and coordinate emergency response efforts (Sanchez et al., 2014a).

If used in combination with open standards, the following model view definitions would be some examples of requirements to support disaster planning and response:

1. Prevention and mitigation:
   a. Selection of internal construction and finishes to reduce the impacts of water ingress;
   b. Selection and design for appropriate wind speeds, flooding conditions;
   c. Design of connections to withstand repetitive load cycles;
   d. Ensuring that the building or structure is maintained at a level so that it can support the design loads.
2. Preparation:
   a. Securing large areas of glass to reduce the risk of breakage and shards;
   b. Clean gutters and stormwater systems;
3. Response:
   a. Access to structural information on damaged structures to assess risk for rescue crews;
   b. Access to data on construction types to assist in selection of equipment to assist in rescue operations;
4. Recovery:
   1. Access to pre-event, "as existing" information on the building for damage assessment;
   2. Access to initial design requirements when considering if a badly damaged structure should be replaced or a new structure built elsewhere.

(Drogemuller, 2013)

In addition to the above, when used during emergency response, responders would require access to the BAS and CMMS linked to the record model, and the ability to understand it (Penn State, 2011). Li et al. (2014) also provide some insight into information requirements for firefighters during emergency events that can be more effectively ensured through this BIM process. During disaster prevention, rule-based analysis of the parametric model and GIS-BIM can help identify the lowest risk area for the construction of new assets. A case study analysing the construction of a school complex in Northern Australia used GIS and BIM models to identify the risk of storm tide and later applied this analysis to a northern Queensland port city (Drogemuller, 2013).

Disaster analysis software can use information from the model to carry out additional analyses. Extreme Loading® for Structures for example is a software platform that can import BIM structural components (ASI, 2015) to carry out:

- physical security assessments;
- advanced structural engineering applications including blast analysis, progressive collapse analysis;
- seismic analysis;
- forensic analysis; and
- demolition planning and prediction.

(*Australian Defence Magazine*, 2015)

Note

- Model View Definitions are 'a subset of the IFC schema that is needed to satisfy one or many Exchange Requirements of the AEC industry' (buildingSMART, 2015).

Field and management tracking

Field and management tracking refers to the process through which

> field management software is used during the construction, commissioning, and handover to manage, track, task, and report on quality (QA/QC), safety, documents to the field, commissioning (Cx), and handover programs, connected to BIM.
>
> (Penn State, 2011)

This includes BIM field management software platforms such as BIM360 Field from Autodesk, former Vela Systems field management and tracking software. This software combines mobile technologies with cloud-based collaboration (Autodesk, 2015e) to synchronise data and communicate it instantly from the work-site to the office. This enabler leverages benefits from lean construction, pre-fabrication, BIM and automated data collection (Moran, 2012).

Mobile construction site management software applications can enable real-time communication of on-site data to improve project efficiency by automating and shortening construction management tasks while enabling fast and reliable communication of accurate data (Moran, 2012). The overall goal of this type of tool is 'to ensure conformance to contract documents, compliance to safety regulations, and performance to owner's project requirements, through BIM-based workflows out in the field and at the point-of-construction' (Penn State, 2011).

In 2012, 15 construction projects carried out in the US and Sweden using a commercial field and management tracking software were analysed through survey data. The aim of the study was to understand the causes of performance variation and highlight factors that influenced these changes. The study analysed 15–20 per cent of all projects carried out by Skanska using this tool, ranging in value from US$15 million to $250 million. This report showed efficiency gains of up to 20 per cent for specific roles such as superintendents. Project-wide efficiency gains were up to 45 per cent, although there was one project that showed a 5 per cent efficiency loss. Variations in efficiency gains were found to be related to the job description of construction management staff (Moran, 2012). A later version of this tool has also been used in other projects in the USA to eliminate significant amounts of data redundancy, paperwork and data entry rework. In these cases, it has also reduced the time required for inspections and increased their scope and effectiveness (Autodesk, 2015i).

This enabler has been integrated experimentally with other technologies such as point cloud-scanning techniques to create real-time monitoring of hazardous engineering procedures.

> Real-time monitoring systems track the resource's (worker, equipment, objects) spatial context variables, such as location and orientation, in real-time using localization and tracking technologies and automatically compare the identified spatial-context with predefined scenarios identified as being dangerous.
>
> (Akula et al., 2013)

In 2013, a group reported such integration for the construction of a railway and bridge project aiming to provide real-time data on whether it was safe to continue drilling. This experiment combined field tracking, point cloud laser-scanning and clash detection to determine whether the drilling personnel were at risk of striking the rebar, utility line or other concealed infrastructure. Its application was, however, restricted to an experimental set-up and is unclear whether it could be used for a normal construction site (Akula et al., 2013). Other scan-to-BIM systems have also been developed for field tracking but have had significant limitations for real-life applications (Bosché et al., 2014). These systems have also been combined with RFID and wireless remote sensors to automate part of the tracking processes during construction within experimental set-ups (Xie et al., 2010; Costin et al., 2012).

Note

- Skanska is one of the largest construction companies in the world with their activities concentrating mostly in Europe and North America (Skanska, 2015).

Front-end planning

The Construction Industry Institute (CII) defines front-end planning in their *Best Practices* as

> the process of developing sufficient strategic information such that owners can address risk and decide to commit resources to maximize the chance for a successful project. Front-End Planning includes putting together the project team, selecting technology, selecting the project site, developing project scope, and developing project alternatives. Front-End Planning is often perceived as synonymous with front-end loading, pre-project planning, feasibility analysis, and conceptual planning.
>
> (Kang et al., 2013)

This is a pre-project planning process that focuses on strategic project issues (Construction Industry Institute, 2014). It is based on the premise that a thorough understanding of project activities, processes and issues that can influence the satisfactory achievement of project outcomes if effectively managed is critical. This applies to pre-project phases and is independent of project size (Faniran et al., 2000). Front-end planning is important not only for new construction projects that wish to reduce issues and rework during the design, construction, handover and operations phases but also for organisations wishing to implement BIM for asset management in the immediate term (Kassem et al., 2015).

An important application of BIM and front-end planning is safety in order to ensure that hazards are not unintentionally created at project planning and execution phases of new built assets. Front-end safety planning can be enhanced by using automated rule-checking based on historical data and BIM to identify corrective actions when it is the least expensive to carry them out (Zhang et al., 2013). This process can also be improved by including digital layouts of temporary structures such as scaffolding systems (Kim and Teizer, 2014).

Front-end planning can also be used with BIM 3D visualisation and 4D modelling to improve productivity and application of lean construction principles. For this process to be most effective, downstream stakeholders should be included through cross-functional teams (Bhatla and Leite, 2012).

Note

- In the UK, the Government Soft Landing (GSL) initiative outlines a policy to use BIM front-end planning to ensure value is achieved during operations (HM Government, 2013).

GIS-BIM

Geographic information system (GIS) is another information system technology which has been developed to use data containing spatial or geographic coordinate references. 'GIS is both a database system with specific capabilities for spatially referenced data, as well as a set of operations for working with the data' (Zhang et al., 2009). Spatial and geographic coordinate systems enable location of objects to be specified by a set of numbers (Wang et al., 2014).

GIS tools are used to visualise and analyse groups of assets based on attributes and their relationship to their spatial location. These systems are widely used in planning, agriculture, infrastructure maintenance, and many other fields. When applied to asset management, GIS can facilitate data collection, processing, and information display. It can also 'integrate asset mapping with project management and budgeting tools so maintenance, inspections, and expenses can be accounted for in the same place' (Zhang et al., 2009). During the construction phase, location-based schedules also present the advantage of integrating 'detailed quantity and position information to represent the true work volume and position to take place during construction' (Gleason, 2013).

GIS is similar to BIM in that it is also used to model spatial information and has been reported to be used for similar purposes such as location-based municipal facilities information queries and management (Kang and Hong, 2015). Linking GIS data to BIM models allows simulating material circulation paths within the site. When this data is further integrated with RFID and GPS technologies for resource tracking, labour-intensive data collection can be reduced or eliminated through automated tracking systems to improve efficiency, reduce data entry errors and reduce labour costs. This kind of integration can also allow real-time transportation information to be compatible with GIS and BIM (Irizarry et al., 2013). Generally this integration is done by extracting information from one system, transforming it to the right format and linking it to the other system (Kang and Hong, 2015).

The buildingSMART Alliance, a council of the National Institute of Building Sciences in the US started the GIS/BIM Information Exchange project in 2010. This project aimed to define best practices for each system and to provide the basis for an information relationship between the two systems (Bush, 2013). However, at the time of writing it is unclear what their latest progress towards those goals has been.

In infrastructure construction, many firms are using systems that integrate BIM and GIS data, as well as other sources of reliable information, to create models that are as complete as possible to reflect existing civil conditions of the work-site. 'Once that model is in place, engineers can leverage it for nearly endless types of analysis, simulation and visualization to optimize design solutions' (Bernstein et al., 2012b). Integration of GIS and BIM data can also be useful for asset management where separate structures are being analysed. Some software developers with BIM applications have started to partner with GIS vendors; Ecodomus, for example, partnered with ESRI ArcGIS and Google Earth to increase integration across platforms (Ecodomus, 2015a). Autodesk's 3D Civil also has functionalities that allow users to directly import GIS data to create surface models (Autodesk, 2015k).

GIS-BIM systems have been proposed to visualise supply chain processes and actual status of material deliveries (Irizarry et al., 2013), to assess energy performance of multiple assets that are geographically disperse (Wu et al., 2014), and for general facilities management (Kang and Hong, 2015).

Note

- Effective interoperability between GIS and BIM is required in order to integrate these systems (Kang and Hong, 2015).

Handheld devices

> The use of mobile hardware in construction has changed the landscape of the modern day project, in both how information is accessed and how it can be added and disseminated. Mobile technology-enabled platforms are now able to communicate between systems and project stakeholders in near real time. One of the major issues in the past in construction has been the inability for feedback and information flow to come from the field fast enough.
>
> (Hardin and McCool, 2015)

In the past, the use of digital formats on construction sites has presented a number of challenges, mainly associated with access to suitable devices with reasonable screen size (Meža et al., 2014). Nowadays, handheld mobile devices such as smartphones and tablets are increasingly used across different industries, including AECO, to allow moving between different spaces while maintaining continuous access to information as required (Williams et al., 2015). These devices provide access to BIM browsers and 2D documents instead of having to carry large amounts of paper documents around the worksite, making information more portable. Mobile devices also provide access to large volumes of detailed information about different types of construction materials, and other relevant information (Yeh et al., 2012). The use of these devices allows teams, for example, to use digital documents and hyperlinks as well as systems that link pdf files of drawings, specifications, submittals, RFIs and other information. 'This all but eliminates version control because a team is only able to look at the "latest and greatest" set of information' (Hardin and McCool, 2015).

In addition to handheld devices, some research has been carried out into the use of wearable mobile devices to collect and access information with limited results (Yeh et al., 2012).

Information delivery manuals

> An Information Delivery Manual is both a product to document information that needs to be exchanged to perform a task in a process, and also a methodology to model and re-engineer the process… The IDM as a methodology utilizes collaborative process re-engineering by involving multiple competencies (such as domain and software experts), as well as knowledge about BIM and the IDM to model or engineer cross-functional processes.
>
> (Berard and Karlshøj, 2012)

IDMs originated in Norway and serve as a framework for multidisciplinary teams to improve communication in the construction processes (Holzer, 2007). They link expert functionalities to a BIM model to provide relevant information, facilitate data exchange and avoid ambiguities by specifying storage, conversion and information exchange, creating links between functional, technical and organisational issues. This is done through process and interaction maps and associated exchange requirement model (ERM). Process maps are used to describe the flow of activities related to a particular topic, the actors' roles and information required, created and consumed. Interaction maps are used to define the roles and transactions related to a specific purpose or functionality. The ERM defines 'a set of information that needs to be exchanged to support a particular business requirement or functionality and is interrelated with Model View Definitions (MVD)' (Volk et al., 2014). Each ERM is described individually in the IDM, which can be divided into requirements for BIM users and requirements for software providers (AEC3, 2012).

The IDM approach allows for a set of functional elements to be defined and be reused to meet different exchange requirements while also accommodating information packages containing multiple sources over time. Attributes of different sets of information required for operation and management can be defined at different stages. The designer can, for example, specify performance requirements of products and systems of the asset, while the actual products, their characteristics and their installation and operation can be defined during the construction phase (Fallon and Palmer, 2007).

Although IDMs are part of the *information exchange framework* for certifying IFC software, their value exceeds this process. They can become legal agreements between multiple stakeholders who aim to improve their digital collaboration (Berard and Karlshøj, 2012). The benefits of IDMs for BIM users include providing easy to understand, plain language descriptions of construction processes, requirements for information to enable them and the expected outcomes. This helps to make information exchange more reliable in projects, improving information quality, decision-making and process effectiveness. For BIM software providers, IDMs identify and describe detailed functional breakdowns of processes and capabilities that need to be supported. This helps to respond better to user needs, guarantee quality of information exchange, and provide reusable software components (Wix and Karlshøj, 2010).

BuildingSMART has developed a detailed IDM guide for IFC, which was last updated in 2010 (Wix and Karlshøj, 2010), and the ISO standard 29481-1:2010: *Building Information Modelling – Information Delivery Manual – Part 1: Methodology and Format* (ISO, 2010; Karlshøj, 2011). They also have about 100 IDM document development projects at different levels of progress for areas across the life-cycle of assets from early design to asset management (Karlshøj, 2013). COBie has also used the IDM approach and simplified the development of specifications for the collection of incremental, process-based information packages (Fallon and Palmer, 2007). The US National BIM Standard Project Committee has been working on IDMs standards for COBie (NIBS-US, 2015).

> The goal is to create an ever growing, internet-accessible and searchable library of use cases and IDMs. In addition, the NBIMS Project Committee is developing end-user templates for defining use cases and their information requirements.
>
> (Fallon and Palmer, 2007)

The latest version of these standards was released by buildingSMART Alliance on 22 July 2015 and is freely available through the National BIM Standards (NBIMS) website (NBIMS-US, 2015b).

Notes

- Model View Definitions (MVD) translate IDMs into a document for software development (Berard and Karlshøj, 2012).
- 'A Model View Definition (MVD) or IFC View Definition defines "a subset of the IFC [...] that is needed to satisfy one or many Exchange Requirements of the AEC industry"' (Volk et al., 2014).

Lean construction principles

Lean construction principles are an adaptation of those that underpinned the Toyota Production System (TPS) first published in the 1980s. These include elements such as waste, variability and cycle time reduction, increasing customer value, output flexibility and process transparency, benchmarks and continuous improvement processes (Gao and Low, 2014). These principles are highly compatible with BIM and integrated project delivery (IPD) principles, which some argue is the delivery method that best integrates both, enhancing processes associated with both approaches when correctly implemented (Sacks et al., 2010). Research has been carried out over the last decade or so about the impact of combining BIM, collaborative delivery approaches and lean construction principles to reduce waste and increase productivity and efficiency (Khanzode, 2010).

A report published in 2010 did a cross-analysis of lean construction principles and BIM functionalities to better understand compatibilities and gaps (Sacks et al., 2010). Another report published that same year by the Stanford University Center for Integrated Facility Engineering (CIFE) attempted to quantify this impact by comparing four case studies focusing on benefits related to better coordination of MEP tasks. This study found that when used together there were grater volumes of pre-fabrication and lower volumes of rework, and proposed an *Integrated Virtual Design and Construction and Lean (IVL)* methodology (Khanzode, 2010). Another case study analysed the use of BIM and lean construction principles by a UK-based architectural design firm focused on social housing and regeneration. This analysis showed efficiency gains across three projects where BIM was piloted with lean principles, these gains were found across eight different areas of waste reduction (Arayici et al., 2011). Other studies have also shown that combining BIM with lean construction has the potential to reduce errors (Hattab and Hamzeh, 2015).

A recent study published in 2015 examined how lean practices can be used as mediators for achieving improved outcomes when implementing BIM. This study compared two construction projects in India using planned percentage complete (PPC) as the main metric. One project was carried out just with BIM, while lean principles were introduced six months prior to the use of BIM. Although the projects presented significant challenges, the weekly PPC levels improved significantly with the introduction of lean principles, achieving a 55 per cent improvement at the end of the six-month evaluation period. Activity alignment-related delays also dropped from 36 per cent to 10 per cent at the end of the six months (Mahalingam et al., 2015).

The integration of BIM and lean construction principles can also be enhanced by using RFIDs to track materials and elements from the design to the detailing phase, through manufacturing and construction into asset management to return cost savings to different project stakeholders (Teichholz,

2013). Object libraries have also been noted to increase efficiencies and create leaner BIM-based social housing construction processes when accompanied by user guidelines (Arayici et al., 2011).

Life-cycle costing

> Life-cycle costing is a valuable technique that is used for predicting and assessing the cost performance of constructed assets. Life-cycle costing is one form of analysis for determining whether a project meets the client's performance requirements.
>
> (ISO, 2008)

Life-cycle costing assessments allow decisions to be made based on the total whole-of-life cost of a constructed asset rather than the immediate cost in order to ensure the lowest cost over the life of the asset while still fulfilling performance requirements. This process can be carried out at any phase starting with generic information such as common statistics and later including increasingly more specific information (Krigsvoll, 2007). When BIM cost planning tools are used for this purpose, an Australian BIM Guide recommends including the following considerations:

- **Passive design analysis:** Model the asset in such a way that specialist applications such as Computer Fluid Design (CFD) software can be used.
- **Mechanical systems analysis:** Carry out virtual testing and balancing.
- **Life cycle analysis (LCA):** Use certified analysis applications; include material quantity measurement methods; assign material codes to model objects as required by applications; use internationally recognised assessment standards such as Building Products Innovation Council's Building Products Life Cycle Inventory; and implement protocols for the correct use of the data published by the source organisation (NATSPEC, 2011).

Life-cycle costing can be calculated as 'net present values, net present costs, annual cost, annual equivalent value, or payback' (Krigsvoll, 2007). The international standard ISO 15686-5 provides extensive detail about how to carry out this analysis in general (ISO, 2008). In order to implement this tool in a BIM environment, a group working at the School of Real Estate and Economics, Dublin Institute of Technology developed a template spreadsheet. This gives the ability to customise calculations using common commercial tools such as CostX and Buildsoft to carry out BIM-based whole life-cycle cost analysis and demonstrated its use in a pilot project (Kehily et al., 2012, 2013).

Note

- The US GSA has expressed that being able to compare design options based on life-cycle cost is a key motivation behind efforts to have simulations that accurately predict heating and cooling (GSA, 2012).

Photogrammetry

Photogrammetry commonly refers to the process of using images and photographs taken from different points of view to automatically or semi-automatically generate a 3D object or model. The process generates a point cloud similar to that from 3D laser scanning with the advantage of it being enriched with colour data for each point and often also being less costly and more flexible than laser scanning (Hichri et al., 2013a; BIMe, 2014; Tuttas et al., 2014). However, this term has also been used to describe the process of including photographs of system components into the model (Thomas et al., 2004). Commercial tools such as Autodesk's BIM360 Field include this functionality so photographs can be taken on the field and directly uploaded to the system for reporting or record-keeping purposes (Autodesk, 2015f). Also from Autodesk, 123D Catch is an app that allows the user to create 3D images and basic 3D models using regular smart mobile devices (Autodesk, 2015l). Trimble has a 3D handheld scanner device specially designed for this purpose which includes a high-resolution camera and android tablet (Trimble, 2015b).

NCC Construction, one of Sweden's largest contractors with more than US$6.5 billion annual sales, have stated that photogrammetry to create BIM models is a promising area especially for civil and infrastructure where traditional surveying techniques are manual and labour-intensive. They claim to have developed an autonomous surveying platform based on photogrammetry that has allowed them to accurately survey larger areas with the same resources (Lindvall, 2015).

Photogrammetry has also been highlighted as an important technique to create as-built models of existing structures such as historical buildings under conservation and other structures being renovated (Hichri et al., 2013b). This technique has also been combined with thermal imaging to automatically and semi-automatically produce as-built models that include thermographic and texture data for energy rehabilitation projects (Lagüela et al., 2013).

RFID

Radio frequency identification (RFID) is a generic term used to describe a system that transmits a unique serial number that identifies an object or person wirelessly, using radio waves (Violino, 2015). This automatic identification technology is used to capture and transmit data, acting as an electronic label for data collection systems to identify and track items (Motamedi and Hammad, 2009). The basic technology consists of a tag which is attached to an object, and a reader that identifies the tag when it is within range (Sattineni and Azhar, 2010).

RFIDs are used in asset construction, operation and maintenance for component tracking, inventory management, and progress, facilities, maintenance, and material management as well as quality control (Motamedi and Hammad, 2009). In previous studies, the use of RFIDs during construction completely eliminated incorrect shipments and missing panels in a project with a large pre-fabrication component. In a 2012 report, delivery and installation of 150 pre-fabricated panels was found to be 93–95 per cent more efficient, and total cost savings achieved were 3 per cent of total project cost thanks to the use of RFID tags (Moran, 2012).

This technology has been available and used in construction for years, RFID-BIM is, however, a relatively new application. When used together, a software application interacts with the BIM database and RFID tag information through an interface; the RFID reader reads the tag, this information is sent to the BIM database which is searched and the object is highlighted with all associated information (Meadati et al., 2010).

RFIDs are then used to facilitate the effective storage and retrieval of critical information needed by management at different project phases. This technology can therefore be used to create a platform upon which stakeholders can organise and share information across phases, providing graphic and non-graphic data related to component geometry, spatial relationships and location, properties and data quality (Cheng and Chang, 2011). Using RFIDs with BIM has also enabled improved safety and productivity, by tracking staff working areas, alerting them when they are leaving safe areas and being used to assign tasks based on proximity to the location. RFIDs are also proposed as a means to monitor safety behaviour and quickly input information about unsafe conditions into the BIM model (Sattineni and Azhar, 2010; Guo et al., 2014).

RFIDs have been proposed to be used 'to store lifecycle and context aware information taken from a BIM during the lifecycle as a distributed database' (Motamedi et al., 2013). Motamedi et al. provide details of conceptual design, data requirements, attribute identification process and new IFC properties (including model code) for this purpose. In the context of asset management, RFID-BIM offers the opportunity for automated data flows, which can improve the ability to address facility needs on a timely and more cost-effective manner (Meadati et al., 2010). RFID-BIM has also been combined with field tracking and 4D modelling to improve quality and efficiency of complex projects with large precast components and lay-down space restrictions (Moran, 2012).

Other special applications of this technology include a 2008 study that proposed the use of RFID, BIM, Wireless LAN (WLAN), ultra-wide-band (WB) and handheld devices to develop a platform for emergency response and recovery way-finding. This technology was used to create an indoor emergency navigation system for firefighters (Rueppel and Stuebbe, 2008). In a different study, RFID and BIM were used for decision-making in structural steel fabrication and erection, developing a portable database to assist fabrication and erection accuracy and efficiency (Shi, 2009; Xie et al., 2010).

Note

- Similar to RFID, projects can use barcodes and quick response (QR) codes.

Space management and tracking

The term 'space management' generally describes the process of managing real-time information about the use of space in an organisation, facility or asset (Reddy, 2011). In a BIM environment, it refers to the process by which BIM models are used to distribute, manage, and track spaces and related resources. BIM is used to allocate space resources during transition planning management in order to define changes required. This applies, for example, during renovation projects where operations and construction activities are paralleled and sections of the asset remain occupied. This application often requires integration with spatial tracking software (Penn State, 2011). Space management and tracking is seen mostly as an operations phase BIM application (Reddy, 2011) and, since BIM for asset management is a relatively recent application, there are few examples of its use (Becerik-Gerber et al., 2012).

The use of BIM asset management systems allows the creation of bi-directional links between the model and CMMS/CAFMs in order to update information such as room IDs, staff room assignments, department locations and work area measurements. Within this context, BIM space management and tracking can be used to 'effectively plan and manage new employee installations as well as move team members and outsourced vendors to new facility locations' through a single interface (Reddy, 2011). The use of models in IFC format can facilitate this data transfer between systems (Sabol, 2008).

A case study based on the renovation of a laboratory space into a set of offices at the Worcester Polytechnic Institute in the UK analysed the application of BIM for space management. In this project, it was suggested that barcodes could help the asset management team better track and manage the use of space within the building. This report concluded that one of the main benefits was being able to coordinate activities and information across space and building management staff. The main barrier for implementing this process across other renovation projects was suggested to be the cost of creating BIM models of existing buildings (Keegan, 2010).

Notes

- This application often requires integration with spatial tracking software (Penn State, 2011).
- In a survey carried out in 2010 in the US, only 6 per cent of participants used BIM for space management and tracking during operations. However, 13 per cent used BIM for this purpose 50–75 per cent of the time and 30 per cent used it at least 25 per cent of the time (this value includes all responses of 25 or higher) (Kreider et al., 2010).

Streamlined logistics

This term refers to the process of integrating the BIM models with Enterprise Resource Planning Systems (ERP) to reduce internal communication costs and errors, increase management control and reduce inventory size of components and finished pieces (Sacks et al., 2005). ERP focus on standardising and synchronising information to integrate industrialised production of construction elements with other business tasks such as logistics and procurement. BIM provides logical and consistent access to information about design elements and their attributes. By linking the two systems 'the model becomes the common denominator which makes the construction process more transparent'. This can lead to more accurate short-term planning and reduce delays and duration of construction processes (Babič et al., 2010). Streamlined logistics has also been proposed as a way of increasing sustainability of the construction process and the finished asset (Ghosh et al., 2011), although no successful examples were found by the authors.

Streamlined logistics can provide precast producers 'the design, production control and financial management tools they need to fully take advantage of the coming revolution in building construction' (Barr, 2014). Enemetric, a small Scottish manufacturer of modular building systems, partnered with the University of Edinburgh to develop new ways of using BIM to automate the design of modular structures. As part of this initiative, they were exploring the development of a BIM-ERP interface to 'permit highly accurate estimation and planning with supplier databases and project management scheduling tools' (Seeam et al., 2013). This was part of a larger strategy to integrate several systems but the level of implementation of this particular system interface was unclear at the time of writing.

Notes

- To avoid traceability problems, the identity of designed objects should be separate from the identity of physical material and/or prefabricated objects. When this is done, the links between these attributes are mapped to provide the basis for the algorithms used for the calculation of material requirements and flows throughout the project (Babič et al., 2010).

- Harty and Koch (2011) offer a comparison of ERP and BIM objects.

Notes

1 These trials have had some limitations, such as requiring the manual calculation of control points, but are steps in the right direction.
2 See definition of an asset in Chapter 1.

Bibliography

3D Print Canal House, 2015. *3D Print Canal House*. Available at: http://3dprintcanalhouse.com [Accessed 1 July 2015].

Aconex, 2015. *Connected BIM, for Better Delivery*. Available at: www.aconex.com/bim-management [Accessed 24 August 2015].

AEC3, 2012. *Information Delivery Manual*. Available at: www.aec3.com/en/5/5_009_IDM.htm [Accessed 27 July 2015].

AEC Magazine, 2013. *BIM to Fabrication*. Available at: http://aecmag.com/57-lead-story/577-bim-to-fabrication [Accessed 1 July 2015].

AGA CAD, 2014. *Smart Assemblies*. Available at: www.aga-cad.com/products/bim-solutions/smart-assemblies [Accessed 6 July 2015].

Ahn, K.-U., Kim, Y.-J., Park, C.-S., Kim, I. and Lee, K., 2014. BIM interface for full vs. semi-automated building energy simulation. *Energy and Buildings*, 68(B), pp. 671–678.

AIA, 2013a. *Guide, Instructions and Commentary to the 2013 AIA Digital Practice Document*, Washington, DC: American Institute of Architects.

AIA, 2013b. *Project Building Information Modelling Protocol Form*, Washington, DC: American Institute of Architects.

AIA, 2013c. *Project Digital Data Protocol Form*, Washington, DC: American Institute of Architects.

Akinci, B., Fischer, M., Levitt, R. and Carlson, R., 2000. *Formalization and Automation of Time-Space Conflict Analysis*, Stanford: Center for Integrated Facility Engineering, Stanford University.

Akula, M., Lipman, R. R., Franaszek, M., Saidi, K. S., Cheok, G. S. and Kamat, V. R., 2013. Real-time drill monitoring and control using building information. *Automation in Construction*, 36, pp. 1–15.

Allsopp, A., 2015. *Google Glass 2 UK Release Date, Price and Specs: No Glass at Google I/O but Smartglasses Team is Growing*. Available at: www.pcadvisor.co.uk/news/wearable-tech/google-glass-2-release-date-price-specs-not-dead-io15-hires-3589338 [Accessed 15 June 2015].

Alwan, Z. and Gledson, B. J., 2015. Towards green building performance evaluation using. *Built Environment Project and Asset Management*, 5(3), pp. 290–303.

Anumba, C., Dubler, C., Goodman, S., Kasprzak, C., Kreider, R., Messner, J., Saluja, C. and Zikic, N., 2010. *Building Information Modelling Execution Planning Guide*, Pennsylvania: Computer Integrated Construction Research Group (CIC), Pennsylvania State University.

Apollonio, F. I., Gaiani, M. and Zheng, S., 2012. BIM-based modelling and data enrichment of classical architectural buildings. *Scientific Research and Information Technology*, 2(2), pp. 41–62.

Arayici, Y., Coates, P., Koskela, L., Kagioglou, M., Usher, C. and O'Reilly, K., 2011. Technology adoption in the BIM implementation for lean architectural practice. *Automation in Construction*, 20(2), p. 189–195.

Arayici, Y., Onyenobi, T. and Egbu, C., 2012. Building information modelling (BIM) for facilities management (FM): the mediacity case study approach. *International Journal of 3D Information Modelling*, 1(1), pp. 55–73.

Archifm, 2015. *Archifm.net Functions*. Available at: www.archifm.net/archifm-net-functionality [Accessed 30 July 2015].

ASHRAE, 2015. *kBIM Template and Library*. Available at: www.ashrae.org/resources–publications/bookstore/kbim-template-and-library [Accessed 24 June 2015].

ASI, 2015. *Extreme Loading for Structures 3.1*. Available at: www.appliedscienceint.com/extreme-loading-for-structures [Accessed 6 July 2015].

Aurasma, 2015. *Aurasma*. Available at: www.aurasma.com [Accessed 15 June 2015].

Australian Defence Magazine, 2015. *ASI Australia/Liberty Industrial*. Available at: www.australiandefence.com.au/99649473-74AA-4116-83D9085273A92AC6 [Accessed 6 July 2015].

Australian Government Productivity Commission, 2014. *Public Infrastructure – Productivity Commission Inquiry Report: Volume 2*, Canberra: Australian Government.

Autodesk, 2013. *Lesson 5: Using BIM for Operations and Facilities Management*. Available at: http://bimcurriculum.autodesk.com/lesson/lesson-5-using-bim-operations-and-facilities-management.html [Accessed 26 May 2015].

Autodesk, 2015a. *Advanced Structural Analysis Software*. Available at: www.autodesk.com/products/robot-structural-analysis/overview [Accessed 6 July 2015].

Autodesk, 2015b. *Autodesk Ecotect Analysis*. Available at: http://usa.autodesk.com/ecotect-analysis [Accessed 24 June 2015].

Autodesk, 2015c. *Autodesk Seek*. Available at: http://seek.autodesk.com [Accessed 28 July 2015].

Autodesk, 2015d. *Autodesk Trust Centre – Overview*. Available at: www.autodesk.com/trust/overview [Accessed 27 July 2015].

Autodesk, 2015e. *BIM 360 Field*. Available at: www.autodesk.com/products/bim-360-field/features [Accessed 6 July 2015].

Autodesk, 2015f. *BIM 360 Field for iPad – Version 4.7*. Available at: http://knowledge.autodesk.com/support/bim-360-field/learn-explore/caas/CloudHelp/cloudhelp/ENU/BIM-360-Field-WhatsNew/files/GUID-8434AC9F-031C-4AD4-9D7D-260E0EAFF0EF-htm.html [Accessed 30 July 2015].

Autodesk, 2015g. *BIM 360 Glue*. Available at: www.autodesk.com/products/bim-360-glue/features [Accessed 30 July 2015].

Autodesk, 2015h. *BIM360 Overview*. Available at: www.autodesk.com/products/bim-360/overview [Accessed 19 March 2015].

Autodesk, 2015i. *Improve Efficiency and Accuracy: Autodesk Helps Construction Firm Implement Cloud-Based BIM Solutions for Better Project Outcomes*, s.l.: Autodesk.

Autodesk, 2015j. *Autodesk and Leica: Transforming the Building Construction Industry, Together*. Available at: http://inthefold.autodesk.com/in_the_fold/2015/06/autodesk-and-leica-transforming-the-building-construction-industry-together.html [Accessed 25 September 2015].

Autodesk, 2015k. *To Create a Surface from GIS Data*. Available at: http://knowledge.autodesk.com/support/autocad-civil-3d/learn-explore/caas/CloudHelp/cloudhelp/2016/ENU/Civil3D-UserGuide/files/GUID-3247F2EA-34D3-41EE-A7D2-DFFA186A43E9-htm.html [Accessed 25 September 2015].

Autodesk, 2015l. *Autodesk 123D*. Available at: www.123dapp.com/Gallery/content/all [Accessed 25 September 2015].

Autodesk, 2015m. *NavisWorks, Features*. Available at: www.autodesk.com/products/navisworks/features/all/gallery-view [Accessed 16 June 2015].

Avsatthi, B., 2015. *BIM-enabled Digital Fabrication of structural Steel – An Advanced Approach. Informed Infrastructure*. Available at: https://informedinfrastructure.com/13449/bim-enabled-digital-fabrication-of-structural-steel-an-advanced-approach [Accessed 5 March 2015].

Azhar, S., 2011. Building information modeling (BIM): trends, benefits, risks, and challenges for the AEC industry. *Leadership and Management in Engineering*, 11(3), pp. 241–252.

Azhar, S., Brown, J. and Farooqui, R., 2009. *BIM-Based Sustainability Analysis: An Evaluation of Building Performance Analysis Software*, paper presented at 45th ASC Annual Conference, Gainesville, FL, 1–4 April.

Azhar, S., Carlton, W. A., Olsen, D. and Ahmad, I., 2011. Building information modeling for sustainable design and LEED® rating analysis. *Automation in Construction*, 20(2), pp. 217–224.

Azhar, S., Khalfan, M. and Maqsood, T., 2012. Building information modelling (BIM): now and beyond. *Australasian Journal of Construction Economics and Building*, 12(4), pp. 15–28.

Babič, N. Č., Podbreznik, P. and Rebolj, D., 2010. Integrating resource production and construction using BIM. *Automation in Construction*, 19, pp. 539–543.

Bahar, Y. N., Pere, C., Landrieu, J. and Nicolle, C., 2013. A thermal simulation tool for building and its interoperability through the building information modelling (BIM) platform. *Buildings*, 3(2), pp. 380–398.

Baharuddin, H. E. A., Wilkinson, S. and Costello, S. B., 2013. *Evaluating Early Stakeholder Engagement (ESE) as a Process for Innovation*, paper presented at *CIB World Building Congress*, Brisbane, Australia, 5–9 May.

Barr, A., 2014. BIM, ERP integration key to future of software for the precast industry. *Concrete Products*, 117(1), pp. 42–45.

Becerik, B. and Pollalis, S. N., 2006. *Computer Aided Collaboration in Managing Construction*, Cambridge: Harvard University Graduate School of Design, Design and Technology Report Series 2006-2.

Becerik-Gerber, B., Jazizadeh, F., Li, N. and Calis, G., 2012. Application areas and data requirements for BIM-enabled facilities management. *Journal of Construction Engineering and Management*, 138(3), pp. 431–442.

Bentley, 2015a. *Bentley for Bridge*. Available at: www.bentley.com/en-US/Products/Bridge+Design+and+Engineering/Index.htm [Accessed 1 July 2015].

Bentley, 2015b. *BIM for Operational Management of Space and Assets*. Available at: www.bentley.com/en-US/Products/Bentley+Facilities [Accessed 1 July 2015].

Bentley, 2015c. *Building Information Modeling (BIM) Software and Integrated Practice*. Available at: www.bentley.com/en-US/Products/Building+Analysis+and+Design [Accessed 1 July 2015].

Bentley, 2015d. *Empowering Intelligent Structural Design Through Integrated Structural Modelling*. Available at: www.bentley.com/en-US/Products/Structural+Analysis+and+Design/ISM/default.htm [Accessed 6 July 2015].

Bentley, 2015e. *Offshore Analysis and Design Software*. Available at: www.bentley.com/en-AU/Products/SACS/Index.htm [Accessed 6 July 2015].

Bentley, 2015f. *Piping Design and Analysis Software for Productivity, Integration, and Nuclear Quality Standards*. Available at: www.bentley.com/en-AU/Products/Bentley+AutoPIPE [Accessed 6 July 2015].

Bentley, 2015g. *Bentley Navigator: Dynamic Project Review and Analysis*. Available at: www.bentley.com/en-AU/Products/ProjectWise+Navigator/Interference+Manager.htm [Accessed 16 June 2015].

Berard, O. and Karlshøj, J., 2012. Information delivery manuals to integrate building product information into design. *Journal of Information Technology in Construction*, 17, pp. 64–74.

Bernstein, H. M., Jones, S. A., Russo, M. A., Laquidara- Carr, D., Messina, F., Partyka, D., Lorenz, A., Buckley, B., Fitch, E. and Gilmore, D., 2010. *Green BIM: How to Building Information Modeling is Contributing to Green Design and Construction*, Bedford, MA: McGraw Hill Construction.

Bernstein, H. M., Jones, S. A., Russo, M. A., Laquidara-Carr, D., Taylor, W., Ramos, J., Lorenz, A., Winn, J., Fujishima, H., Fitch, E., Buckley, B. and Gilmore, D., 2012a. *The Business Value of BIM in North America: Multi-Year Trend Analysis and User Ratings (2007–2012)*, Bedford, MA: McGraw Hill Construction.

Bernstein, H. M., Jones, S. A., Russo, M. A., Laquidara-Carr, D., Taylor, W., Ramos, J., Healy, M., Lorenz, A., Fujishima, H., Buckley, B., Fitch, E. and Gilmore, D., 2012b. *The Business Value of BIM for Infrastructure: Addressing America's Infrastructure Challenges with Collaboration and Technology SmartMarket Report*, Bedford, MA, US: McGraw Hill Construction.

Bhatla, A. and Leite, F., 2012. *Integration Framework of BIM with the Last Planner System*, paper presented at the 20th Annual Conference of the International Group for Lean Construction San Diego, CA, 18–20 July.

BIM Task Group, 2012. *COBie UK 2012*. Available at: www.bimtaskgroup.org/cobie-uk-2012 [Accessed 25 September 2015].

BIM Task Group, 2013. *Uniclass2 Beta Classification Tables*. Available at: www.bimtaskgroup.org/uniclass2 [Accessed 30 July 2015].

BIM Task Group, 2014. *BS 1192–4:2014 Collaborative Production of Information Part 4: Fulfilling Employers Information Exchange Requirements Using COBie – Code of Practice*. Available at: www.bimtaskgroup.org/bs-1192-42014-collaborative-production-of-information-part-4-fulfilling-employers-information-exchange-requirements-using-cobie-code-of-practice [Accessed 27 July 2015].

BIMe, 2014. *BIM Dictionary – Photogrammetry*. Available at: http://bimexcellence.com/dictionary/photogrammetry [Accessed 21 November 2014].

BIMEX, 2015. *Building Maintenance Scheduling*. Available at: http://bimex.wikispaces.com/Building+Maintenance+Scheduling [Accessed 24 June 2015].

BIM-MEP AUS, 2014. *BIM-MEP AUS*. Available at: www.bimmepaus.com.au/home_page.html [Accessed 24 August 2015].

Blippar, 2015. *Blippar*. Available at: https://blippar.com/en [Accessed 15 June 2015].

Bloomberg, M. R., Burney, D. J. and Resnick, D., 2012. *BIM Guidelines*, New York: New York City Department of Design + Construction.

Bosché, F., 2010. Automated recognition of 3D CAD model objects in laser scans and calculation of as-built dimensions for dimensional compliance control in construction. *Advanced Engineering Informatics*, 24(1), pp. 107–118.

Bosché, F., Guillemet, A., Turkan, Y., Haas, C. T. and Haas, R., 2014. Tracking the built status of MEP works: assessing the value of a Scan-vs-BIM system. *Journal of Computing in Civil Engineering*, 28(4), pp. 1–28.

Botta, S., Comoglio, C. and Petrosillo, I., 2013. Implementing the environmental and an integrated management system: social policies of a municipality through theoretical framework and case study. *Journal of Environmental Planning and Management*, 56(7), pp. 1073–1095.

Brilakis, I., Lourakis, M., Sacks, R., Savarese, S., Christodoulou, S., Teizer, J. and Makhmalbaf, A., 2010. Toward automated generation of parametric BIMs based on hybrid video and laser scanning data. *Advanced Engineering Informatics*, 24(a), pp. 456–465.

Bryde, D., Broquetas, M. and Volm, J. M., 2013. The project benefits of Building Information Modelling (BIM). *International Journal of Project Management*, 31, pp. 971–980.

BSI, 2013. *PAS 1192-2:2013 Specification for Information Management for the Capital/Delivery Phase of Construction Projects Using Building Information Modelling*. Available at: http://shop.bsigroup.com/Navigate-by/PAS/PAS-1192–22013 [Accessed 27 July 2015].

BSI, 2014. *PAS 1192–3:2014 Specification for Information Management for the Operational Phase of as Sets Using Building Information Modelling*. Available at: http://shop.bsigroup.com/forms/PASs/PAS-1192–3 [Accessed 27 July 2015].

buildingSMART, 2010. *Investors Report: Building Information Modelling (BIM)*, London: UK Department of Business Innovation and Skills and buildingSMART.

buildingSMART, 2015. *Model View Definition Summary*. Available at: www.buildingsmart-tech.org/specifications/ifc-view-definition [Accessed 6 July 2015].

Bush, L., 2013. GIS/BIM Information Exchange. Available at: www.nibs.org/?page=bsa_gisbimie [Accessed 23 July 2015].

Caldas, C. H. and O'Brien, W. J. (eds.), 2009. *A Research Outlook for Real-Time Project Information Management by Integrating Advanced Field Data Acquisition Systems and Building Information Modelling*, paper presented at Computing in Civil Engineering, Austin, TX, 24–27 June.

Chappell, E., 2014. *AutoCAD Civil 3D 2014 Essentials*. Indianapolis: Wiley Brand.

Charlesraj, V. P. C., 2014. *Knowledge-Based Building Information Modelling (K-BIM)*, paper presented at the 31st International Symposium on Automation and Robotics in Construction and Mining, ISARC 2014, Sydney, Australia, 9–11 July.

Chen, S.-M., Griffis, F. D., Chen, P.-H. and Chang, L.-M., 2013. A framework for an automated and integrated project scheduling and management system. *Automation in Construction*, 35, pp. 89–110.

Cheng, M.-Y. and Chang, N.-W., 2011. *Radio Frequency Identification (RFID) Integrated with Building Information Model (BIM) for Open-Building Life Cycle Information Management*, paper presented at 28th International

Symposium on Automation and Robotics in Construction, Seoul, Korea, 29 June–2 July.

Chi, H.-L., Chen, Y.-C., Kang, S.-C. and Hsieh, S.-H., 2012. Development of user interface for tele-operated cranes. *Advanced Engineering Informatics*, 26(3), pp. 641–652.

Chi, H.-L., Kang, S.-C. and Wang, X., 2013. Research trends and opportunities of augmented reality applications in architecture, engineering, and construction. *Automation in Construction*, 33, pp. 116–122.

CIOB, 2013a. *CIOB Contract for Use with Complex Projects*, Berkshire: Chartered Institute of Building.

CIOB, 2013b. *CIOB Contract for Use with Complex Projects*. Berkshire: Chartered Institute of Building.

CodeBook, 2015. *CodeBook*. Available at: www.codebookinternational.com/about [Accessed 27 July 2015].

Conover, D., Crawley, D., Hagan, S., Knight, D., Barnaby, C. S., Gulledge, C., Hitchcock, R., Rosen, S., Emtman, B., Holness, G., Iverson, D., Palmer, M. and Wilkins, C., 2009. *An Introduction to Building Information Modelling (BIM)*, Atlanta: American Society of Heating Refrigerating and Air-Conditioning Engineers.

Construction Industry Institute, 2014. *Front End Planning, Including the Project Definition Rating Index (PDRI)*. Available at: www.construction-institute.org/kd/pdc/clee1-3.cfm?section=prodev [Accessed 14 July 2015].

Coombes, A., Senior, C., Ross, G., Skidmore, G., John, I., Austin, J., Stott, J., Brett, M., Bartyzel, M., Farmer, M., Johnson, M., Grant, S. and Wright, S., 2012. *AEC (UK) BIM Protocol V2.0*, s.l.: AEC (UK).

Costa, A., Keane, M. M., Torrens, J. I. and Corry, E., 2013. Building operation and energy performance: Monitoring, analysis and optimisation toolkit. *Applied Energy*, 101, pp. 310–316.

Costin, A., Pradhananga, N., Teizer, J. and Marks, E., 2012. *Real-Time Resource Location Tracking in Building Information Models (BIM)*, paper presented at *9th International Conference on Cooperative Design, Visualisation and Engineering*, Osaka, Japan, 2–5 September.

Costrel, F. and Rega, S., 2015. The first 3D printed house is coming, and the construction industry will never be the same. *Business Insider Australia*. Available at: www.businessinsider.com.au/3d-printed-houses-construction-industry-neighborhoods-2015-3 [Accessed 23 March 2015].

CRC for Construction Innovation, 2006. *Code Checking*. Available at: www.construction-innovation.info/indexf5c2.html?id=772 [Accessed 22 June 2015].

CRC for Construction Innovation, 2009. *National Guidelines for Digital Modelling*, Brisbane: Cooperative Research Centre for Construction Innovation.

Darling Geomatics, 2015. *3D Laser Scanning – Building Information Modelling (BIM)*. Available at: www.darlingltd.com/3d_laser_scanning/building_information_modeling.html [Accessed 29 May 2015].

Dash, J., Steinle, E., Singh, R. P. and Bahr, H. P., 2004. Automatic building extraction from laser scanning data: An input tool for disaster management. *Advances in Space Research*, 33(3), pp. 317–322.

Davies, R. and Harty, C., 2013. Implementing 'siteBIM': a case study of ICT innovation on a large hospital project. *Automation in Construction*, 30, pp. 15–24.

Department of Ecology, State of Washington, 2015. *The Integrated Process*. Available at: www.ecy.wa.gov/programs/swfa/greenbuilding/Charrettes.html [Accessed 1 July 2015].

Dimyadi, J. and Amor, R., 2013. *Automated Building Code Compliance Checking – Where is it at?* Brisbane: CIB World Building Congress.

Ding, L. and Xu, X., 2014. Application of cloud storage on BIM life-cycle management. *International Journal of Advanced Robotic Systems*, 11(129), pp. 1–10.

Dodge Data and Analytics, 2015. *Sweets*. Available at: http://sweets.construction.com/QuickLinks/building-information-modeling-bim [Accessed 28 July 2015].

Downton, P., 2011. Building environmental performance assessment: methods and tools. *Environmental Design Guide*, 70, pp. 1–8.

dRofus, 2015. *About dRofus*. Available at: www.drofus.no/en/product [Accessed 19 March 2015].

Drogemuller, R., 2013. *BIM Support for Disaster Response*, paper presented at 9th Annual International Conference of the International Institute for Infrastructure Renewal and Reconstruction, Risk-informed Disaster Management: Planning for Response, Recovery and Resilience, Brisbane, Australia, 8–10 July.

Duddy, K., Beazley, S., Drogemuller, R. and Kiegeland, J., 2013. A platform – independent product library for BIM, in Z. Ma, J. Zhang, Z. Hu and H. Guo (eds.), *Proceedings of the 30th CIB W78 International Conference*, Beijing: WQBook, pp. 389–399.

Dunston, P. S., Arns, L. L. and McGlothlin, J. D., 2010. Virtual reality mock-ups for healthcare facility design and a model for technology hub collaboration. *Journal of Building Performance Simulation*, 3(3), pp. 185–195.

East, B., 2012. *Construction Operations Building Information Exchange (COBie)*, Washington, DC: buildingSMART alliance, National Institute of building Sciences.

East, B., 2015. *Construction Operations Building Information Exchange (COBie)*. Available at: www.nibs.org/?page=bsa_cobie [Accessed 23 June 2015].

Eastman, C., Lee, J.-M., Jeong, Y.-S. and Lee, J.-K., 2009. Automatic rule-based checking of building designs. *Automation in Construction*, 18(8), pp. 1011–1033.

Ecodomus, 2015a. *BIM GIS*. Available at: www.ecodomus.com/index.php/bim-gis [Accessed 23 July 2015].

Ecodomus, 2015b. *Ecodomus FM/BIM Software for Lifecycle Facilities Management*. Available at: www.ecodomus.com/index.php/ecodomus-fm [Accessed 23 June 2015].

Elmualim, A. and Pelumi-Johnson, A., 2009. Application of computer-aided facilities management (CAFM) for intelligent buildings operation. *Facilities*, 27(11/12), pp. 421–428.

Exactal, 2015a. *CostX® is Fully Integrated 2D and 3D Estimating Software*. Available at: www.exactal.com.au/products/costx-bim [Accessed 29 June 2015].

Exactal, 2015b. *CostX Takeoff*. Available at: www.exactal.com.au/products/costx-takeoff [Accessed 29 June 2015].

Eynon, J., 2013. *Design Manager's Handbook*, 4th edn, Somerset, NJ: John Wiley & Sons.

Fallon, K. K. and Palmer, M. E., 2007. *General Buildings Information Handover Guide: Principles, Methodology and Case Studies*, Gaithersburg: US Department of Commerce, National Institute of Standards and Technology.

Faniran, O. O., Love, P. and Smith, J., 2000. *Effective Front-End Project Management: A Key Element in Achieving Project Success in Developing Countries*, paper presented at the 2nd International Conference on Construction in Developing Countries: Challenges Facing the Construction Industry in Developing Countries, Gabarone, Botswa, 15–17 November.

Fathi, H. and Brilakis, I., 2013. A videogrammetric as-built data collection method for digital fabrication. *Advanced Engineering Informatics*, 27(4), pp. 466–476.

Filkins, B., 2012. *Leveraging Virtual Mockups for Design Verification and Construction Understanding*. Available at: http://bimforum.org/wp-content/uploads/2012/05/Leveraging-Virtual-Mock-Ups-for-Design-Verification-and-Construction.pdf [Accessed 29 June 2015].

Fillingham, V., Malone, A. and Gulliver, S. R., 2014. *Birmingahm City University: City Centre Campus Development*, Reading: Faithful+Gould.

Fischer, M., 2013. *Keynote Address*, presented at the CIB World Building Congress 2013 – Day 1, Brisbane, Australia, 5–9 May.

Fussell, T., Beazley, S., Aranda-Mena, G., Chevez, A., Crawford, J., Succar, B., Hainsworth, J., Hardy, S., McAtee, S., McCann, G., Rizzalli, R., Akhurst, P., Linning, C., Marchant, D., Law, J., Lord, P., Morse, D., Crapper, P. and Spathonis, J., 2009. *National Guidelines for Digital Modelling: Case Studies*, Brisbane: CRC for Construction Innovation.

Gao, J. and Fischer, M., 2008. *Framework & Case Studies Comparing Implementations & Impacts of 3D/4D Modeling Across Projects*, Stanford: Center for Integrated Facility Engineering (CIFE), Stanford University.

Gao, S. and Low, S. P., 2014. *Lean Construction Management: The Toyota Way*, Singapore: Springer.

Gerbov, A., 2014. *Process Improvement and BIM in Infrastructure Design Projects – Findings from 4 Case Studies in Finland*, Aalto: Aalto University.

Ghosh, S., Negahban, S., Kwak, Y. H. and Skibniewski, M. J., 2011. *Impact of Sustainability on Integration and Interoperability between BIM and ERP: A Governance Framework*, paper presented at the Technology Management Conference (ITMC), 2011 IEEE International, San Jose, CA, 27–30 June.

Gijezen, S., Hartmann, T., Buursema, N. and Hendriks, H., 2009. *Organizing 3D Building Information Models with the Help of Work Breakdown Structures to Improve the Clash Detection Process*, Twente: Visico – Centre for Visualization and Simulation in Construction, University of Twente.

Gilligan, B. and Kunz, J., 2007. *VDC Use in 2007: Significant Value, Dramatic Growth, and Apparent Business Opportunity*, Stanford: Center for Integrated Facility Engineering, Stanford University.

Gleason, D., 2013. *Laser Scanning for an Integrated BIM*, paper presented at the Lake Constance 5D Conference, Constance, Germany, 28–29 October.

Gleeson, J., 2005. *Computer-Aided Green Design*. Available at: www.architectureweek.com/2005/0330/tools_1-2.html [Accessed 20 July 2015].

Good, N., 2015. *Eco-Charrette*. Available at: www.betterbricks.com/design-construction/reading/eco-charrette [Accessed 1 July 2015].

Google, 2014. *Google Goggles*. Available at: https://play.google.com/store/apps/details?id=com.google.android.apps.unveil&hl=en [Accessed 15 June 2015].

GPS World staff, 2015. *Google Patent Seeks to Link Robots via Smartphones*. Available at: http://gpsworld.com/google-patent-seeks-to-link-robots-via-smartphones [Accessed 8 June 2015].

Grading and Excavation Contractor, 2007. *PennDOT About to Embrace GPS Technology*. Available at: www.gradingandexcavation.com/GX/Articles/PennDOT_About_to_Embrace_GPS_Technology_2554.aspx [Accessed 4 June 2015].

Greenwood, D., Lockley, S., Malsane, S. and Matthews, J., 2010. *Automated Compliance Checking Using Building Information Models*, paper presented at the Construction, Building and Real Estate Research Conference of the Royal Institution of Chartered Surveyors, Paris, France, 2–3 September.

Grilo, A. and Jardim-Goncalves, R., 2010. Value proposition on interoperability of BIM and collaborative working environments. *Automation in Construction*, 19, pp. 522–530.

GSA, 2009. *BIM Guide for 3D Imaging*, Washington, DC: US General Services Administration.

GSA, 2011. *GSA BIM Guide for Facility Management*, Washington, DC: US General Services Adminnistration.

GSA, 2012. *GSA BIM Guide Series 05*, Washington, DC: US General Services Administration.

GSA, 2015. *3D Laser Scanning*. Available at: www.gsa.gov/portal/content/102282 [Accessed 29 May 2015].

Guo, H., Liu, W., Zhang, W. and Skitmor, M., 2014. *A BIM-RFID Unsafe On-Site Behavior Warning System*, Kunming: ICCREM 2014, Smart Construction and Management in the Context of New Technology.

Hajian, H. and Becerik-Gerber, B., 2009. *A Research Outlook for Real-Time Project Information Management by Integrating Advanced Field Data Acquisition Systems and Building Information Modelling*, paper presented at Computing in Civil Engineering, Austin, TX, 24–27 June.

Hakkarainen, M., Woodward, C. and Rainio, K., 2009. *Software Architecture for Mobile Mixed Reality and 4D BIM Interaction*, paper presented at 26th CIB W78 Conference, Istanbul, Turkey, 15–17 July.

Hannon, J. J., 2007. *National Coorperation Highway Research Program (NCHRP) Synthesis 372: Emerging Technologies for Construction Delivery*, Washington, US: Transportation Research Board.

Hardin, B. and McCool, D., 2015. *BIM and Construction Management: Proven Tools, Methods, and Workflows*, 2nd edn, Indianapolis: Wiley.

Hartmann, T., van Meerveld, H., Vossebeld, N. and Adriaanse, A., 2012. Aligning building information model tools and construction management methods. *Automation in Construction*, 22, pp. 605–613.

Harty, C. and Koch, C., 2011. *Revisiting Boundary Objects: ERP and BIM Systems as Multi-Community Artefacts*, paper presented at the 6th Nordic Conference on Construction Economics and Organisation – Shaping the Construction/Society Nexus, Copenhagen, Denmark, 13–15 April.

Hattab, M. A. and Hamzeh, F., 2015. Using social network theory and simulation to compare traditional versus BIM – lean practice for design error management. *Automation in Construction*, 52, pp. 59–69.

Heidari, M., Allameh, E., de Vries, B., Timmermans, H., Jessurun, J. and Mozaffar, F., 2014. Smart-BIM virtual prototype implementation. *Automation in Construction*, 39, pp. 134–144.

Hergunsel, M. F., 2011. *Benefits of Building Information Modelling for Construction Managers and BIM Based Scheduling*, Worcester: Worcester Polytechnic Institute.

Herring, S., 2013. *GPS Machine Control Grading & BIM (Building Information Modelling)*. Available at: www.ucls.org/assets/documents/1_machine-control-grading%20with%20notes.pdf [Accessed 8 June 2015].

Hichri, N., Stefani, C., De Luca, L., Veron, P. and Hamon, G., 2013a. From point cloud to BIM: A survey of existing approaches. *International Archives of the Photogrammetry, Remote Sensing and Spatial Information Sciences*, XL-5/W2, pp. 343–348.

Hichri, N., Stefani, C., De Luca, L. and Veron, P., 2013b. Review of the 'as-built BIM' approaches. *International Archives of the Photogrammetry, Remote Sensing and Spatial Information Science*, XL-5/W1, pp. 107–112.

HM Government, 2013. *Government soft landings (GSL): GSL Powered by BIM Summary*. Available at: www.bimtaskgroup.org/wp-content/uploads/2013/02/GSL-Summary-170213.pdf [Accessed 25 September 2015].

Holzer, D., 2007. *Are You Talking To Me? Why BIM Alone is Not the Answer*, paper presented at 4th International Conference of the Association of Architecture Schools of Australasia, Sydney, Australia, 27–29 September.

Huang, T., Kong, C. W., Guo, H. L., Baldwin, A. and Li, H., 2007. A virtual prototyping system for simulating construction processes. *Automation in Construction*, 16, p. 576–585.

Hunt, G., 2013. *Comprehensive Facility Operation & Maintenance Manual*. Available at: www.wbdg.org/om/om_manual.php [Accessed 24 June 2015].

Irizarry, J., Karam, E. P. and Jalaei, F., 2013. Integrating BIM and GIS to improve the visual monitoring of construction supply. *Automation in Construction*, 31, pp. 241–254.

ISO, 2008. *ISO15686-5: Buildings and Constructed Assets – Part 5: Life-Cycle Costing*, Geneva, Switzerland: International Organization for Standardization.

ISO, 2010. *ISO 29481-1:2010 – Building Information Modelling – Information Delivery Manual – Part 1: Methodology and Format*. Available at: www.iso.org/iso/catalogue_detail.htm?csnumber=45501 [Accessed 25 September 2015].

Jackson, T., 2014. *Topcon Unveils the LN-100: World's First 3D Positioning Solution Designed for BIM and Construction Layout*. Available at: www.equipmentworld.com/topcon-unveils-the-ln-100-worlds-first-3d-positioning-solution-designed-for-bim-and-construction-layout [Accessed 8 June 2015].

Jenkins, B., 2004. *Laser Scanning Saves $2 Million for Boston's Big Dig*. Available at: www.sparpointgroup.com/news/asset-management/laser-scanning-saves-2-million-for-bostons-big-dig#sthash.pyff2cLN.dpuf [Accessed 29 May 2015].

Jenkins, B., 2007. Utah DOT uses accelerated bridge construction to slash road closure times. *SparView*, 3 October 2007. [Online] Available at: www.sparpointgroup.com/news/mobile-scanning-and-mapping/utah-dot-uses-accelerated-bridge-construction-to-slash-road-closure-times [Accessed 9 December 2015].

Jiang, L. and Leicht, R. M., 2015. Automated rule-based constructability checking: case study of formwork. *Journal of Management in Engineering*, 31(1), pp. 1–10.

Jiao, Y., Zhang, S., Li, Y., Wang, Y. and Yang, B., 2013. Towards cloud augmented reality for construction application by BIM and SNS integration. *Automation in Construction*, 33, pp. 37–47.

Jones, K. and Collis, S., 1996. Computerized maintenance management systems. *Property Management*, 14(4), pp. 33–37.

Jongeling, R. and Olofsson, T., 2007. A method for planning of work-flow by combined use of location-based scheduling and 4D CAD. *Automation in Construction*, 16(2), pp. 189–198.

Jongeling, R., Kim, J., Fischer, M., Mourgues, C. and Olofsson, T., 2008. Quantitative analysis of workflow, temporary structure usage, and productivity using 4D models. *Automation in Construction*, 17(6), pp. 780–791.

Jotne IT, 2015. *BIM / VDC*. Available at: www.epmtech.jotne.com/solutions/bim [Accessed 24 June 2015].

Jung, J., Hong, S., Jeong, S., Kim, S. and Cho, H., 2014. Productive modelling for development of as-built BIM of existing indoor structures. *Automation in Construction*, 42, pp. 68–77.

Kamat, V. R. and El-Tawil, S., 2007. Evaluation of augmented reality for rapid assessment of earthquake-induced building damage. *Journal of Computing in Civil Engineering*, 21(5), pp. 303–310.

Kaner, I., Sacks, R., Kassian, W. and Quitt, T., 2008. Case studies of BIM adoption for precast concrete design by mid-sized structural engineering firms. *ITcon*, 13, pp. 303–323.

Kang, J. H., Anderson, S. D. and Clayton, M. J., 2007. Empirical study on the merit of web-based 4D visualization in collaborative construction planning and scheduling. *Journal of Construction Engineering and Management*, 133(6), pp. 447–461.

Kang, T. W. and Hong, C. H., 2015. A study on software architecture for effective BIM/GIS-based facility management data integration. *Automation in Construction*, 54, pp. 25–38.

Kang, Y., O'Brien, W. J. and Mulva, S. P., 2013. Value of IT: indirect impact of IT on construction project performance via best practices. *Automation in Construction*, 35, pp. 383–396.

Karlshøj, J., 2011. *Information Delivery Manuals*. Available at: http://iug.buildingsmart.org/idms [Accessed 25 September 2015].

Karlshøj, J., 2013. *Overview of Information Delivery Manuals Independent of Their Status*. Available at: http://iug.buildingsmart.org/idms/overview [Accessed 27 July 2015].

Kassem, M., Kelly, G., Dawood, N., Serginson, M. and Lockley, S., 2015. BIM in facilities management applications: a case study of a large university complex. *Built Environment Project and Asset Management*, 5(3), pp. 261–277.

Keegan, C. J., 2010. *Building Information Modelling in Support of Space Planning and Renovation in Colleges and Universities*, MSc thesis report, Worcester, Worcester Polytechnic Institute.

Kehily, D., McAuley, B. and Hore, A., 2012. Leveraging whole life cycle costs when utilising building information modelling technologies. *International Journal of 3-D Information Modelling*, 1(4), pp. 40–49.

Kehily, D., Woods, T. and McDonnell, F., 2013. Linking effective whole life cycle cost data requirements to parametric building information models using BIM technologies. *International Journal of 3-D Information Modeling*, 2(4), pp. 1–11.

Khanzode, A., 2010. *An Integrated, Virtual Design and Construction and Lean (IVL) Method for Coordination of MEP*, Stanford: Center for Integrated Facility Engineering (CIFE), Stanford University.

Kim, J. B., Jeong, W. S., Clayton, M. J., Haberl, J. S. and Yan, W., 2015. Developing a physical BIM library for building thermal energy simulation. *Automation in Construction*, 50, pp. 16–28.

Kim, K. and Teizer, J., 2014. Automatic design and planning of scaffolding systems using building information modeling. *Advanced Engineering Informatics*, 28(1), pp. 66–80.

Kivits, R. A. and Furneaux, C., 2013. BIM: Enabling sustainability and asset management through knowledge management. *The Scientific World Journal*, Article ID 983721.

Klein, L., Li, N. and Becerik-Gerber, B., 2012. Imaged-based verification of as-built documentation of operational buildings. *Automation in Construction*, 21, pp. 161–171.

KMC Controls, 2015. *Understanding Building Automation and Control Systems*. Available at: www.kmccontrols.com/products/Understanding_Building_Automation_and_Control_Systems.aspx [Accessed 24 June 2015].

Kreider, R., Messner, J. and Dubler, C., 2010. *Determining the Frequency and Impact of Applying BIM for Different Purposes on Projects*, paper presented at the 6th International Conference on Innovation in Architecture, Engineering & Construction (AEC), University Park, PA, 9–11 June.

Krigsvoll, G., 2007. Life cycle costing as part of decision making-use of building information models, in L. Bragança, Pinheiro, M., Jalali, S., Mateus, R., Amoêda, R. and Correia Guedes, M. (eds.), *Portugal SB07 Sustainable Construction, Materials and Practices: Challenge of the Industry for the New Millennium*. Amsterdam: IOS Press, pp. 433–440.

Krygiel, E. and Nies, B., 2008. *Green BIM: Successful Sustainable Design with Building Information Modeling*. Hoboken: Wiley.

Ku, K. and Mills, T., 2010. *Research Needs for Building Information Modeling for Construction Safety*, paper presented *at International Proceedings of Associated Schools of Construction 45nd Annual Conference*, Boston, MA.

Kunz, J. and Fischer,. M., 2012. *Virtual Design and Construction: Themes, Case Studies and Implementation Suggestions*, CIFE Working Paper #097, Stanford: Sanford University.

Kwon, O.-S., Park, C.-S. and Lim, C.-R., 2014. A defect management system for reinforced concrete work utilizing BIM, image-matching and augmented reality. *Automation in Construction*, 46, pp. 74–81.

Lagüela, S., Díaz-Vilariño, L., Martínez, J. and Armesto, J., 2013. Automatic thermographic and RGB texture of as-built BIM for energy rehabilitation purposes. *Automation in Construction*, 31, pp. 230–240.

LandTech Consultants, 2014. *3D Modeling – GPS Machine Control*. Available at: www.ltc3d.com/3D_modeling-GPS_machine_control.php [Accessed 8 June 2015].

Lee, G., Lee, J., Jones, S. A., Uhm, M., Won, J., Ham, S. and Park, Y., 2012a. *The Business Value of BIM in South Korea: How Building Information Modeling is Driving Positive Change in the South Korean Construction Industry*, Bedford: McGraw Hill Construction.

Lee, J.-K. Lee, J., Jeong, Y., Sheward, H., Sanguinetti, P., Abdelmohsen, S. and Eastman, C. M., 2012b. Development space database for automated building design review systems. *Automation in Construction*, 24, pp. 203–212.

Lee, J. M., 2010. *Automated Checking of Building Requirements on Circulation over a Range of Design Phases*, Atlanta: Georgia Institute of Technology.

Levy, F., 2012. *BIM in Small-Scale Sustainable Design*. Hoboken: John Wiley & Son.

Lindvall, P., 2015. *BIM Show Live – Stage 2: Business Value of Photogrammetry in Infrastructure Projects*. Available at: www.bimshowlive.co.uk/seminar/stage-2-business-value-of-photogrammetry-in-infrastructure-projects [Accessed 26 February 2015].

Lin, Y.-C., 2014. Construction 3D BIM-based knowledge management system: a case study. *Journal of Civil Engineering and Management*, 20(2), pp. 186–200.

Li, N., Yang, Z., Ghahramani, A., Becerik-Gerber, B. and Soibelman, L., 2014. Situational awareness for supporting building fire emergency response: information needs, information sources, and implementation requirements. *Fire Safety Journal*, 63, pp. 17–28.

Liu, X., Akinci, B., Berges, M. and Garrett, J. H., 2012. *Exploration and Comparison of Approaches for Integrating Heterogeneous Information Sources to Support Performance Analysis and HVAC Systems*, paper presented at the International Conference on Computing in Civil Engineering, ASCE, Clearwater Beach, FL, 17–20 June.

Mahalingam, A., Yadav, A. K. and Varaprasad, J., 2015. Investigating the role of lean practices in enabling BIM adoption: evidence from two Indian cases. *Journal of Construction Engineering and Management*, 141(7), pp. 5001–5006.

Mäki, T. and Kerosuo, H., 2015. Site managers' daily work and the uses of building information modelling in construction site management. *Construction Management and Economics*, 33(3), pp. 163–175.

Malkin, R., 2010. BIM for efficient sustainable design. *Architecture Australia*, 99(5), pp. 105–107.

Manchester City Council, 2013. *BIM Case Study: Using BIM to Enhance On-site Health & Safety*, Manchester: Manchester City Council.

Manning, R. and Messner, J. L., 2008. Case studies in BIM implementation for programming of healthcare facilities. *ITCon*, 13, pp. 446–457.

Marzouk, M. and Abdelaty, A., 2014. BIM-based framework for managing performance of subway stations. *Automation in Construction*, 41, pp. 70–77.

Marzouk, M., Hisham, M., Ismail, S., Youssef, M. and Seif, O., 2010. *On the Use of Building Information Modeling in Infrastructure Bridges*, paper presented at the 27th International Conference – Applications of IT in the AEC Industry (CIB W78), Cairo, Egypt, 16–19 November.

Mathur, V. N., Price, A. D. and Austin, S., 2008. Conceptualizing stakeholder engagement in the context of sustainability and its assessment. *Construction Management and Economics*, 26(6), pp. 601–609.

Matta, C., Zahm, F., Rocha, F., Hellmann, I., Tinsley, M., Cheng, P., Koenig, P., Gee, R. and Fallon, K., 2015. *BIM Guide 02 Spatial Program Validation*, Washington, DC: United States General Service Administration.

McGraw Hill Construction, 2014a. *The Business Value of BIM for Construction in Major Global Markets: How Contractors aaround the World are Driving Innovation with Building Information Modeling*, Bedford, MA: McGraw Hill Construction.

McGraw Hill Construction, 2014b. *The Business Value of BIM in Australia and New Zealand: How Building Information Modeling is Transforming the Design and Construction Industry SmartMarket Report*, Bedford, MA: McGraw Hill Construction.

Meadati, P., Irizarry, J. and Akhnoukh, A. K., 2010. *BIM and RFID Integration: A Pilot Study*, paper presented at the Second International Conference on Construction in Developing Countries (ICCIDC-II), Cairo, Egypt, 3–5 August.

Merschbrock, C. and Munkvold, B. E., 2014. *Succeeding with Building Information Modeling: A Case Study of BIM Diffusion in a Healthcare Construction Project*, paper presented at the 47th Hawaii International Conference on System Science, Hawaii Island, 6–9 January.

Meža, S., Turk, Ž. and Dolenc, M., 2014. Component based engineering of a mobile BIM-based augmented reality system. *Automation in Construction*, 42, pp. 1–12.

Microsoft, 2015. *Microsoft HoloLens: A New Way to See the World*. Available at: www.microsoft.com/microsoft-hololens/en-us/hardware [Accessed 15 June 2015].

Migilinskas, D., Popov, V., Juocevicius, V. and Ustinovichius, L., 2013. The benefits, obstacles and problems of practical BIM implementation. *Procedia Engineering*, 57, pp. 767–774.

Moran, M. S., 2012. *Assessing the Benefits of a Field Data Management Tool*, Delft: Delft University of Technology.

Motamedi, A. and Hammad, A., 2009. Life-cycle management of facilities components using radio frequency identification and building information modelling. *Journal of Information Technology in Construction*, 14, pp. 238–262.

Motamedi, A., Setayeshgar, S., Soltani, M. M. and Hammad, A., 2013. *Extending BIM to Incorporate Information of RFID Tags Attached to Building Assets*, paper presented at the 4th Construction Specialty Conference, Montreal, Canada, 29 May–1 June.

Motamedi, A., Hammad, A. and Asen, Y., 2014. Knowledge-assisted BIM-based visual analytics for failure root cause detection in facilities management. *Automation in Construction*, 43, pp. 73–83.

Motawa, I. and Almarshad, A., 2013. A knowledge-based BIM system for building maintenance. *Automation in Construction*, 29, pp. 173–182.

NATSPEC, 2011. *NATSPEC National BIM Guide*, s.l.: Construction Information Systems Limited.

NATSPEC, 2014a. *BIM R&D Projects*. Available at: http://bim.natspec.org/index.php/research-development/bim-r-d-projects [Accessed 29 July 2015].

NATSPEC, 2014b. *Introduction to BIM*. Available at: http://bim.natspec.org/index.php/resources/introduction-to-bim [Accessed 11 December 2015].

NATSPEC BIM, 2015. *National Object Library Object Creation Standard*. NATSPEC BIM Team Update, 29 July.

Nawari, N. O., 2012. *Automated Code Checking in BIM Environments*, paper presented at the 14th International Conference of Computing in Civil and Building Engineering, Moscow, Russia, 27–29 June.

NBIMS-US, 2015. *buildingSMART Alliance Releseases NBIMS-US TM Version 3*. Available at: https://www.nationalbimstandard.org/buildingSMART-alliance-Releases-NBIMS-US-Version-3 [Accessed 27 July 2015].

NBS, 2014. *NBS Pioneering Automated Checking of Building Regulations*. Available at: www.thenbs.com/corporate/press/14-02-18.asp [Accessed 22 June 2015].

NBS National BIM Library, 2015. *NBS National BIM Library*. Available at: www.nationalbimlibrary.com [Accessed 28 July 2015].

Nemetschek Group, 2014. *Nemetschek Advances to Become Leading BIM 5D Provider*. Available at: www.nemetschek.com/en/presse/press-releases/detail/nemetschek-advances-to-become-leading-bim-5d-provider [Accessed 29 June 2015].

Newforma Inc., 2015. *Newforma Building Information Management*. Available at: www.newforma.com.au/products/newforma-building-information-management [Accessed 24 September 2015].

Nguyen, T.-H. and Kim, J.-L., 2011. *Building Code Compliance Checking Using BIM Technology*, paper presented at the Winter Simulation Conference, Phoenix, AZ, 11–14 December.

NIBS-US, 2015. *Frequently Asked Questions About the National BIM Standard-United States*. Available at: www.nationalbimstandard.org/faqs#faq3 [Accessed 27 July 2015].

Oculus VR, 2015. *Rift: Next-Generation Virtual Reality*. Available at: www.oculus.com/en-us/rift [Accessed 25 September 2015].

O'Donoghue, C. D. and Prendergast, J. G., 2004. Implementation and benefits of introducing a computerised maintenance management system into a textile manufacturing company. *Journal of Materials Processing Technology*, 153–154, pp. 226–232.

Olbrich, M., Graf, H., Kahn, S., Engelke, T., Keil, J., Riess, P., Webel, S., Bockholt, U. and Picinbono, G., 2013. Augmented reality supporting user-centric building information management. *The Visual Computer*, 29(10), pp. 1093–1105.

OmniClass™, 2006. *OmniClass™: A Strategy for Classifying the Built Environment. Introduction and User's Guide*, Toronto: OmniClass.

Onuma, 2011. *Products*. Available at: http://onuma.com/products [Accessed 30 July 2015].

OpenBIM, 2012. *BIM for Facilities Management – University Campus*. Available at: www.openbim.org/case-studies/university-campus-facilities-management-bim-model [Accessed 26 May 2015].

Page, S., 2012. *3D Laser Scanning: As-Built Reality Capture for BIM*. Available at: www.aecbytes.com/viewpoint/2012/issue_66.html [Accessed 29 May 2015].

Park, C.-S., Lee, D.-Y., Kwon, O.-S. and Wang, X., 2013. A framework for proactive construction defect management using BIM, augmented reality and ontology-based data collection template. *Automation in Construction*, 33, pp. 61–71.

Peavey, E. K., Zoss, J. and Watkins, N., 2012. Simulation and mock-up research methods to enhance design decision making. *HERD*, 5(3), pp. 133–144.

Penn State, 2011. *BIM Execution Plan*. Available at: http://bim.psu.edu/Uses/Resources/default.aspx [Accessed 5 September 2014].

Peterson, F., Fischer, M. and Tutti, T., 2009. *Integrated Scope-Cost-Schedule Model System for Civil Works*, Stanford: Center for Integrated Facility Engineering, Stanford University.

Reddy, K. P., 2011. *BIM for Building Owners and Developers: Making a Business Case for Using BIM on Projects*, Chichester: John Wiley & Sons.

Redmond, A., Hore, A., Alshawi, M. and West, R., 2012. Exploring how information exchanges can be enhanced through cloud BIM. *Automation in Construction*, 24, pp. 175–183.

Rueppel, U. and Stuebbe, K. M., 2008. BIM-based indoor-emergency-navigation-system for complex buildings. *Tsinghua Science & Technology*, 13, pp. 362–637.

Rundell, R., 2008. *BIM and Digital Fabrication*, s.l.: Autodesk.

Sabol, L., 2008. *Building Information Modeling & Facility Management*, paper presented at *IFMA World Workplace*, Dallas, 15–17 October.

Sacks, R. and Barak, R., 2008. Impact of three-dimensional parametric modeling of buildings on productivity in structural engineering practice. *Automation in Construction*, 17, pp. 439–449.

Sacks, R., Eastman, C. M., Lee, G. and Orndorff, D., 2005. A target benchmark of the impact of three-dimensional parametric modeling in precast construction. *PCI Journal*, 50(4), pp. 126–138.

Sacks, R., Koskela, L., Dave, B. A. and Owen, R., 2010a. Interaction of lean and building information modeling in construction. *Journal of Construction Engineering and Management*, 136(9), pp. 968–980.

Sacks, R., Radosavljevic, M. and Barak, R., 2010b. Requirements for building information modelling based lean production management systems for construction. *Automation in Construction*, 19, pp. 641–655.

Sahlman, W., 2015. *An Australian First for BIM Capability in Bridge Maintenance*, paper presented at Road and Rail Structures 2015, Sydney, Australia, 28–29 April.

Sanchez, A. X., Kraatz, J. A., Hampson, K. D. and Loganathan, S., 2014a. *BIM for Sustainable Whole-of-Life Transport Infrastructure Asset Management*, paper presented at IPWEA Sustainability in Public Works Conference, Tweed Heads, Australia, 27–29 July.

Sanchez, A. X., Kraatz, J. A. and Hampson, K. D., 2014b. *Document Review. Research Report* 2, Perth, WA: Sustainable Built Environment National Research Centre.

Sanchez, A. X., Hampson, K. D. and Mohamed, S., 2015a. *Case Study Report – Sydney Opera House*, Perth, Australia: Sustainable Built Environment National Research Centre (SBEnrc).

Sanchez, A. X., Hampson, K. D. and Mohamed, S., 2015b. *Perth Children's Hospital Case Study Report*, Perth: Sustainable Built Environment National Research Centre.

Sattineni, A. and Azhar, S., 2010. *Techniques for Tracking RFID in a BIM Model*, paper presented at the 27th International Symposium on Automation and Robotics in Construction (ISARC 2010), Slovakia, 25–27 June.

SBEnrc, 2012. *Collaborative Object Libraries- Project 3.1 Final Report*, Brisbane: Sustainable Built Environment National Research Centre.

Schall, G., Mendez, E., Kruijff, E., Veas, E., Junghanns, S., Reitinger, B. and Schmalstieg, D., 2009. Handheld augmented reality for underground infrastructure visualization. *Personal and Ubiquitous Computing*, 13(4), pp. 281–291.

Schlueter, A. and Thesseling, F., 2009. Building information model based energy/exergy performance assessment in early design stages. *Automation in Construction*, 18, pp. 153–163.

Scott, T., Barda, P., Marshall, C., Kane, C., Eynon, D., Schuck, R., Burt, N., Canham, C., Mitchell, J., Collard, S. and Jurgens, D., 2014. *A Framework for the Adoption of Project Team Integration and Building Information Modelling*, Canberra: Australian Construction Industry Forum and Australasian Procurement and Construction Council.

Seeam, A., Zheng, T., Lu, Y., Usmani, A. and Laurenson, D., 2013. BIM integrated workflow management and monitoring system for modular buildings. *International Journal of 3D Information Modeling*, 2(1), pp. 17–28.

Seely, J. C., 2004. *Digital Fabrication in the Architectural Design Process*, Cambridge, MA: Massachusetts Institute of Technology.

Sherman, H., 2013. *BIM: The Way Forward*. Available at: www.minterellison.com/publications/BIM-the-way-forward-CLA201303 [Accessed 31 May 2013].

Shi, W., 2009. *Framework for Integration of BIM and RFID in Steel Construction*, doctoral dissertation, Gainesville, FL: University of Florida.

Singh, V., Gu, N. and Wang, X., 2011. A theoretical framework of a BIM-based multi-disciplinary collaboration platform. *Automation in Construction*, 20, pp. 134–144.

Skanska, 2015. Skanska in Brief. Available at: http://group.skanska.com/about-us/skanska-in-brief [Accessed 30 July 2015].

SmartBIM, 2015. SMARTBIM Library. Available at: http://library.smartbim.com/#area=3 [Accessed 28 July 2015].

Smith, P., 2014. BIM & the 5D Project Cost Manager. *Procedia – Social Behavioural Science*, 119, pp. 475–484.

Solibri, 2015. *Solibri Model Checker*. Available at: http://solibri.com.au/products/solibri-model-checker [Accessed 24 June 2015].

Solihin, W. and Eastman, C., 2015. Classification of rules for automated BIM rule checking development. *Automation in Construction*, 53, pp. 69–82.

SPAR Point Group Staff, 2005. 3D laser scanning shortens repair schedule for Petronius. *SparView*, 17 May.

Stanley, R. and Thurnell, D., 2014. The benefits of, and barriers to, implementation of 5D BIM for quantity surveying in New Zealand. *Australasian Journal of Construction Economics and Building*, 14(1), pp. 105–117.

Starr, M., 2015. World's first 3D-printed apartment building constructed in China. *CNET*, 20 January. Available at: www.cnet.com/au/news/worlds-first-3d-printed-apartment-building-constructed-in-china.

Staub, A., 2014. Germany – researching sustainability, in K.D. Hampson, J.A. Kraatz and A.X. Sanchez (eds.), *R&D Investment and Impact in the Global Construction Industry*. London: Routledge, pp. 135–156.

Staub-French, S. and Khanzode, A., 2007. 3D and 4D modeling for design and construction. *ITCON*, 12, pp. 381–407.

Stone, A., 2013. *Preparing for the Next Superstorm: BIM for Natural Disaster Preparedness and Recovery*. Available at: www.revitcommunity.com/feature_article.php?read=1&g=8c4070de-0b67-11e3-8a5e-d4ae52bb7f09&cpfeatureid=77251&page=all [Accessed 11 April 2014].

Su, Y. C., Lee, C. Y. and Lin, Y. C., 2011. *Enhancing Maintenance Management Using Building Information Modelling in Facilities Management*, paper presented at the 28th International Symposium on Automation and Robotics in Construction, Seoul, South Korea, 29 June–2 July.

Sulankivi, K., Makela, T. and Kiviniemi, M., 2009. *BIM-Based Site Layout and Safety Planning*, paper presented at the First International Conference on Improving Construction and Use through Integrated Design Solutions, Espoo, Finland, 10–12 June.

Sullivan, C. C., 2007. Integrated BIM and design review for safer, better buildings: how project teams using collaborative design reduce risk, creating better health and safety in projects. *AIA/Architectural Record Continuing Education Series*, 6, pp. 191–199.

Sullivan, G. P., Pugh, R., Melendez, A. P. and Hunt, W. D., 2010. *Operations & Maintenance Best Practices: A Guide for Achieving Operational Efficiency*, Washington, DC: US Department of Energy.

Tang, A., Owen, C., Biocca, F. and Mou, W., 2003. *Comparative Effectiveness of Augmented Reality in Object Assembly*, paper presented at the SIGCHI Conference on Human Factors in Computing Systems, Ft. Lauderdale, FL, 5–10 April.

Tang, P., Huber, D., Akinci, B., Lipman, R., and Lytle, A., 2010. Automatic reconstruction of as-built building information models from laser-scanned point clouds: a review of related techniques. *Automation in Construction*, 19, pp. 829–843.

Tang, P., Anil, E. B., Akinci, B. and Huber, D., 2011. *Efficient and Effective Quality Assessment of As-Is Building Information Models and 3D Laser-Scanned Data*, paper presented at the International Workshop on Computing in Civil Engineering, American Society of Civil Engineers, Miami, FL, 19–22 June.

Teichholz, P., 2013. *BIM for Facility Managers*. Hoboken: Wiley.

Thomas, S. R., Lee, S.-H., Spencer, J. D. and Tucker, R. L., 2004. Impacts of design information technology on project outcomes. *Journal of Construction Engineering and Management*, 130(4), pp. 586–597.

Topcon Positioning Systems, 2008. *Automatic Stakeless Grading*, Livermore: Topcon Positioning Systems.

Trafikverket, 2015. *Beställa förvaltningsdata (Order Management Data)*. Available at: www.trafikverket.se/Foretag/Bygga-och-underhalla/Jarnvag/Forvaltningsdata/Bestalla-forvaltningsdata [Accessed 28 July 2015].

Trimble, 2015a. *3D Warehouse*. Available at: https://3dwarehouse.sketchup.com/index.html [Accessed 28 July 2015].

Trimble, 2015b. *DPI-8 Handheld 3D Scanner for Building Construction*. Available at: www.trimble.com/construction/DPI-8-Handheld-Scanner [Accessed 25 September 2015].

Tutt, D. and Harty, C., 2013. *Journeys Through the CAVE: The Use of 3D Immersive Environments for Client Engagement Practices in Hospital Design*, paper presented at 29th Annual ARCOM Conference, Association of Researchers in Construction Management, Reading, UK, 2–4 September.

Tuttas, S., Braun, A., Borrmann, A. and Stilla, U., 2014. *Comparison of Photogrammetric Point Clouds with BIM Building Elements for Construction Progress Monitoring*, paper presented at ISPRS-International Archives of the Photogrammetry, Remote Sensing and Spatial Information Sciences, Zurich, Switzerland, 5–7 September.

Union Square, 2015a. *What We Offer*. Available at: www.unionsquaresoftware.com.au/solutions/union-square-for-construction/what-we-offer [Accessed 24 September 2015].

UnionSquare, 2015b. *The Open BIM Alliance*. Available at: www.unionsquaresoftware.com.au/the-open-bim-alliance [Accessed 24 September 2015].

US Army Corps of Engineers, 2012. *The US Army Corps of Engineers Roadmap for Life-Cycle Building Information Modeling (BIM)*, Washington, DC: US Army Corps of Engineers.

USGBC, 2014. *LEED v4 User Guide*, Washington, DC: US Green Building Council.

Vico Software, 2013. *The Mission Hospital – St. Joseph Health System*. Available at: www.vicosoftware.com/Portals/658/docs/Case%20Study_Mission%20Hospital.pdf [Accessed 29 June 2015].

Vico Software, 2015a. *The 5D BIM Workflow in Vico Office*. Available at: www.vicosoftware.com/5D-BIM-Workflow-in-Vico-Office/tabid/223771/Default.aspx [Accessed 29 June 2015].

Vico Software, 2015b. *Vico Takeoff Manager*. Available at: www.vicosoftware.com/products/vico-office-quantity-takeoff-manager/tabid/85287/Default.aspx [Accessed 29 June 2015].

Vico Software, 2015c. *CSI Uniformat and Masterformat*. Available at: www.vicosoftware.com/csi-standards-uniformat-quantities/tabid/87560 [Accessed 30 July 2015].

Vico Software, 2015d. *Virtual Mock-Up*. Available at: www.vicosoftware.com/virtual-mock-up-construction-services/tabid/84370 [Accessed 29 June 2015].

Violino, B., 2015. *RFID Journal – What is RFID?* Available at: www.rfidjournal.com/articles/view?1339 [Accessed 21 November 2014].

Volk, R., Stengel, J. and Schultmann, F., 2014. Building Information Modeling (BIM) for existing buildings – literature review and future needs. *Automation in Construction*, 38, pp. 109–127.

Vorakulpipat, C. and Rezgui, Y., 2008. Value creation: the future of knowledge management. *The Knowledge Engineering Review*, 23(3), pp. 283–294.

Wang, B., Li, H., Rezgui, Y., Bradley, A. and Ong, H. N., 2014a. BIM based virtual environment for fire emergency evacuation. *The Scientific World Journal*, Article ID 589016, pp. 1–22.

Wang, X., 2009. Augmented reality in architecture and design: potentials and challenges for application. *International Journal of Architectural Computing*, 7(2), pp. 309–326.

Wang, X. and Dunston, P. S., 2006. Potential of augmented reality as an assistant viewer for computer-aided drawing. *Journal of Computing in Civil Engineering*, 20(6), pp. 437–441.

Wang, X. and Dunston, P. S., 2007. Design, strategies, and issues towards an augmented reality-based construction training platform. *ITcon*, 12, pp. 363–380.

Wang, X., Love, P E., Kim, M. J., Park, C.-S., Sing, C.-P. and Hou, L., 2013a. A conceptual framework for integrating building information modeling. *Automation in Construction*, 34, pp. 37–44.

Wang, X., Truijens, M., Hou, L., Wang, Y. and Zhou, Y., 2014b. Integrating augmented reality with building information modeling: Onsite construction process controlling for liquefied natural gas industry. *Automation in Construction*, 40, pp. 96–105.

Wang, Y., Wang, X., Wang, J., Yung, P. and Jun, G., 2013b. Engagement of facilities management in design stage through BIM: framework and a case study. *Advances in Civil Engineering*, Article ID 189105, pp. 1–8.

Waterhouse, R., Chapman, I., Monswhite, D., Rutland, C., Duncan, A., Malleson, A., Hamil, S., Reeves, J., Poulter, J., Munkley, J., Anwyl, J., Ball, T., Allen, N., Annable, R. and Moorhouse, J., 2014. *NBS National BIM Report*, Newcastle upon Tyne: NBS.

Weygant, R. S., 2011. *BIM Content Development: Standards, Strategies, and Best Practices*, Hoboken: Wiley and Sons.

Wikitude GmbH, 2015. *Wikitude: See More*. Available at: www.wikitude.com [Accessed 15 June 2015].

Williams, G., Gheisari, M., Chen, P.-J. and Irizarry, J., 2015. An efficient BIM translation to mobile augmented reality applications. *Journal of Management in Engineering*, 31, pp. 1–8.

Williams, J., Amor, R., Apleby, S., Boyden, G., Davis, S., Greenstreet, N., Hawkins, J., Jowett, G., Read, H., Hunter, F. and Reding, A., 2014. *New Zealand BIM Handbook: A Guide to Enabling BIM on Building Projects*, s.l.: Building and Construction Productivity Partnership.

Winke, J., 2014. Project profile: river shore restoration tech. Erosion control, *Forester Daily News*, 8 April 2014. Available at: http://foresternetwork.com/daily/soil/project-profile-river-shore-restoration-tech [Accessed 9 December 2015].

Wix, J. and Karlshøj, J., 2010. *Information Delivery Manual: Guide to Components and Development Methods*, s.l.: buildingSMART.

Wong, J. K.-W. and Kuan, K.-L., 2014. Implementing 'BEAM Plus' for BIM-based sustainability analysis. *Automation in Construction*, 44, pp. 163–175.

Wong, J. K. W. and Zhou, J., 2015. Enhancing environmental sustainability over building life cycles through green BIM: a review. *Automation in Construction*, 57, pp. 156–165.

Wong, J., Wang, X., Li, H., Chan, G. and Li, H., 2014. A review of cloud-based BIM technology in the construction sector. *Journal of Information Technology in Construction*, 19, pp. 281–291.

Wong, K.-D. and Fan, Q., 2013. Building information modelling (BIM) for sustainable building design. *Facilities*, 31(3/4), pp. 138–157.

Woo, J., Wilsmann, J. and Kang, D., 2010. *Use of As-Built Building Information Modeling*, paper presented at Construction Research Congress, Banff, Alberta, Canada, 8–10 May.

Wu, W., Yang, X. and Fan, Q., 2014. *GIS-BIM Based Virtual Facility Energy Assessment (VFEA) – Framework Development and Use Case of California State University, Fresno*, paper presented at Computing in Civil and Building Engineering (ASCE), Orlando, FL, 23–25 June.

Xie, H., Shi, W. and Issa, R., 2010. *Implementation of BIM/RFID in Computer-Aided Design-Manufacturing-Installation Process*, paper presented at 3rd IEEE International Conference on Computer Science and Information Technology (ICCSIT), Chengdu, China, 9–11 July.

Xiong, X., Adan, A., Akinci, B. and Huber, D., 2013. Automatic creation of semantically rich 3D building models from laser scanner data. *Automation in Construction*, 31, pp. 325–337.

Yeh, K.-C., Tsai, M.-H. and Kang, S.-C., 2012. On-site building information retrieval by using projection-based augmented reality. *Journal of Computing in Civil Engineering*, 26(3), pp. 342–355.

Zhang, L. and Issa, R. R., 2012. *Comparison of BIM Cloud Computing Frameworks*, paper presented at the International Conference on Computing in Civil Engineering, Clearwater Beach, FL, 17–20 June.

Zhang, S., Teizer, J., Lee, J.-K., Eastman, C. M., Venugopal, M., 2013. Building information modeling (BIM) and safety: automatic safety checking of construction models and schedules. *Automation in Construction*, 29, pp. 183–195.

Zhang, S., Sulankivi, K., Kiviniemi, M., Romo, I., Eastman, C. M. and Teizer, J., 2015. BIM-based fall hazard identification and prevention in construction safety planning. *Safety Science*, 72, pp. 31–45.

Zhang, X., Arayici, Y., Wu, S., Abbott, C. and Aouad, G., 2009. *Integrating BIM and GIS for Large Scale (Building) Asset Management: A Critical Review*, paper presented at the Twelfth International Conference on Civil, Structural and Environmental Engineering Computing, Funchal, Madeira, Portugal, 1–4 September.

ZSL, 2012. *Fornax – Plan Check Expert*. Available at: www.zsl.com/egovernance-solutions/fornax-plan-check-expert [Accessed 24 June 2015].

Zuuse, 2015. *Zuuse*. Available at: www.zuuse.com/operate [Accessed 25 September 2015].

Metrics dictionary

Adriana X. Sanchez and Will Joske

This *Metrics dictionary* provides an overview of metrics that can be associated with benefits to measure the progress towards achieving project and programme goals. The benefits and failures of BIM need to be measured throughout the life-cycle of the asset to guarantee a continual improvement process (Eadie et al., 2013) and support the adoption by new users that depends on how the real benefits of the transition are perceived (Lu et al., 2012).

Different stakeholders in different locations with different levels of expertise tend to measure the value achieved from implementing BIM in different ways. In a global survey published by McGraw Hill (2014a), metrics such as cost savings, profitability and productivity (labour intensity) were the most commonly used indicators to measure benefits from BIM by contractors, especially those with high BIM engagement levels. Process metrics related to schedule such as overall time for project delivery were the second most preferred indicators. However, Japanese contractors rated safety metrics among the top most important measures while South Korean and US firms rated them lower. Subcontractors also highlighted safety and off-site manufacturing (pre-fabrication) significantly more frequently than head contractors (McGraw Hill Construction, 2014a).

It is recommended that internal metrics dictionaries are created as well as a repository where all the metrics which are to be measured reside and an expanding history of actual values can be recorded. Dictionaries such as this provide a practical way of avoiding wasted efforts (Bradley, 2010) and the one in this book, in particular, can be used as a starting point or reference document to develop those internal dictionaries. The following sections provide a set of metrics from which each team can select those that are most appropriate to their goals and needs. These metrics are mostly based on literature but also include key performance indicators (KPIs) proposed by the authors based on professional experience.

Each metric profile has aimed to provide the definition of the KPI and information about challenges and reported use of these metrics when available. These have been categorised into:

1. **People** – serve to monitor benefits achieved through changes in behavioural patterns or that directly affect staff.
2. **Processes** – monitor benefits achieved through changes to general process improvement and generally aim to measure the efficiency of these processes.
3. **Procurement** – monitor benefits achieved during or through procurement and asset management processes.
4. **Sustainability and future-proofing** – monitor benefits achieved in terms of better environmental sustainability outcomes and improved emergency management.

People

Satisfaction

This metric can refer to client, staff or end-user satisfaction and is usually measured qualitatively through surveys and satisfaction scales feedback. Rankin et al. (2008), for example, suggest a seven-point scale for client satisfaction where one is extremely dissatisfied and seven is extremely satisfied or a five-point scale where one is low satisfaction and five is high satisfaction/delighted. In this last case, the success criterion could be to achieve a rating of four or more.

When applied specifically to client satisfaction, this metric can be divided into product, service and other client-specified criteria (KPI Working Group, 2000). Table MD.1 provides a more detailed description of these metrics.

Also to measure client satisfaction Rankin et al. (2008) propose to monitor: (1) quality issues – available for use; and (2) quality issues – warranty. The first measures 'the level of client satisfaction with the product at the time the product is considered available for use as measured by the number of open (outstanding) non-conformances when product was available for use'; on a scale from one to seven, where seven is highly satisfied. Similarly, the latter measures the level of satisfaction at the end of warranty based on outstanding non-conformances.

On the other hand, if the goal is associated with improving staff satisfaction, this can be estimated through associated performance indicators such as staff turnover rate, absenteeism and motivation (based on a qualitative feedback scale) (Cox et al., 2003).

End-user satisfaction can be similarly monitored through a seven-point scale in customer feedback surveys, logs of customer feedback and post-occupancy evaluations (Chan and Chan, 2004; CURT, 2005; Kunz and Fischer, 2012).

Table MD.1 Client satisfaction indicators (KPI Working Group, 2000)

Indicator	*Definition*
Client satisfaction product – standard criteria	In a scale of 1 to 10, how satisfied was the client with the finished output/product? Where 1 is extremely dissatisfied and 10 is extremely satisfied/exceeded expectations.
Client satisfaction service – standard criteria	In a scale of 1 to 10, how satisfied was the client with the service of the advisor, suppliers and contractors? Where 1 is extremely dissatisfied and 10 is extremely satisfied/exceeded expectations.
Client satisfaction – client-specified criteria	In a scale of 1 to 10, how satisfied was the client with certain client-specified criteria? Where 1 is extremely dissatisfied and 10 is extremely satisfied/exceeded expectations – weighted together to determine level of importance.

Satisfaction metrics can also be applied to different stakeholders during project delivery to measure satisfaction with communication between team members. This can be expressed as a percentage of stakeholders who are satisfied with a meeting or current communication style and frequency (Kam et al., 2014). In this case, this metric can provide an indicator of communication improvement or lack thereof.

Safety

Safety is commonly a major concern in construction projects and asset management. Metrics associated with the goal of *improved safety* commonly aim to measure the degree to which changes and processes promote conditions that lead to accident/injury-free delivery and operations (Cox et al., 2003; Chan and Chan, 2004). There are a number of indicators that can be used for quantitative measures such as accident rate (Equation 1); experience modification ratings (EMRs); exposure hours (work hours); lost man-hours due to accident or injury; number of recordable incidents/accidents (including fatal and non-fatal); number of doctor and first-aid cases; number of near-miss incidents, and the ratio between the number of lost time accidents and hours worked (Cox et al., 2003; Chan and Chan, 2004; Suermann, 2009; CURT, 2005).

Equation 1 Accident rate (Chan and Chan, 2004)

$$Accident\ Rate = \frac{Total\ no.\ of\ reportable\ construction\ site\ accidents}{Total\ no: of\ workers\ employed\ or\ man-hours\ worked} \times 100\%$$

Equation 2 Lost time accidents ratio (Suermann, 2009)

$$Lost\ time\ accidents\ ratio = \frac{Number\ of\ lost\ time\ accidents \times 20{,}000}{Total\ hours\ worked}$$

Reportable lost-time accidents (including fatal and non-fatal) can be calculated, as a total or segregated into fatal and non-fatal, per 100,000 hours worked (KPI Working Group, 2000). This can be measured in terms of the amount of time lost to incidents or the number of incidents per 100,000 hours worked (Rankin et al., 2008). EMR calculations are relatively more complicated but Hinze et al. (1995) provide a good description of how to use this indicator to monitor safety performance improvement in construction projects.

BIM has been proposed to improve safety statistics by reducing risk through modelling the egress, safety, hazards and simulating safety training (Kam et al., 2014). Kunz and Fischer (2012) suggest that the success criteria for BIM-enabled projects may be that BIM processes for safety significantly and measurably (through some of the indicators mentioned above) improve overall safety.

This metric can provide the basis for BIM-based safety improvement programmes that aim to 'eliminate losses due to poor working practices that could impact workforce well-being', productivity and legal liability (Cox et al., 2003). However, some of these statistics can be difficult to obtain for benchmarking purposes due to complicated subcontracting systems, rapid flow of labour (Chan and Chan, 2004) and lack of integration across the supply chain.

Examples of reported improvement of safety related metrics due to the use of BIM include a survey carried out by Suermann (2009). Here, safety was among the six highest rated KPIs when implementing BIM, with 53.7 per cent of respondents reporting that BIM improves this KPI. This study, however, did not report on specific levels of improvement.

Meetings

Metrics related to meeting communication improvement due to the use of BIM and enabling tools can include effectiveness, efficiency and appropriateness.

Effectiveness

This metric is semi-quantitative and is calculated as the fraction of stakeholders who self-report that they had timely and meaningful participation in project meetings. A potential success criterion could be achieving 90 per cent or higher. 'Meeting effectiveness requires careful attention to meeting participation, excellent attendance, and highly relevant meeting content so that appropriate stakeholders can have timely and meaningful participation in project design decisions' (Kunz and Fischer, 2012). Although it has an element of subjectivity introduced by self-reporting, it provides a hard ratio that can be compared across projects.

Efficiency

Similarly to meeting effectiveness, this is a semi-quantitative metric. It is the fraction of actions taken during a meeting that relate to value-adding activities such as evaluation, prediction, alternative formulation or decision-making. Non-value-adding activities include description, explanation and negotiation (Kunz and Fischer, 2012). This is considered semi-quantitative because it depends on the judgement of the reporting staff on whether activities were value-adding or not.

A potential success criterion could be ≥70 per cent (Kunz and Fischer, 2012).

Agenda appropriateness

This indicator is similar to meeting efficiency but only applies to the agenda. It is the fraction of items in the meeting agenda that are acceptable as value-adding topics for a majority of meeting participants. The success criterion could be ≥ 70 per cent (Kunz and Fischer, 2012).

Stakeholder involvement

This metric aims to measure the degree to which stakeholders have timely and significant participation in review and approval processes. It is based on a self-assessment scale and can be applied to specific tasks within a life-cycle phase or overall across that phase. The success criterion could be to achieve 90 per cent (Kunz and Fischer, 2012).

Learning curve

Learning curves are mathematical models that can be used to analyse the impact of technology implementation on the learning processes of workers (Anzanello and Fogliatto, 2011). These models are based on the assumption that workers and teams learn by doing, and that the frequency of repetition increases efficiency in carrying out a specific task (Lu et al., 2012).

Lu et al. (2012) developed a method that can be used to measure the learning effects of using BIM-based visualisations on repetitive construction tasks such as raising safety screens, erecting column formwork, fixing precast edge beams and raising Holland hoists. This method is somewhat more complex than common indicators but can be used to justify initial investments and budgets for BIM implementation (Lu et al., 2012).

Reported examples of the use of this metric to measure the aggregated learning effect of using BIM in construction include a 66-floor high-rise building with a gross-floor area of standard floors equal to 3,200m^2. In this study, performance benchmark data such as staff-hours/m^2 was collected from previous non BIM-enabled projects from site logs, employee timesheets and progress reports, and compared to the same indicators in their current work using BIM. This case study showed that although the total variation in productivity was similar (1.51 staff-hours/m^2 improvement without BIM and 1.44 staff-hours/m^2 with BIM), the aggregate effect of using BIM produced a significant performance improvement on a per cycle[1] basis. This method allowed the researchers to estimate an associated financial benefit of US$55,920 during the construction of just five of the 66 floors (Lu et al., 2012).

Conflict

Conflict metrics can include the number and cost of claims or the time required to solve contractor claims and conflicts between trades (Suermann, 2009; Khanzode, 2010). The success criteria can be the complete absence of legal claims (Chan and Chan, 2004) or a lower number of conflicts than in similar projects. Improvement in these metrics can be an indication of improved coordination and more efficient communications (Khanzode, 2010).

In order to measure different benefits, this indicator can be measured as a total or separately for specific causes such as issues related to space management, changes, errors and omissions, and misunderstandings, or for specific tasks such as construction site issues.

Labour intensity

Labour intensity is commonly seen as a measure of productivity and efficiency. It can be monitored per task by asking employees to record the hours worked each week for listed activity and compare to project estimates (Suermann, 2009; Cox et al., 2003; Allen Consulting Group, 2010; Khanzode, 2010). Examples of these include labour intensity required to complete cost estimates, carrying out scenario and alternative analysis or other tasks that were not automated. During the life-cycle of an asset, this can also be measured in terms of financial value by directly multiplying the number of hours worked by the salaries of those employees. However, this type of currency-based measurement unit provides little opportunity to easily compare across projects, assets or organisations due to differences in employee salaries (Sacks et al., 2005).

The use of *man hours per units* (Equation 3) can provide an organisation with more neutral and consistent statistics (Sacks et al., 2005; Cox et al., 2003). This method measures the completed units of output per individual man-hour of work. Data to calculate this can be quickly gathered and be used for any basic task or activity (Cox et al., 2003). This could, for example, be applied as labour intensity per drawing, scenario or report.

Equation 3 Labour intensity per unit (Suermann, 2009)

$$Labour\,intensity\,per\,unit = \frac{percentage\,completed \times total\,square\,feet}{Man-hours\,to\,date}$$

Another popular approach is the earned man-hours baseline method for measuring performance. To calculate it, the estimated unit rates are multiplied by the amount of work completed (units) to date. The actual number of man-hours charged to a task can then be subtracted from the number of earned man-hours to provide an indicator of job productivity (Cox et al., 2003). This can be used as an indicator to different asset types by changing the unit; kilometre of road for linear infrastructure, square metre of floor for buildings, etc.

Lost-time accounting can also provide a metric related to labour intensity. Similarly to rework, this metric measures the time lost in wasted work hours by monitoring the man-hours lost due to idle time due to delays in either material delivery, instructions or work orders (Cox et al., 2003). This can be a good metric when trying to measure improvements on construction programming and sequencing.

Finally, multifactor productivity combines labour intensity with financial investments by calculating the ratio of the value of an output, such as project cost, to the combined input of labour and capital investment (Allen Consulting Group, 2010). This can be calculated per unit of output or per cycle. However, this data can be difficult to obtain for benchmarking efforts.

Note

- In a study by Eadie et al. (2013), 35 per cent of respondents reported using person-hours worked as a KPI to measure productivity improvements achieved from using BIM.

Knowledge management

> Knowledge management refers to the processes of creation, storage/retrieval, transferring/sharing, and applying knowledge to fulfill organizational aims and objectives. Benefits gained through such strategies are diverse and can include: improved decision making, support for knowledge retrieval, improved communication between staff, improved information about and for customers; benefits in workflow, quality and productivity; improvements in innovation and business opportunities; expansion in organizational and personal knowledge bases and finally financial outcomes in volume of sales, profit margin, competitive advantage and improvements in cash flow.
>
> (Zyngier and Burstein, 2012)

There have been some books written about what knowledge management is and how to measure it, the most recent perhaps being Geisler and Wickramasinghe (2015), which also discusses the role of technology as an enabler of knowledge management. This profile will therefore only provide a short introduction to the topic. Some methods currently used to measure knowledge management include balance scorecard, cost savings and uniform business processes (Zyngier and Burstein, 2012). The balance scorecard is a somewhat complex method designed to translate organisational objectives and strategy into specific, quantifiable goals while monitoring performance towards achieving those goals. This is measured across five dimensions; financial, client, internal processes, learning and growth (Bose, 2004).

These and other approaches are highly dependent on the organisation type and objectives. One published in 2014 in the *Business Information Review* based the selection of the specific metrics on the following series of steps:

- Understand the business problem you are trying to solve.
- Identify use cases that have meaningful impact.

- Determine the metrics that align with each use case.
- Understand your baseline and establish a target.
- Measure and monitor: be prepared to change.

(Hanley, 2014)

Other methods traditionally used include economic value-added and intellectual capital (Bontis et al., 1999). However, perhaps a more practical set of indicators for measuring knowledge management when implementing BIM is:

- Hours of training per employee.
- Hours spent in briefings.
- Hours spent by senior staff explaining strategy and actions.
- Savings from implementing team suggestions.
- Number of new solutions suggested.
- Processes completed without errors.
- Training efforts in terms of expenses per employee.

(Liebowitz and Suen, 2000)

Another approach relates to the time required to access and accessibility of knowledge about specific issues. This type of practical system audit requires answers to the questions: Can knowledge be quickly accessed by people who need it and when they need it? (Zyngier and Burstein, 2012). It could, for example, measure the amount of time required to access lessons learned about a specific issue, availability of knowledge on common asset management issues (e.g., out of a list of common issues, what percentage is easily accessible to new staff?) or even the number of clicks required to find information about a specific issue.

There are also survey-based approaches such as that applied by Lin (2014) in the evaluation of their BIM-based knowledge management system. This evaluation was made by system users who were asked to rate specific indicators in a scale from one to five, where five reflects strong agreement. These indicators included: functionality issues such as ease of knowledge-sharing and illustration; utility issues such as user interface and information sufficiency; and capability issues such as reduction of unnecessary cost and error rate as well as improvement of knowledge communication, sharing and location. This approach has been used in other BIM-based knowledge sharing systems such as that developed by Ho et al. (2013) with similar survey results.

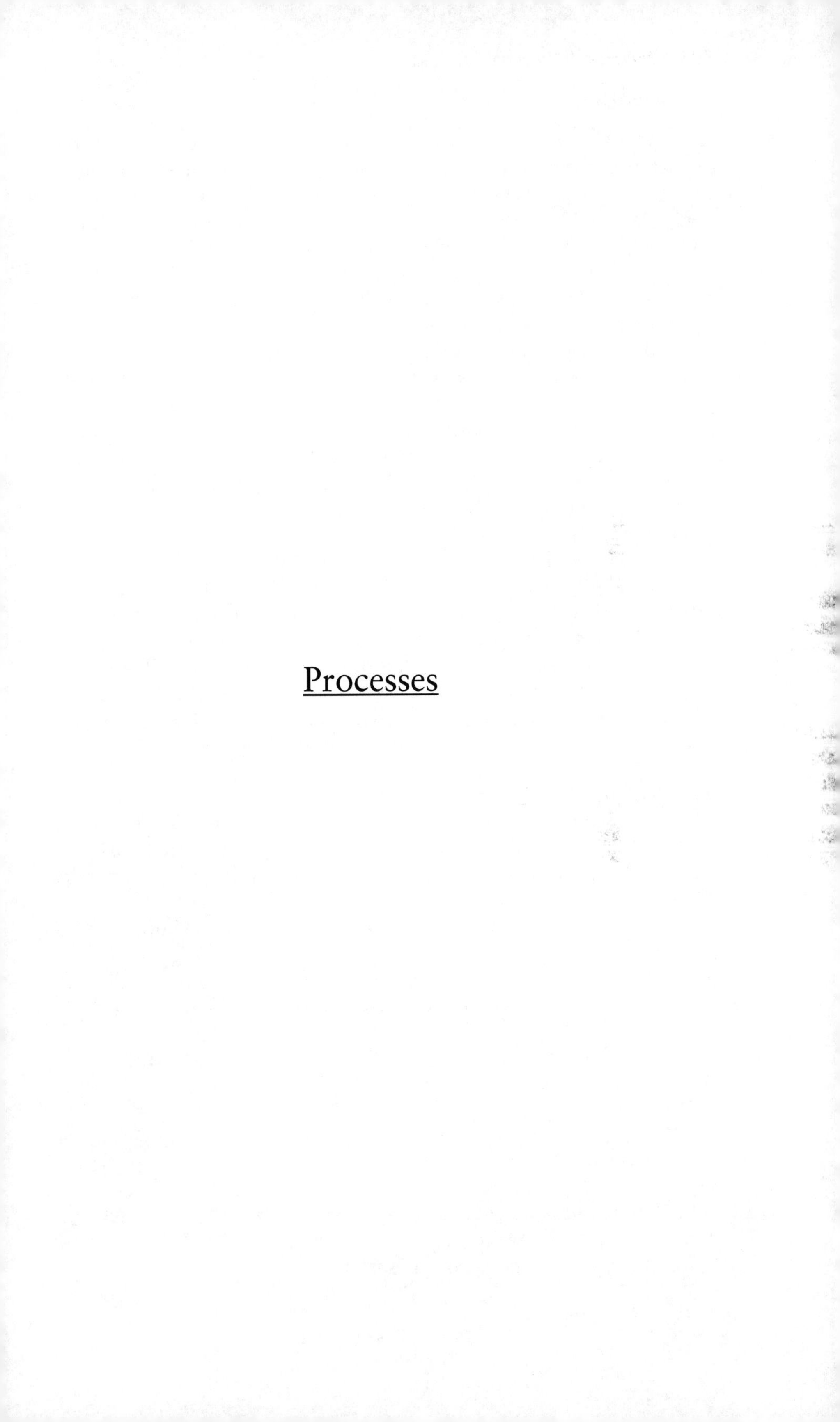

Processes

Requests for information

Requests for information (RFI) are often cited as one of the most useful indicators of the value achieved by implementing BIM and are a tool used to clarify unresolved issues (BSI and buildingSMART, 2010; Khanzode, 2010; Leite et al., 2011; Eadie et al., 2013). The reason for this is that most companies keep RFIs logs, providing a benchmark (Giel and Issa, 2013).

RFIs can be used as a metric by monitoring the number as a total or by type based on the cause, such as RFIs related to ambiguous information, errors and omissions or coordination issues. They can also be measured in terms of the cost associated with the creation and resolution of each RFI, the RFI turnaround time or cost avoidance by comparing against historic benchmarks in similar projects (BSI and buildingSMART, 2010; Barlish and Sullivan, 2012; Suermann, 2009; buildingSMART UK, 2010; Gao and Fischer, 2008; Gilligan and Kunz, 2007; Becerik and Pollalis, 2006). In the case of cost avoidance, it can also be retroactively calculated by reviewing RFI logs and focusing on RFIs related to issues that could have been discovered by BIM (Giel and Issa, 2013).

The number of RFIs has been suggested as a metric of coordination, productivity and process automation (Khanzode, 2010; BSI and buildingSMART, 2010; Becerik and Pollalis, 2006). BIM has been argued to reduce the number of RFIs by allowing the team to identify more conflict-related issues and completely resolve issued RFIs in less time (Khanzode, 2010). The suggested success criterion could be to have no RFIs related to issues that could have been identified before construction (Kunz and Fischer, 2012).

Common reasons for RFIs are (Becerik and Pollalis, 2006):

- Inquiring about design intent and requiring clarifications due to insufficient or missing information, or errors, omissions and conflicts.
- Accessing information about subcontractors and their activities.
- Accessing information related to specifications, contract or shop drawings, or personnel.
- Accessing work package information, such as scope of works, materials, and other.
- Inquiring about means and methods.
- Requesting guidance regarding errors made or problems during delivery.

Giel and Issa (2013) suggested the following types of RFIs can be avoided through the use of BIM:

- Dimensional inconsistencies due to rounding errors and discrepancies between documents.
- Document discrepancies between disciplines due to, for example, the use of segregated 2D document preparation by different disciplines.
- 2D errors and omissions, often driven by a lack of information needed and drafting errors in annotations.

- Grid or column alignment issues and discrepancies between floors.
- Direct clashes due to direct conflicts between systems or disciplines.

Notes

- In a survey carried out by Eadie et al. (2013), 39 per cent of respondents often use number of RFIs as a KPI for BIM-enabled projects.
- Research carried out by Becerik and Pollalis (2006) showed that integrated project environments can reduce the RFI turnaround time from an average of 14 days to 5.4 days and produced savings of up to US$536,500 at the organisation level, depending on the portfolio and organisation characteristics.

Time predictability

This metric refers to the change between actual time required for a cycle, task or phase and the estimated time for that phase at *commit to invest* stage, expressed as a percentage of actual time invested (Rankin et al., 2008).

Equation 4 Time predictability (Rankin et al., 2008)

$$Time\,Predictability = \frac{actual\,time - estimated\,time}{actual\,time} \times 100$$

It can also be measured as the time growth with respect to construction contract scheduled date or occupancy date, when completed and signed, or funded but pending (Suermann, 2009). A special application of this metric could be time predictability of evacuation during an emergency drill or event.

Time per unit

This metric refers to the average time required to produce a unit of output. Examples can be months/days/hours per kilometre of pipe or road, months/days/hours per square metre of floor space or hours per modelled room (Rankin et al., 2008; Sacks et al., 2005). Equation 5 shows how to measure the overall time used for the construction phase as a fraction of a capacity measure such as square metres of floor.

Equation 5 Time per unit (Rankin et al., 2008)

$$Time\,per\,unit = \frac{contract\,time\,for\,construction}{capacity\,measurement}$$

Time per unit can be a good indicator to benchmark outputs that depend on wages, such as hours required by drafters to produce drawings and other documents. In this case, the cost per unit would be highly dependent on contextual issues such as corporate salary policies and location, reducing the capacity to screen out the impact of non-BIM related factors (Sacks et al., 2005).

This data can often be derived from project documentation such as site logs, timesheets, and progress reports (Lu et al., 2012). Time per unit can be, for example, used to measure improved efficiency and coordination when using BIM (Porwal and Hewage, 2013).

Speed of production

This can be seen as the inverse metric of time per unit. It measures relative time by dividing a capacity or output measurement by the time invested in producing it (Chan and Chan, 2004). Equation 6 provides an example for the speed of construction. This metric may be more useful when monitoring benefits resulting from task automation than time per unit as in some of these cases the new process may produce a number of outputs (e.g., drawings, views or as-built documentation) in a few minutes rather than hours, days or months.

Equation 6 Speed of production (Chan and Chan, 2004)

$$Speed\,of\,Production = \frac{Gross\,floor\,area\ (m^2)}{Construction\,time\ (\frac{days}{weeks})}$$

The speed of production can also be measured through the percentage completed method, which can be useful for relatively minor tasks (Cox et al., 2003).

Overall time

This refers to the absolute time required for a specific phase or cycle (such as planning, design or construction) or overall project by practical completion date (GSA, 2007; de Souza, 2011). This can be calculated as the number of days or weeks from start to practical completion (Equation 7) (Chan and Chan, 2004).

Equation 7 Construction time (Chan and Chan, 2004)

$$Construction\ time = Practical\ completion\ date - Project\ commencement\ date$$

It can also be calculated as a percentage of standard duration if benchmarks for similar projects are available (Barlish and Sullivan, 2012). This metric can also be applied to accumulated time invested due to specific issues or tasks. Examples of applications to issues are overall time due to changes, rework, data-capturing or to resolve conflicts, and examples of applications to specific tasks include overall time investment in field coordination or compliance tasks.

Time for defects-warranty is a particular application of this metric to measure the efficiency of this process. It is calculated based on the time taken by the contractor to rectify all defects during the operations and maintenance phase, between available for use date and the end of the contractually agreed period for rectifying defects. This metric is normally expressed in weeks (Rankin et al., 2008).

Other special applications of this metric include:

- overall equipment or staff idle/stand-by time;
- overall lost time (normally calculated in days) due to accidents (this can also be calculated as a rate when expressed as a percentage of total number of hours/days worked by all employees in a year); and
- overall time required to access information during emergency drills and events.

This metric can also be useful to measure improved access to information and knowledge by measuring the overall time invested in training.

Note

- In a survey carried out by Becerik-Gerber and Rice (2009), 50 per cent of the respondents reported up to 25 per cent overall time reduction and 58 per cent reported reduction of overall construction time. In another survey, 43 per cent of respondents reported using overall programme duration as a KPI for BIM (Eadie et al., 2013).

Schedule conformance

This metric compares estimated vs actual schedule (Khanzode, 2010). As a lagging indicator, it can refer to the fraction of activities that start and finish within one day of schedule as shown in Equation 8. A suggested success criterion for projects using BIM is to achieve ≥ 95 per cent 'of all design and construction activities started and completed within one day of their planned start and finish dates, usually based on a 2–3 week look ahead schedule' (Kunz and Fischer, 2012).

Equation 8 Schedule conformance lagging indicator (Kunz and Fischer, 2012)

$$Schedule\,conformance(lagging) = \frac{\begin{array}{c} activities\,completed\,within \\ 1\,day\,from\,schedule \end{array}}{total\,activities} \times 100$$

This can also be applied to specific milestones to calculate on-time milestone completion to determine if, for example, the design or construction process is proceeding according to schedule (Cox et al., 2003). Alternatively, it can be calculated as a time variation discounting the effect of extension of time (EOT) granted by the client as shown in Equation 9.

Equation 9 Time variation (Chan and Chan, 2004)

$$Time\,variation = \frac{construction\,time - original\,contract\,period - EOT}{original\,contract\,period + EOT} \times 100\,per\,cent$$

These metrics can also be used at an organisational level to monitor schedule variance across the entire portfolio (Kunz and Fischer, 2012).

As a leading indicator, it can be calculated by subtracting the current forecasted completion duration from the originally planned completion duration, where the unit is days. This can then be evaluated against the remaining duration of the project to provide a better evaluation tool as shown in Equation 10 (Roper and McLin, 2005). By doing so, acceptable productivity goals can be determined (Cox et al., 2003). In this case, the success criterion can be to have a positive schedule conformance (leading), and thus having a completion date significantly earlier than planned (O'Connor and Yang, 2004).

Equation 10 Schedule conformance leading indicator (Roper and McLin, 2005)

$$Schedule\,conformance(leading) = \frac{original\,completion - current\,forecasted\,completion}{remaining\,project\,duration} \times 100$$

Improvements in the measured schedule conformance in BIM-enabled projects can be due to reduction of uncertainty, faster iterations to find solutions and fewer conflicts (Kam et al., 2014).

Note

- In a survey carried out by Suermann (2009), 82.8 per cent of respondents reported using schedule conformance as a KPI for BIM and in a survey by Gilligan and Kunz (2007) 20 per cent of BIM users reported tracking final schedule conformance quantitatively.

Cost of change

Cost of change is often used as a KPI for BIM (Eadie et al., 2013) and is probably most applicable to planning, design and construction phases. This cost can include: daily cost of the contractor's general conditions; the developer's administrative fees; the designer's contract administration fee; and the cost of interest on the owner's construction loan (Giel and Issa, 2013). The cost of change can be calculated directly through the accumulated value of variation orders (buildingSMART UK, 2010). It can also be calculated in relative terms, either with respect to the total project cost (Rankin et al., 2008) or to the cost of change in non-BIM projects that are similar in scope and size (Gilligan and Kunz, 2007).

This indicator can also be divided into cost of change – demand and cost of change – supply. The first (Equation 11) refers to changes attributable to client approved and generated change orders between the construction cost at available for use and the originally estimated cost at commitment to construct. The latter (Equation 12) refers to changes attributable to orders originated by the contractor (Rankin et al., 2008).

Equation 11 Cost of change – demand (Rankin et al., 2008)

$$Cost\,of\,change - demand = \frac{\begin{array}{c}approved\,cost\,for\,change\\ originating\,from\,client\end{array}}{total\,project\,cost} \times 100$$

Equation 12 Cost of change – supply (Rankin et al., 2008)

$$Cost\,of\,change - supply = \frac{\begin{array}{c}approved\,cost\,for\,change\,originating\\ from\,contractor\end{array}}{total\,project\,cost} \times 100$$

The indirect cost of changes can also be calculated retrospectively by multiplying the project delay (in days) due to BIM-discoverable issues by the daily cost penalties outlined in the contract (Giel and Issa, 2013).

Notes

- Giel and Issa (2013) assumed a 5 per cent capitalisation (CAP) rate for their cost of change calculations.
- Delays in processing, meeting notification requirements, and confronting changes causes significant challenges to the successful negotiation of changed conditions. Lack of detailed point-of-impact documentation on schedules and costs (contemporaneous records, in legal terms) weakens a general contractor's legal position. Owners prefer to do business with contractors who can be trusted to provide timely information and who show supporting analysis for all schedule and cost implications on their projects (Roper and McLin, 2005).
- In a case study carried out by Barlish and Sullivan (2012), change orders saw savings of 42 per cent of standard cost.

Variations and change orders

'The goal of the change-order process is to inform the owner of changing conditions that impact cost and schedule' (Roper and McLin, 2005). Changes can be used as a metric directly by recording the total number of changes or variation/change orders and comparing to similar non-BIM projects (Suermann, 2009; buildingSMART UK, 2010; GSA, 2007; Leite et al., 2011).

The number of change orders can be used as a measure of efficient coordination (Khanzode, 2010) and can be derived from the change order logs for each project (Giel and Issa, 2013). This metric can also be divided into change orders – client and change orders – project manager. The first refers to the number of individual change orders approved by the client between commit to construct and available for use, originating from the client. The latter refers to those originating from the project manager (KPI Working Group, 2000).

This indicator can also be used to monitor specific issues such as change orders due to errors and omissions or coordination issues.

Note

- In a survey reported by Eadie et al. (2013), 38 per cent of respondents said to measure change orders to monitor benefits from BIM.

Time for change

As with cost of change, time for change can be used as a metric in absolute terms (total number of additional project days due to change orders) or as a percentage of total project time (Equation 13). It can also be divided into time for change – demand and time for change – supply (Rankin et al., 2008; Suermann, 2009).

Equation 13 Time for change (Rankin et al., 2008)

$$Time\,for\,change = \frac{actual\,construction\,time - estimated\,construction\,time}{total\,project\,time} \times 100$$

Latency

Latency can be divided between decision latency and response latency. The first refers to the time between the moment information is made available and the time when the decision is announced. The latter refers to the time between asking a question or issuing an RFI, and the moment when a complete response is received (Kunz and Fischer, 2012). Latency can also be applied to change orders, that is the time between issuing a change order and the time when it is resolved.

Latency can be used as a quantitative metric of stakeholder participation and communication (Gilligan and Kunz, 2007). A potential success criterion could be *≤ 2 minutes during critical design and construction activities* and > *98 per cent reliability* (Kunz and Fischer, 2012) or less than five days (Kam et al., 2014).

A special case is emergency latency, which will be discussed in the sustainability and future-proofing indicators section of this dictionary.

Notes

- In a survey carried out by Gilligan and Kunz (2007), 40 per cent of respondents reported some improvement in latency, and 60–80 per cent of BIM veteran users reported improvements of two days or better.
- Latency of structural permitting was observed to be reduced from similar projects' 39 months to 15 months for a BIM-enabled project (Kam et al., 2013).
- In 2005, the typical practice in the US was to have latency periods of two days to a month or more (Kunz and Fischer, 2012).
- Interviewees of a study by Gilligan and Kunz (2007) tied improved latency to: (1) less room for interpretation; (2) use of a common model which reduces potential for discrepancies; and (3) improved communication medium combined with collaborative contracts. These were reported to provide the environment for more trust and cooperation driving lower latencies.

Model (or drawing) coordination consistency

This metric records the number of multidisciplinary models/drawings that contain conflicts, interferences or inconsistencies at major project milestones. The success criterion could, for example, be to have zero 'coordination inconsistencies at Construction Document review, during construction' (Kunz and Fischer, 2012). This metric can also be applied to other project documentation.

Visualisation

This is a qualitative metric that aims to measure how satisfied stakeholders are with the level of clarity and accessibility of models and analyses of outputs, organisations and processes. It can be reported on a scale from one to five, where one represents low level of clarity and accessibility and five represents a high level. The potential success criterion could be 'good visualizations as part of VDC methods and stakeholder engagement dramatically improve client and all stakeholder satisfaction' (Kunz and Fischer, 2012). Alternatively, 80 per cent of stakeholders report four or higher could be used as a success criterion.

Volume of rework

This is the volume of work that is required to be redone due to unanticipated conditions and errors. This is a commonly used KPI for BIM (Suermann, 2009; Kunz and Fischer, 2012). This can be measured in terms of time spent by staff in rectifying issues (Khanzode, 2010). For example, by 'calculating the change in the number of man-hours and material cost for repairing work in place or rehandling materials' (Cox et al., 2003).

This metric can be used to monitor improved coordination and process efficiency (Khanzode, 2010). Potential success criteria could be to have zero rework during field work and 20–40 per cent of virtual work (Kunz and Fischer, 2012).

The volume of rework can also be calculated relative to bid estimates based on traditional methods by dividing labour-hours in the field or total labour hours of rework to the original estimates for that task (Kunz and Fischer, 2012).

This metric can also be further subdivided by cause of rework to gain insight into the specific processes that are generating it. These issues can, for example, be rework due to data capturing/re-entry or programming errors.

Notes

- In 2003, rework was traditionally responsible for 6–12 per cent of the overall expenditure in construction projects (Cox et al., 2003).

Clashes

This metric refers to geometric clashes found in the model. Teams can track the number of clashes or issues that were found using BIM that would normally go undetected until they reached the field in traditional projects. This can then be compared to the number of actual clashes in non-BIM projects that are similar in size and scope (Gilligan and Kunz, 2007). Alternatively, the absolute number of smart clashes found during the design phase in a BIM project can be compared to the absolute number of clashes found in a non-BIM project of similar characteristics during the design phase (GSA, 2007).

This metric can also be applied to reported field/work-site clashes.

Off-site manufacturing

This metric aims to measure the amount of pre-fabrication and off-site manufacturing enabled by BIM (Khanzode, 2010). It can be calculated by monitoring the number of off-site pre-fabrication man-hours and reporting them as a percentage of all project man-hours (Barlish and Sullivan, 2012), the volume of pre-fabrication by each trade and the labour intensity per unit of pre-fabricated assembly (Khanzode, 2010).

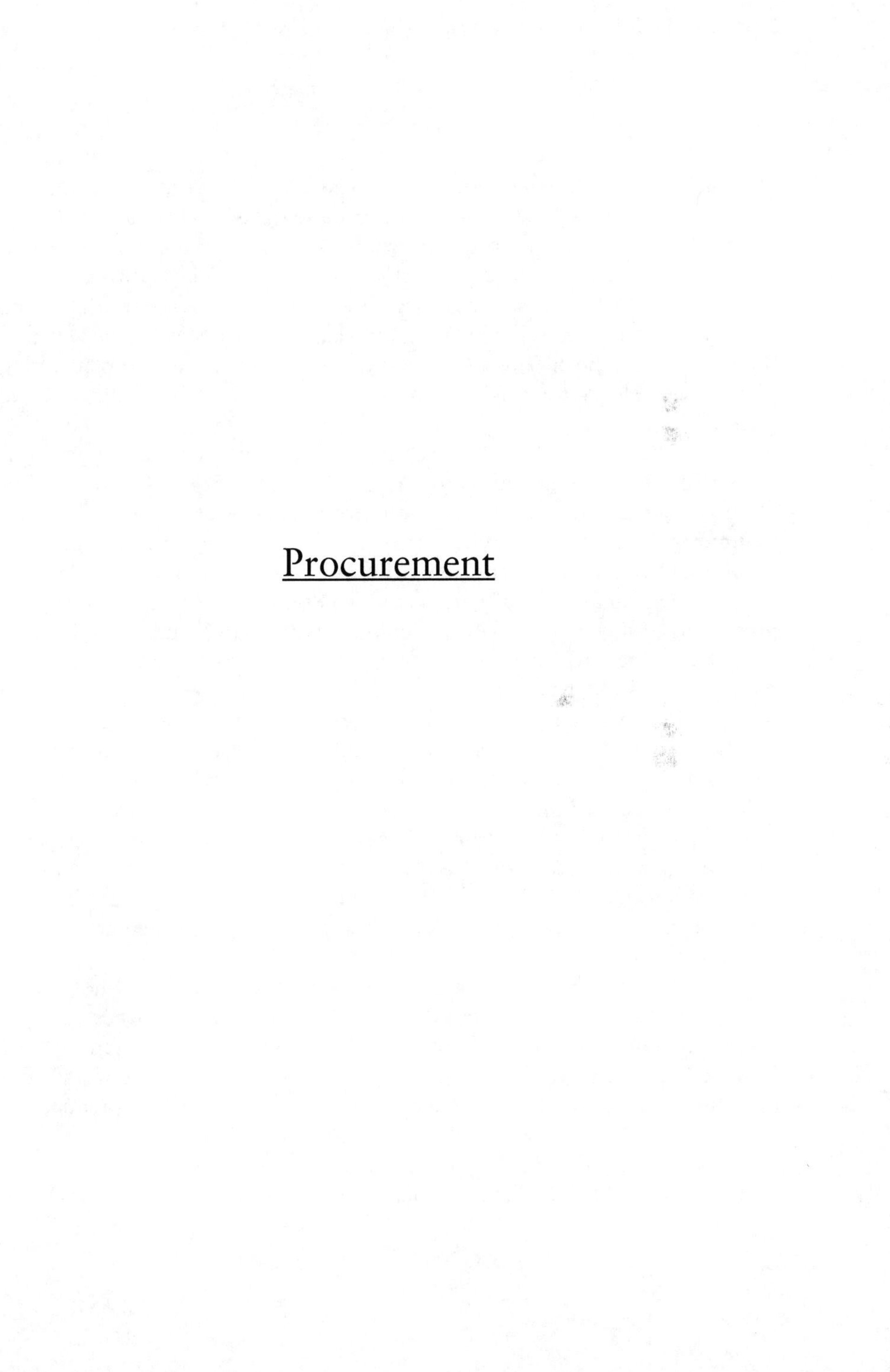

Procurement

Return on investment (ROI)

ROI compares potential benefits or gains of an investment against its cost. Giel and Issa (2013) suggest the following generic equation (Equation 14):

Equation 14 Return on investment (Giel and Issa, 2013)

$$ROI = \frac{Gain\ from\ Investment - Cost\ of\ investment}{Cost\ of\ investment}$$

One of the main challenges with this approach is estimating the return on investment due to avoided cost (BSI and buildingSMART, 2010). Giel and Issa (2013) attempted to do this by studying projects that could have used BIM. Through this approach, they added the cost of delays, change orders and rework attributed to issues that could have been discovered by using BIM in the design phase. Then, they calculated the net BIM savings by subtracting the cost the organisations would have incurred due to implementing BIM. Please refer to the original reference to access the complete model and methodology.

ROI can be estimated in a number of ways. Some users, for example, focus on project outcome improvements such as 'reduced RFIs and field coordination problems, better communication through 3D visualization, productivity improvement of personnel, positive impact of winning projects, lifecycle values of BIM, and initial cost of staff training' (Giel and Issa, 2013).

Barlish and Sullivan (2012) calculated the ROI by comparing BIM-enabled projects against similar non-BIM projects that formed the benchmark based on data extracted from historic records. The limitation of this approach is that they could only compare those metrics that were already being monitored and recorded in previous projects. Additionally, surveys and studies carried out by Becerik-Gerber and Rice (2010) suggest that to effectively measure the ROI from BIM, a minimum timeframe of five to eight years is required.

Many projects monitoring ROI from BIM estimate the return on investment through savings based on estimates of staff-hours and material costs that would have been required if all (or a portion of) conflicts detected through automated clash detection would have happened (Giel and Issa, 2013). This approach, however, does not account for other savings achieved indirectly (Azhar, 2011).

Some organisations have also estimated the unit-cost and staff-hours that would have had to be invested to find different types of clashes through traditional means (Giel and Issa, 2013). However, as pointed out by Becerik-Gerber and Rice (2009), a 'comprehensive study that covers a full spectrum of hard and soft costs, as well as the concept of cost avoidance has to be conducted in order to evaluate the real value of BIM'.

Users have also reported measuring ROI indirectly through processes such as fewer RFIs and change orders, and shorter delivery time as well as financial/procurement measures such as cost savings and profit (Bernstein et al., 2014). Other metrics also used to measure ROI indirectly relate to people indicators such as safety, client satisfaction and conflicts that lead to project disruption (Bernstein et al., 2014).

Notes

- The ROI from using BIM has been reported to be overall positive and up to 50 per cent. For BIM (M) it has been estimated to be greater than 60 per cent. Surveys have shown that ROI from BIM for owners can be between 10 and 25 per cent (BIM Industry Working Group, 2011; Tsai et al., 2014; Giel and Issa, 2013; Becerik-Gerber and Rice, 2010; Allen Consulting Group, 2010).
- In a series of case studies carried out for online collaboration and project management, the ROI was found to be between US$42,000 and $500,000 per year per asset owner (Becerik and Pollalis, 2006).
- Azhar (2011) analysed the detailed cost data from ten projects. Through this analysis they found ROI from BIM varied from 140 per cent to 39,900 per cent, averaging at 1,633 per cent for all projects and 634 per cent for those that did not have a planning or value analysis phase.
- A 2012 survey by McGraw Hill found that the leading factor influencing positive ROI estimates in infrastructure is improved process outcomes. This includes fewer RFIs and field coordination problems, and is particularly true for contractors. Other influencing factors included improved communication and sustainability measures (Bernstein et al., 2012).

Cost predictability

Cost predictability is mostly a lagging indicator described as the change between the actual costs at output handover and the estimated cost at the commit to invest phase. This is normally expressed as a percentage of the estimated cost as shown in Equation 15 but can also be calculated as a percentage of client management cost with respect to base estimates (KPI Working Group, 2000; Rankin et al., 2008).

Equation 15 Cost predictability (Rankin et al., 2008)

$$Cost\ Predictability = \frac{actual\ cost - estimated\ cost}{estimated\ cost} \times 100$$

This can be calculated for the design, construction or operations phases, or, for example, based only on client initiated change orders during the construction phase (KPI Working Group, 2000). It can also be expressed as detailed cost conformance calculated as a fraction of estimated cost items whose actual cost was within 2 per cent of the budgeted cost. Kunz and Fischer (2012) suggest the following success criterion when using this metric in BIM-enabled projects: '≥ 95 per cent of budgeted items cost within 2 per cent of their budgeted costs'.

Cost predictability can also be measured through budget estimate conformance metrics. This is the fraction of budgeted items within 5 per cent of budget cost in the guaranteed maximum price estimate with zero dollars contingency. Kunz and Fischer (2012) suggest the follow metric: during design, 95 per cent of budgeted items are within 5 per cent of budgeted cost at relevant milestones; and during construction, 95 per cent of budgeted items are within 2 per cent of budgeted cost. It can also be measured as a percentage of corporate management approved budget, where the suggested success criterion is that companies deliver 98 per cent of the projects with less than 2 per cent of unbudgeted changes (Kunz and Fischer, 2012).

This metric can also be used continually as the ratio between the project current estimate and the programmed amount (Suermann, 2009). Cost predictability can also be measured through the cost growth, which is the variance between budgeted cost and actual cost. In procurement, this can be expressed as shown in Equation 16 and can be calculated based on completed and signed work, funded but pending, and unfunded and pending (Suermann, 2009; buildingSMART UK, 2010).

Equation 16 Cost growth (Suermann, 2009)

$$Cost\,growth = \frac{Tender\,value}{Outcome\,value} \times 100$$

The percentage of net variation over final cost is another way to measure cost predictability by calculating the ratio of net variations of final contract cost (Equation 17). This metric provides an indication of cost overruns or underruns (Chan and Chan, 2004).

Equation 17 Percentage net variation (Chan and Chan, 2004)

$$\%NETVAR = \frac{\left\langle \begin{matrix} Final \\ contract \\ sum \end{matrix} - \begin{matrix} original \\ contract \\ sum \end{matrix} - \begin{matrix} final \\ rise\,and \\ fall \end{matrix} - \begin{matrix} contingency \\ allowance \\ \end{matrix} \right\rangle}{final\,contract\,sum} \times 100\%$$

During operations, cost predictability can be applied to asset operation estimates. In buildings, for example, this could be applied to energy consumption versus energy models used during the design phase (GSA, 2012).

BIMe (2014) provides a leading indicator for cost predictability based on cost certainty metrics that provide the level of confidence in cost (low, medium or high).

Cost per unit

This metric is centred on the value of the output represented as currency unit per capacity units such as dollars per square metre of floor space or dollars per kilometre of road. The dollar value can refer either to the contract value (Equation 18) or the dollar value associated with completing $1m^2$ of floor space, including materials costs, labour costs, waste and equipment costs, among other (Rankin et al., 2008; Chan and Chan, 2004; Suermann, 2009; Cox et al., 2003). The benefit of the latter is that it can provide a breakdown of production elements at a measurable scale (Cox et al., 2003).

Equation 18 Unit cost (Chan and Chan, 2004)

$$Unit\,cost = \frac{Final\,contract\,sum}{Gross\,floor\,area\ (m^2)}$$

This metric is one of the most commonly used KPIs for productivity in the construction industry (Suermann, 2009; Cox et al., 2003). For linear infrastructure construction projects, it can also be expressed in terms of cost per lane kilometre, tunnel lane kilometre or square metre of bridge. When applied to tender cost, it can be expressed as shown in Equation 19 (Rankin et al., 2008).

Equation 19 Cost per unit in relation to tender cost (Rankin et al., 2008)

$$Cost\,per\,unit = \frac{tender\,cost}{capacity\,measurement}$$

In general, cost per unit measures can provide a more neutral measure of productivity as they normalise the value across different projects and companies. However, productivity will differ substantially in relation to the type, size and complexity of a project. Therefore, the type of project should be considered when using this metric as a way of benchmarking (Sacks et al., 2005).

Kunz and Fischer (2012) propose reducing these values by 20 per cent for similar or improved functions, quality and schedule as a breakthrough success criterion or goal.

Capital productivity can be seen as a special case of cost per unit. This metric then measures the ratio between the value of the output and that of the input of capital, and can be used to analyse efficiency changes as well as productivity (Allen Consulting Group, 2010).

At the company level, this metric can be used to measure the value added per employee. This is the ratio of turnover minus all costs subcontracted to or supplied by other parties to number of employees (de Souza, 2011).

Cost per defects – warranty

Cost per defects – warranty is the cost needed to rectify all defects during the operations and maintenance period, between practical completion and the end of the contractually agreed period for rectifying defects. This is normally calculated as a percentage of the actual construction cost at agreed period for rectifying defects as per Equation 20 (Rankin et al., 2008; KPI Working Group, 2000).

Equation 20 Cost for defects-warranty (Rankin et al., 2008)

$$Cost\,for\,defects - warranty = \frac{construction\,cost\,of\,rectifying\;all\,defects}{final\,cost\,of\,construction} \times 100$$

This metric can be used to monitor benefits such as fewer errors and better change management.

Cost avoidance or savings

This metric aims to estimate the cost that would have been incurred if BIM had not been used. It can be done by assigning estimated costs to clashes found based on past cost or delays generated by similar issues (Gilligan and Kunz, 2007). An avoidance cost log may be kept to track this (Barlish and Sullivan, 2012). This indicator can be further detailed into cost avoided by identifying errors and omissions at earlier stages; for example, by identifying those errors found during design that if BIM would not have been used, had not been caught until the construction phase and how much it would have cost to repair the error then. A special application of cost savings is cost recovery during decommissioning. These are not savings per se but income from identifying valuable materials that can be recovered and repurposed or sold (Kivits and Furneaux, 2013).

Note

- Azhar (2011) reported a project in which clashes resolved by using BIM were tracked and associated savings calculated based on estimates of the cost of making design changes or field modifications if the clash would have gone undetected. The net cost was calculated assuming that 75 per cent of clashes detected through BIM would have also been detected through conventional practices. In this case, the net adjusted cost savings were estimated to be US$200,392 for a project with a contract value of US$46 million, where the cost associated to implementing BIM was estimated to be US$90,000.

Overall cost

Overall cost is also one of the most common KPIs used in construction to measure benefits from BIM (Suermann, 2009). This metric can refer to overall cost of:

- a cycle such as design, construction or operations (Lu et al., 2012; GSA, 2007);
- a functional area such as value engineering or assembly (Barlish and Sullivan, 2012; GSA, 2007);
- a process such as documentation (NATSPEC, 2014); or
- a whole project (Suermann and Issa, 2007).

Ideally, the cost of BIM will only be factored in for those areas that use the system directly (Barlish and Sullivan, 2012). When applied to cost per cycle, data can be derived from project documentation such as site logs, timesheets, progress reports, inventory records or supplier material forms (Lu et al., 2012).

Overall cost for the operations cycle, for example, can be calculated following Equation 21 on annual terms, expressed as a percentage of the actual design and construction costs (Rankin et al., 2008). This could also include maintenance (GSA, 2007; buildingSMART UK, 2010). The cost of preparing the asset management database can also be used as a functional area overall cost during operations (buildingSMART UK, 2010).

Equation 21 Operating cost (Rankin et al., 2008)

$$Operating\,cost = \frac{annual\,operating\,cost\,arranged\,over\,years}{final\,cost\,of\,construction\,and\,design} \times 100$$

The success criterion for this metric can be to achieve a total installed cost significantly lower than authorised budget (O'Connor and Yang, 2004).

Challenges to using overall cost as a benchmarking metric are: (1) it is highly project-type dependent (Sacks et al., 2005); and (2) the proprietary nature of bidding information limiting the accessibility to this data due to confidentiality issues (Barlish and Sullivan, 2012; GSA, 2007).

Notes

- Overall cost has been reported to be reduced by 8–10 per cent for BIM-enabled projects during pre-construction and construction phases, and 8–18 per cent for the design phase. These reductions have been attributed to improved understanding, spatial coordination, clash detection and construction planning (buildingSMART UK, 2010).
- In a survey carried out by Suermann and Issa (2007), 84 per cent of respondents reported improved overall cost when using BIM and 50

per cent of respondents to a survey carried out by Eadie et al. (2013) reported using overall cost as a metric for BIM.

Profit

Competitive advantage has been reported to be gained through increased profits when using BIM (NBS, 2014). This can refer to tender profit or output profit through higher fees for new services or more efficient use of resources (buildingSMART UK, 2010; NATSPEC, 2014), or marketing profitability (Kam et al., 2014).

It can be calculated through net present value (Equation 22; where *t* is construction time interval).

Equation 22 Net present value (Chan and Chan, 2004)

$$NPV = \sum_{t=0}^{n} \frac{(Net\,Cash\,FLow)_t}{(1 + discount\,rate)^t}$$

However, this metric poses similar challenges to overall cost for benchmarking purposes; the data is difficult to obtain due to confidentiality issues. Profit may also not be appropriate to measure benefits for public clients (Chan and Chan, 2004).

Quality

Quality of output can be assessed based on the number of open (outstanding) non-conformances at the moment the output is available for use or at the end of the warranty period (Rankin et al., 2008; KPI Working Group, 2000). Other indicators of quality are number of quality assurance (QA) and quality control (QC) issues (Suermann, 2009) such as: QC tests (number performed, frequency, percentage passed/failed); number of change requests and root causes; number of exceptions at turnover; and cost of quality (cost of QA resources, QA cost as a percentage of construction, or QA cost + cost of rework) (CURT, 2005).

Quality can also be measured through qualitative methods by using a seven or ten point scale based either on: (1) quality, defined as 'the totality of features required by a product or services to satisfy a given need; fitness for purpose; meeting technical speciation' (Chan and Chan, 2004); or (2) defects, defined as the impact, at the time of handover, caused by the condition of the asset with respect to defects (KPI Working Group, 2000).

Note

- In a survey carried out by Suermann (2009) quality control/rework was one of the highest rated KPIs used for BIM-enabled projects, where 87.7 per cent of respondents said that BIM improves this KPI.

Accuracy and number of errors/omissions

Accuracy can refer to that of the 2D drawing and other design deliverables produced from the BIM model (Porwal and Hewage, 2013; NATSPEC, 2014), of the cost estimates and design coordination (Leite et al., 2011), and that of the as-built model and energy models (GSA, 2012).

The number of errors and omissions can also be used as a lagging metric of accuracy (GSA, 2007) and can be further divided by tasks such as quantity take-offs or project set-up.

Field material delivery

Field material delivery is mainly a construction and operations metric. It is calculated by dividing all material deliveries made 24 or fewer hours ahead of scheduled use by the total number of material deliveries over a period of time or cycle. A potential success criterion could be achieving ≥ 95 per cent of all field material deliveries within 24 hours or less from scheduled use (Kunz and Fischer, 2012).

Programme capacity

Programme capacity is an organisational metric that can be used when using BIM across different projects. It is expressed as a percentage of the total number of ongoing projects in programme years that meet the growth goals of the organisation (Suermann, 2009).

Risk

There are a number of well-established and emerging methodologies to assess risk associated with project delivery (Cretu et al., 2011; Dikmen et al., 2007; Zavadskas et al., 2010). This is normally represented by the size of the contingency budget, which can also serve as a metric for comparison between similar projects (Suermann, 2009).

Globalisation

Kunz and Fischer (2012) from the American Centre for Integrated Facility Engineering also suggest monitoring the level of materials and services procured from and provided to global supply chains. This is a relevant metric for organisations that aim to remain globally competitive and take advantage of BIM's benefits related to improved communication across geographically sparse supply chains. This indicator can have two success criteria: (1) '≥ 50 per cent of components and services potentially obtained from global supply chains'; or (2) '≥ 50 per cent of products and services sold to global market' (Kunz and Fischer, 2012).

Equipment asset remaining useful life

Equipment assets' remaining useful life refers to the length from the current time to the end of the 'period during which an asset or property is expected to be usable for the purpose it was acquired' (Si et al., 2011). This is a common indicator used by asset managers to help identify infrastructure needs and make informed decisions during operations (Koronios et al., 2005). Si et al. (2011) provide a good review of different alternatives to model and measure this indicator.

The use of computerised preventive maintenance systems, such as BIM-based asset (preventative) maintenance scheduling, have been shown to significantly extend average useful life by up to 53 per cent, providing estimated savings to owners of up to 1.12 per cent of the current replacement value (Teichholz, 2013).

Sustainability and future-proofing

Sustainability and environmental performance scores

Sustainability of the design, construction and/or operations can be measured through the use of checklists of standard practices or sustainability rating tools such as Leadership in Energy & Environmental Design (LEED) (Rankin et al., 2008). These tools are useful because they provide a historic benchmark for similar projects that were not carried out using BIM, while also potentially providing a marketing tool. Tools such as the Australian Infrastructure Sustainability (IS) and the American Institute for Sustainable Infrastructure Envision rating systems can provide a similar benchmark for infrastructure projects (Infrastructure Sustainability Council of Australia, 2014; Institute for Sustainable Infrastructure, 2015).

Environmental performance can also be measured through Environmental Impact Assessment scores, the application of international standards such as ISO14000, and the total number of environmental performance complaints received either at the end of the warranty period or annually during operations (Chan and Chan, 2004).

Note

- Although many factors can influence the score achieved, according to Autodesk, by 2005 the use of BIM could already facilitate up to 20 points for LEED (Autodesk, 2005).

Resources use and management

BIM has been argued to improve the resource use of the final facility through a number of ways such as better design outcomes and use of life-cycle simulations (Kam et al., 2014; Kunz and Fischer, 2012). Cost per unit of resource use reduction is a special type of cost per unit that can be applied for long-term sustainability benefits measurement. Examples can be energy or water use, or CO_2 emissions reduction per year as a ratio of hours spent in value engineering and life-cycle simulations (NATSPEC, 2011). The University of South Carolina, for example, used the energy consumption savings averaged over ten years as a metric to measure sustainability benefits from implementing BIM in their Columbia Campus (approximately US$900,000) (Azhar and Brown, 2009).

This metric can also be used directly by recording resource consumption such as water and energy use from utility bills and comparing this to either similar projects or to the original design (Kunz and Fischer, 2012). Another criterion could be recyclability of the final asset. This would be the percentage of the materials that can be recycled during decommissioning (Volk et al., 2014).

During construction, resource management can be a useful tool for monitoring the changes in material waste by calculating the amount of material and resources used during the construction operations to produce one unit of output (Cox et al., 2003). One unit can be a capacity measure (kilometre of road or square metre of floor), or a value measure based on the asset value.

Carbon emissions and footprint

Carbon emissions (CO_2 equivalents) and carbon footprint are two commonly used measures for sustainability of built assets. Carbon footprint measures the total amount of CO_2 emissions that is 'directly and indirectly caused by an activity or is accumulated over the life stages of a product' (Wiedmann and Minx, 2008). Standard methodologies to calculate it include the international standard ISO/TS 14067:2013 (ISO, 2013).

Annual or cycle (construction or operations) CO_2 emissions caused by, for example, energy use can also be used in absolute numbers or as a ratio of 100m^2 of gross floor area of the asset (kg CO_2/100m^2) (de Souza, 2011). Embodied carbon in materials used can also be applied as a metric for improved environmental performance and calculated according to international standards such as ISO14020, ISO14040 or ISO14025 (Airaksinen and Matilainen, 2011).

Although tools such as LEED include carbon footprint assessment, this is often done only during the design phase based on estimates as a leading indicator to compare different design options. However, this process can be extended to the construction process to compare different construction methods and make decisions about materials and services. Finally, original estimates can be later compared to the actual carbon footprint and CO_2 emissions based on the record model.

There have been several efforts to use carbon footprint and emissions as a leading indicator during construction. Memarzadeh and Golparvar-Fard (2012), for example, developed a method that combines BIM phase planning with carbon footprint measures. This way they were able to not only actively use this metric to monitor the impact of schedule and material changes on the emissions of projects but also facilitated its visualisation for review processes with a colour-coded view. A Canadian group developed a similar system and applied it to housing construction, which allowed them to identify specific areas of improvement in the construction process that could lower the asset's footprint (Mah et al., 2011).

Emergency latency

As highlighted in the latency profile, this metric can be divided between decision latency and response latency. The first refers to the time between the moment information is made available and the time when the decision is announced. The latter refers to the time between requiring the information and the moment when a complete response is received (Kunz and Fischer, 2012).

This metric can be applied to emergency management efforts. In this case, the emergency decision latency would refer to the time between when emergency responders have access to information about the asset and the event in progress, and then defining a course of action. Emergency response latency would refer to the time between the emergency responders requiring information about the asset's conditions and the moment when they receive all the information needed to take action. This information could be retrieved from emergency logs/systems.

Emergency plan and response effectiveness

There are a number of ways of assessing the effectiveness of emergency plans and response. A group which developed a BIM-based tool to improve emergency plan effectiveness during construction works, for example, included five aspects: excavation risk management plan, crane management plan, fall protection plan for leading edges, fall protection plan for roofers and emergency response plan. The latter included five sub-plans: construction crew entry/exit; construction equipment and deliveries routes; temporary facilities and job trailer locations; emergency vehicles route; and severe weather shelter. The effectiveness of this system was evaluated through a questionnaire that asked users to rate it as compared to traditional planning methods. The scale used was –2 for highly ineffective/inaccurate, zero for no difference and +2 for highly effective/accurate. The indicators rated were: ease in hazard assessment, ease in communication, and accuracy and incident and accidents control. For this specific case, the results pointed to moderate to high effectiveness (Azhar and Behringer, 2013).

The US Environmental Protection Agency has established a series of emergency response effectiveness indicators such as community involvement and satisfaction (expressed satisfaction by public and state and local governments) and operational activities indicators (no deaths or injuries during response, no work days lost due to attending the response, completion within schedule, and lack of technical errors) (Amarakoon et al., 2008). Other indicators that have been suggested include:

- Damages caused by a specific type and level of emergency.
- Number of victims that were located and received assistance from responders in a timely manner.
- Time from the onset of the event until the time when the scene is declared as being 'under control'.
- Responders injured and severity of injures.
- Total area affected by emergency event.
- Percentage of victims who sustained further injuries or whose condition worsened after being identified by responder (Brown and Robinson, 2005).

Quantitative indicators that may also be applicable to a BIM environment and evaluate the effectiveness of the plan development effort include number of staff-hours spent in: developing the emergency plans and training programmes, collecting hazard data, and conducting hazard analyses (Lindell and Whitney, 1995). The International Association of Emergency Managers also includes indicators such as the level of inclusiveness of the planning process through, for example, the number of stakeholders involved in the development of the plan (Jensen and Duncan, 2011).

Fire safety

Fire safety is often a special application of risk metrics. Common indicators for asset fire safety are expected risk to life (ERL), expected risk of injury (EROI) and expected annual fire loss (AFL). ERL is defined as the expected death frequency per year per individual user of an asset. This risk equation is a function of the annual fire frequency and user population size of the asset, and the sum of the probability and number of deaths of different scenarios. EORI is similar to ERL but is based on individual scenario calculations and AFL is a function of probability of a scenario and its total cost to the asset (Li et al., 2013). Fridrich and Kubečka (2014) provide a methodology to use BIM to carry out and improve fire risk assessment.

Other fire safety indicators include fire death rate per 100,000 population, annual mortality rate (Xin and Huang, 2013). Another indicator used within fire safety is asset evacuation capability, which can be evaluated by comparing required safety egress time to the available safety egress time. This is the difference between the time required by users to reach a safe location and the time the asset will reach fatal environmental conditions (Wang et al., 2015). A Taiwanese group, for example, integrated a BIM model with a fire dynamics simulator to evaluate personnel safety and escape time during fire emergency events (Wang et al., 2014).

Event-specific fire safety indicators can also be modelled and monitored during fire emergency events. BIM has been used to automatically calculate fluid dynamics of emissions during fire events in the Manchester Town Hall Complex refurbishment project (Codinhoto et al., 2013).

Note

1 A cycle is defined in this study as a quarter of an entire floor.

Bibliography

Airaksinen, M. and Matilainen, P., 2011. A carbon footprint of an office building. *Energies*, 4(8), pp. 1197–1210.

Allen Consulting Group, 2010. *Productivity in the Buildings Network: Assessing the Impacts of Building Information Models*, s.l.: BEIIC.

Amarakoon, S., Kemp, D., Finan, B., Byrd, D. and Watson, Y., 2008. *Performance Indicators for EPA Emergency Response and Removal Actions*, s.l.: EPA.

Anzanello, M. J. and Fogliatto, F. S., 2011. Learning curve models and applications: Literature review and research directions. *International Journal of Industrial Ergonomics*, 41(5), pp. 573–583.

Autodesk, 2005. *Building Information Modeling for Sustainable Design*. Available at: http://images.autodesk.com/latin_am_main/files/bim_for_sustainable_design_oct08.pdf [Accessed 7 July 2015].

Azhar, S., 2011. Building information modelling (BIM): trends, benefits, risks, and challenges for the AEC industry. *Leadership and Management in Engineering*, 11(3), pp. 241–252.

Azhar, S. and Behringer, A., 2013. *A BIM-Based Approach for Communicating and Implementing a Construction Site Safety Plan*, paper presented at the 49th ASC Annual International Conference Proceedings, Associated Schools of Construction, San Luis Obispo, CA, 10–13 April.

Azhar, S. and Brown, J., 2009. BIM for sustainability analyses. *International Journal of Construction Education and Research*, 5(4), pp. 276–292.

Barlish, K. and Sullivan, K., 2012. How to measure the benefits of BIM – a case study approach. *Automation in Construction*, 24, pp. 149–159.

Becerik, B. and Pollalis, S. N., 2006. *Computer Aided Collaboration in Managing Construction*, Cambridge, UK: Harvard University Graduate School of Design, Design and Technology Report Series 2006-2.

Becerik-Gerber, B. and Rice, S., 2009. *The Value of Building Information Modelling: Can We Measure the ROI of BIM?* Available at: www.aecbytes.com/viewpoint/2009/issue_47_pr.html [Accessed 20 May 2014].

Becerik-Gerber, B. and Rice, S., 2010. The perceived value of building information modelling in the US building industry. *Journal of Information Technology in Construction*, 15, pp. 185–201.

Bernstein, H. M., Jones, S.A., Russo, M.A., Laquidara-Carr, D., Taylor, W., Ramos, J., Healy, M., Lorenz, A., Fujishima, H., Buckley, B., Fitch, E. and Gilmore, D., 2012. *The Business Value of BIM for Infrastructure: Addressing America's Infrastructure Challenges with Collaboration and Technology SmartMarket Report*, Bedford, MA: McGraw Hill Construction.

BIM Industry Working Group, 2011. *A Report for the Government Construction Client Group Building Information Modelling (BIM) Working Party*, strategy paper, s.l.: Department of Business, Innovation and Skills.

BIMe, 2014. *Project Assessment: The BIMe Approach*. Available at: http://bimexcellence.net/projects [Accessed 1 September 2014].

Bontis, N., Dragonetti, N. C., Jacobsen, K. and Roos, G., 1999. The knowledge toolbox: a review of the tools available to measure and manage intangible resources. *European Management Journal*, 17(4), pp. 391–406.

Bose, R., 2004. Knowledge management metrics. *Industrial Management & Data Systems*, 104(6), pp. 457–468.

Bradley, G., 2010. *Benefit Realisation Management: A Practical Guide to Achieving Benefits through Change*, 2nd edn, Surrey: Gower.

Brown, D. E. and Robinson, D., 2005. *Development of Metrics to Evaluate Effectiveness or Emergency Response Operations*, paper presented at the 10th

International Command and Control Research and Technology Symposium, Charlottesville, 13–16 June.

BSI and buildingSMART, 2010. *Constructing the Business Case: Building Information Modelling*, London: British Standards Institution.

buildingSMART UK, 2010. *Investing in BIM Competence*, London: buildingSMART UK.

Chan, A. P. and Chan, A. P., 2004. Key performance indicators for measuring construction success. *Benchmarking: An International Journal*, 11(2), pp. 203–221.

Codinhoto, R., Kiviniemi, A., Kemmer, S., Essiet, U. B., Donato, V. and Tonso, L. G., 2013. *BIM-FM: Manchester Town Hall Complex Research Report*, Manchester: University of Salford and Manchester City Council.

Cox, R. F., Issa, R. R. and Ahrens, D., 2003. Management's perception of key performance indicators for construction. *Journal of Construction Engineering and Management*, 129(2), pp. 142–151.

Cretu, O., Stewart, R. and Berends, T., 2011. *Risk Management for Design and Construction*. Hoboken, NJ: Wiley.

CURT, 2005. *Construction Measures: Key Performance Indicators*, s.l.: Construction Users Roundtable.

de Souza, J., 2011. Key performance indicators and benchmarking, in *Construction Statistics, No. 11*, Newport: Office for National Statistics.

Dikmen, I., Birgonul, M. T. and Han, S., 2007. Using fuzzy risk assessment to rate cost overrun risk in international construction projects. *International Journal of Project Management*, 25(5), pp. 494–505.

Eadie, R., Browne, M., Odeyinka, H., McKeown, C. and McNiff, S., 2013. BIM implementation throughout the UK construction project lifecycle: an analysis. *Automation in Construction*, 36, pp. 145–151.

Fridrich, J. and Kubečka, K., 2014. Fire risk in relation to BIM. *Advanced Materials Research*, 899, pp. 552–555.

Gao, J. and Fischer, M., 2008. *Framework & Case Studies Comparing Implementations & Impacts of 3D/4D Modeling Across Projects*, Stanford: Center for Integrated Facility Engineering (CIFE), Stanford University.

Geisler, E. and Wickramasinghe, N., 2015. *Principles of Knowledge Management: Theory, Practice, and Cases*. London: Routledge.

Giel, B. K. and Issa, R. R., 2013. Return on investment analysis of using building information modelling in construction. *Journal of Computing in Civil Engineering*, 27(5), pp. 511–521.

Gilligan, B. and Kunz, J., 2007. *VDC Use in 2007: Significant Value, Dramatic Growth, and Apparent Business Opportunity*, Stanford: Center for Integrated Facility Engineering, Stanford University.

GSA, 2007. *GSA Building Information Modelling Guide Series 01 – Overview*, Washington, DC: US General Services Administration.

GSA, 2012. *GSA BIM Guide Series 05*, Washington, DC: US General Services Administration.

Hanley, S., 2014. Measure what matters: a practical approach to knowledge management metrics. *Business Information Review*, 31(3), pp. 154–159.

Hinze, J., Bren, D. C. and Piepho, N., 1995. Experience modification rating as measure of safety performance. *Journal of Construction Engineering and Management*, 121(4), pp. 455–458.

Ho, S.-P., Tserng, H.-P. and Jan, S.-H., 2013. Enhancing knowledge sharing management using BIM technology in construction. *Scientific World Journal*, Article ID 170498, pp. 1–10.

Infrastructure Sustainability Council of Australia, 2014. *IS Overview*. Available at: http://182.160.161.41/is-rating-scheme/is-overview/is-rating-tool [Accessed 10 July 2015].

Institute for Sustainable Infrastructure, 2015. *Envision® Sustainable Infrastructure Rating System*. Available at: www.sustainableinfrastructure.org/rating [Accessed 7 July 2015].

ISO, 2013. *ISO/TS 14067:2013 – Greenhouse Gases – Carbon Footprint of Products – Requirements and Guidelines for Quantification and Communication*. Available at: www.iso.org/iso/home/store/catalogue_tc/catalogue_detail.htm?csnumber=59521 [Accessed 7 July 2015].

Jensen, J. and Duncan, R. C., 2011. *Preparedness: A Principled Approach to Return on Investment*, Falls Church, VA: International Association of Emergency Managers.

Kam, C., Senaratna, D., Xiao, Y. and McKinney, B., 2013. *The VDC Scorecard: Evaluation of AEC Projects and Industry Trends*, Stanford: Center for Integrated Facility Engineering (CIFE), Stanford University.

Kam, C. Senaratna, D., McKinney, B., Xiao, Y. and Song, M., 2014. *The VDC Scorecard: Formulation and Validation*, Stanford: Center for Integrated Facility Engineering (CIFE), Stanford University.

Khanzode, A., 2010. *An Integrated, Virtual Design and Construction and Lean (IVL) Method for Coordination of MEP*, Stanford: Centre for Integrated Facility Engineering, Stanford University.

Kivits, R. A. and Furneaux, C., 2013. BIM: enabling sustainability and asset management through knowledge management. *The Scientific World Journal*, Article ID 983721, pp. 1–14.

Koronios, A., Lin, S. and Gao, J., 2005. *A Data Quality Model for Asset Management in Engineering Organisations*, paper presented at the International Conference on Information Quality Conference, Cambridge, MA, 4–6 November.

KPI Working Group, 2000. *KPI Report for the Minister for Construction*, London: UK Department of Environment, Transport and Regions.

Kunz, J. and Fischer,. M., 2012. *Virtual Design and Construction: Themes, Case Studies and Implementation Suggestions*, CIFE Working Paper #097, Stanford: Sanford University.

Leite, F., Akcamete, A., Akinci, B., Atasoy, G. and Kiziltas, S., 2011. Analysis of modelling effort and impact of different levels of detail in building information models. *Automation in Construction*, 20(5), pp. 601–609.

Liebowitz, J. and Suen, C. Y., 2000. Developing knowledge management metrics for measuring intellectual. *Journal of Intellectual Capital*, 1(1), pp. 54–67.

Li, X., Zhang, X. and Hadjisophocleous, G., 2013. Fire risk analysis of a 6-storey residential building using CUrisk. *Procedia Engineering*, 62, pp. 609–617.

Lin, Y.-C., 2014. Construction 3D BIM-based knowledge management system: a case study. *Journal of Civil Engineering and Management*, 20(2), pp. 186–200.

Lindell, M. K. and Whitney, D. J., 1995. Effects of organizational environment, internal structure, and team climate on the effectiveness of local emergency planning committees. *Risk Analysis*, 15(4), pp. 439–447.

Lu, W., Peng, Y., Shen, Q. and Li, H., 2012. Generic model for measuring benefits of BIM as a learning tool in construction tasks. *Journal of Construction Engineering and Management*, 139(2), pp. 195–203.

Mah, D., Manrique, J. D., Yu, H., Al-Hussein, M. and Nasseri, R., 2011. House construction CO_2 footprint quantification: a BIM approach. *Construction Innovation*, 11(2), pp. 161–178.

McGraw Hill Construction, 2014a. *The Business Value of BIM for Construction in Major Global Markets: How Contractors Around the World Are Driving Innovation With Building Information Modelling*, Bedford, MA: McGraw Hill Construction.

McGraw Hill Construction, 2014b. *The Business Value of BIM in Australia and New Zealand: How Building Information Modelling is Transforming the Design and Construction Industry SmartMarket Report*, Bedford, MA: McGraw Hill Construction.

Memarzadeh, M. and Golparvar-Fard, M., 2012. *Monitoring and Visualization of Building Construction Embodied Carbon Footprint Using DnAR – N-Dimensional Augmented Reality Models*, paper presented at the Construction Research Congress (ASCE), West Lafayette, Indiana, 21–23 May.

NATSPEC, 2011. *NATSPEC National BIM Guide*, Sydney: Construction Information Systems Limited.

NATSPEC, 2014. *Getting Started with BIM*, Sydney: NATSPEC.

NBS, 2014. *NBS National BIM Report*, Newcastle upon Tyne: RIBA.

O'Connor, J. T. and Yang, L.-R., 2004. Project performance versus use of technologies at project and phase levels. *Journal of Construction Engineering and Management*, 130, pp. 322–329.

Porwal, A. and Hewage, K. N., 2013. Building information modelling (BIM) partnering framework for public construction projects. *Automation in Construction*, 31, pp. 204–214.

Rankin, J., Robinson Fayek, A., Meade, G., Haas, C., and Manseau, A., 2008. Initial metrics and pilot program results for measuring the performance of the Canadian construction industry. *Canadian Journal of Civil Engineering*, 35(9), pp. 894–907.

Roper, K. and McLin, M., 2005. *Key Performance Indicators Drive Best Practices for General Contractors*, Raleigh, NC: FMI, Management Consulting, Investment Banking for the Construction Industry, Microsoft Corporation.

Sacks, R., Eastman, C. M., Lee, G. and Orndorff, D., 2005. A target benchmark of the impact of three-dimensional parametric modelling in precast construction. *PCI Journal*, pp. 126–138.

Si, X.-S., Wang, W., Hu, C.-H. and Zhou, D.-H., 2011. Remaining useful life estimation – a review on the statistical data. *European Journal of Operational Research*, 213(1), pp. 1–14.

Suermann, P. C., 2009. *Evaluating the Impact of Building Information Modelling (BIM) on Construction*, doctoral thesis, Gainesville, FL: University of Florida.

Suermann, P. C. and Issa, R. R., 2007. *Evaluating the Impact of Building Information Modelling (BIM) on Construction*. University Park, PA: CONVR, pp. 206–2015.

Teichholz, P., 2013. *BIM for Facility Managers*, Hoboken: Wiley.

Tsai, M.-H., Mom, M. and Hsieh, S.-H., 2014. Developing critical success factors for the assessment of BIM technology adoption: Part I. Methodology and survey. *Journal of the Chinese Institute of Engineers*, pp. 1–14.

Volk, R., Stengel, J. and Schultmann, F., 2014. Building Information Modelling (BIM) for existing buildings – literature review and future needs. *Automation in Construction*, 38, pp. 109–127.

Wang, K.-C., Shih, S.-Y., Chan, W.-S., Wang, W.-C., Wang, S.-H., Gansonre, A.-A., Liu, J.-J., Lee, M.-T., Cheng, Y.-Y. and Yeh, M.-F., 2014. *Application of Building Information Modelling in Designing Fire Evacuation – A Case Study*, paper presented at 31st International Symposium on Automation and Robotics in Construction and Mining, Sydney, Australia, 9–11 July.

Wang, S.-H., Wang, W.-C., Wang, K.-C. and Shih, S.-Y., 2015. Applying building information modelling to support fire safety management. *Automation in Construction*, 59, pp. 158–167.

Wiedmann, T. and Minx, J., 2008. A definition of 'carbon footprint', in C. C. Pertsova (ed.) *Ecological Economics Research Trends*, Hauppauge, NY: Nova Science Publishers, pp. 1–11.

Xin, J. and Huang, C., 2013. Fire risk analysis of residential buildings based on scenario clusters. *Fire Safety Journal*, 62(A), pp. 72–78.

Zavadskas, E. K., Turskis, Z. and Tamošaitiene, J., 2010. Risk assessment of construction projects. *Journal of Civil Engineering and Management*, 16(1), pp. 33–46.

Zyngier, S. and Burstein, F., 2012. Knowledge management governance: the road to continuous benefits realization. *Journal of Information Technology*, 27(2), pp. 140–155.

Index

Benefits dictionary index

Enablers dictionary index

Metrics dictionary index